AF361648

Exoskeleton Robotic Systems

Exoskeleton Robotic Systems

Editors

Ionuţ Daniel Geonea
Cristian Petre Copilusi

Basel • Beijing • Wuhan • Barcelona • Belgrade • Novi Sad • Cluj • Manchester

Guest Editors

Ionuț Daniel Geonea	Cristian Petre Copilusi
University of Craiova	University of Craiova
Craiova	Craiova
Romania	Romania

Editorial Office
MDPI AG
Grosspeteranlage 5
4052 Basel, Switzerland

This is a reprint of the Special Issue, published open access by the journal *Applied Sciences* (ISSN 2076-3417), freely accessible at: https://www.mdpi.com/journal/applsci/special_issues/Exoskeleton_Robotic_Systems.

For citation purposes, cite each article independently as indicated on the article page online and as indicated below:

Lastname, A.A.; Lastname, B.B. Article Title. *Journal Name* **Year**, *Volume Number*, Page Range.

ISBN 978-3-7258-2469-4 (Hbk)
ISBN 978-3-7258-2470-0 (PDF)
https://doi.org/10.3390/books978-3-7258-2470-0

Contents

About the Editors

Ionuţ Daniel Geonea

Ionut Daniel Geonea, Associate Professor, PhD. Eng.: Ionut Daniel Geonea graduated in 2005 from the Faculty of Mechanics of the University of Craiova with a degree in Agricultural Machinery and Installations. He completed a master's degree in the specialization "modeling and simulation in mechanical engineering" in 2007, and in "automotive engineering—manufacture and design" in 2017. He completed his doctoral studies between 2005 and 2009 at the "Academician-Radu Voinea" Doctoral School of the University of Craiova under Professor Nicolae Dumitru on the topic: "Research on the Analysis and Synthesis of Some Mechanisms of Agricultural Machines". He has published 170 research articles, 91 of which are indexed in Scopus and 68 in Web of Knowledge. Of these, several articles on the design of systems for rehabilitation are in journals with high impact factors. His research interests include mechanical design, mechanisms and transmissions, machine components and tribology, and experimental measurements. From 2005 to 2006, he worked for Agricultural Machines and Tractors Company, in Craiova-Romania, as a design engineer within the constructive design workshop (tractors). He continued his professional career in the Faculty of Mechanics, Craiova, Romania, as a university lecturer (2006–2011), university assistant (2011–2012), and lecturer (2012–2018). Since 2018, he has worked as an assistant professor at the Faculty of Mechanics, University of Craiova, Romania. Since 2014, he has served as the president of the Dolj branch of the Romanian Tribology Association.

Cristian Petre Copilusi

Cristian Petre Copilusi, Associate Professor, PhD. Eng.: Cristian Petre Copilusi graduated from the Faculty of Mechanics, Road Vehicles at the University of Craiova in 2005. He then completed a master's degree in "Modeling and simulation in mechanical engineering" in 2007. In tandem with this master's degree, he undertook doctoral studies between 2005 and 2009 at the Doctoral School "Acad. Radu P. Voinea" of the University of Craiova under the supervision of the distinguished professor, Dr. Eng. Nicolae Dumitru, with a public defense of the doctoral thesis entitled "Research on mechanical systems applicable in medicine". Between 2009 and 2013, he undertook postdoctoral studies in a program at the University of Craiova. He has published 63 research articles, of which 28 are indexed in Web of Science and the rest in other international databases. Among these, two articles dedicated to human gait recovery systems were published in journals with notable impact factors. His primary research interests are machine parts, biomechanics, and biomedical applied research. His academic development is as follows: 2006–2012, Assistant Professor; 2012–2016, Lecturer; and 2016–present, Associate Professor in the Faculty of Mechanics, University of Craiova. Since 2018 and until now, he holds the position of Vice-Dean of Teaching, Quality, and Students in the Faculty of Mechanics.

Editorial

Special Issue on Exoskeleton Robotic Systems

Ionut Daniel Geonea * and Cristian Petre Copilusi

Faculty of Mechanics, University of Craiova, 200512 Craiova, Romania; cristian.copilusi@edu.ucv.ro
* Correspondence: ionut.geonea@edu.ucv.ro; Tel.: +40-727779866

Citation: Geonea, I.D.; Copilusi, C.P. Special Issue on Exoskeleton Robotic Systems. *Appl. Sci.* **2023**, *13*, 9788. https://doi.org/10.3390/app13179788

Received: 26 July 2023
Accepted: 28 August 2023
Published: 30 August 2023

Over the past 20 years, the therapies available in rehabilitation clinics have improved substantially, aided by robotic rehabilitation systems, which have seen rapid progress and increased performance. This progress has contributed to the diversification of the applications of exoskeleton robotic systems, namely, to provide assistance: in climbing up and down stairs, to workers in various activities (carrying weights, working with raised arms or in a bent position), and even in executing sit-to-stand movements. Exoskeleton robotic systems are used in a wide range of fields, both medical (rehabilitation, paraplegics, and amputees) or non-medical (rescue operations, construction work).

The progress of exoskeleton robotic systems is notable, and is visible in both lower-limb and upper-limb exoskeletons. However, there remain challenges, such as optimizing control systems and human–machine interaction.

This Special Issue presents a collection of articles highlighting recent achievements, applications, studies, and review articles within the field of exoskeleton robotic systems. These contributions cover a wide range of the applications of these systems, from the rehabilitation of the upper and lower limbs to assisting human gait when ascending and descending stairs, actuator selection criteria, and the implementation of actuation moment control algorithms.

The topics addressed within this Special Issue—"Exoskeleton Robotic Systems"—are varied. One such topic is the investigation of daily variations in the physical effects of using exoskeletons with the upper limbs [1]. The study aimed to evaluate the long-term biomechanics of wearing an exoskeleton at work to assist the upper limbs. Four activities are proposed for assistance, namely, lifting and lowering a box, moving while holding a box, moving without the box, and static bending during mechanical work. The data collected from the wearer while performing these activities are kinematic (movements using a full-body inertial motion capture system) and electromyographic (with sEMG sensors placed on the subject).

Review articles such as [2] aim to study the impact of COVID-19 on the evolution of patent proposals for lower limb exoskeleton robotic systems. The authors conducted a comparative analysis of the situation during the pandemic compared to the previous decade. The main purpose of the study was to provide a comprehensive overview of the efforts to innovate exoskeleton robotic systems during the pandemic. The results of the study confirm the trend of a decreasing number of patent proposals during the pandemic.

Another review study [3] provides a summary and overview of the use of the Hybrid Assistive Limb (HAL) robotic system in patients with spinal cord injuries. The conclusion of this study is that the use of this robotic system demonstrates a wide variety of applications, in both acute to chronic patients. Its use is generally safe, with no serious user injuries reported. Additionally, the vast majority of users experienced functional improvements, showing increased walking distances, gait speeds, and pain reduction.

Recent developments in exoskeleton robotic systems are presented in [4]. Namely, an exoskeleton robotic system has been developed which provides support for all three joints of the human lower limb. In order to design the motion laws of the drive motors, an experimental analysis of human gait with the Vicon equipment was used as a starting point, and a database of joint angle variation laws for a group of 30 people was created. A dynamic study of the exoskeleton was carried out using the Newton Euler method supplemented with Lagrange multipliers, in conjunction with a dynamic study of the virtual model using the ADAMS software for dynamic analysis of multibody systems. Experimental tests of the physical model were also carried out, demonstrating the feasibility of the prototype.

In this collection of Special Issue articles, there is also one [5] which proposes a new selection criterion for linear pneumatic motors used to drive wearable assistive devices, namely, the energy-to-mass ratio. The requirements for a wearable assistive device are compactness, lightweightness, and energy efficiency. The authors propose a criterion for choosing an actuator that meets the above-mentioned requirements, namely, the energy-to-mass ratio.

In the field of locomotor rehabilitation, prostheses are commonly required by people with trans tibial amputations. The authors of [6] propose the constructive solution of an ankle prosthesis based on an intelligent fluid joint (magneto rheological). Additionally, the sole of the prosthesis is equipped with eight pressure sensors which register the pressure between the sole and the ground, which is necessary to determine the kinematic and dynamic parameters of the person's gait.

Robotic rehabilitation systems also involve the upper limbs. Upper limb exoskeletons are used for two types of activity: as an assistive system that provides physical support and reduces the wearer's strain at work, and as a robotic system for the rehabilitation of stroke patients.

Thus, in [7], a dynamic analysis of a mechanism in the structure of an upper limb rehabilitation robot, called ParReEx, is presented. The dynamic analysis was performed in the first phase, considering the rigid kinematic elements, and later, the elasticity of the elements, as well as the friction in the kinematic couplings. A method of structural optimization of the constructive shape of the mechanism's elements is also presented, with the aim of reducing mechanical stress concentrators as well as minimizing the mass of the rehabilitation device.

A similar system designed for upper limb rehabilitation is the ASPIRE robot, a dynamic study of which is presented in [8]. The purpose of this system is the rehabilitation of brachial monoparesis. The dynamic analysis was performed using MSC.ADAMS software, a powerful tool for the study of mobile multibody systems. The robot is composed of three modules: the first providing the flexion/extension movement for the shoulder, the second providing the adduction/abduction movement, and the third providing forearm pronation/supination. The results obtained from the ADAMS simulation computed torque for the flexion/extension module and the adduction/abduction module are compared with those obtained experimentally. The similarity of these results validates the dynamic model, confirming that the connection forces determined from the dynamic simulation are valid. These data are used to study the mechanical strength of the flexion/extension modulus using the finite element method.

The outstanding performance of exoskeleton robotic systems is evidenced by their use as assistive systems for stair ascent/descent, as presented in [9]. Stair climbing involves slow movements within the joints but with greater amplitudes than in normal walking. The objective of this study was to evaluate the biomechanics of a new robotic exoskeleton system intended for stair climbing. The results of the study show that an individual application of assistive force is required for adequate assistance when climbing stairs.

Exoskeletons have also been developed that provide rehabilitation of a specific joint of the human lower limb. Thus, in [10], the energy consumption and torque control of such an exoskeleton are studied. The study focuses on sit-to-stand and walking activities. The constructive design of the exoskeleton hip robot was performed using Autodesk Inventor software, and the kinematic analysis was based on Denavit–Hartenberg parameters inputted into Matlab computational software. The Linear Quadratic Regulator (LQR) method was considered in order to obtain the optimum controller for energy consumption and walking activity.

Author Contributions: I.D.G. and C.P.C. are the Guest Editors of the Special Issue. They conceived and wrote this editorial article. All authors have read and agreed to the published version of the manuscript.

Acknowledgments: Thanks to all authors and peer reviewers for their valuable contributions to the Special Issue, "Exoskeleton Robotic Systems".

Conflicts of Interest: The authors declare no conflict of interest.

References

1. Riemer, J.; Jaitner, T.; Wischniewski, S. Investigating Inter-Day Variations in the Physical Effects of Exoskeletons: Requirements for Long-Term Biomechanical Studies. *Appl. Sci.* **2023**, *13*, 6483. [CrossRef]
2. Pană, C.F.; Rădulescu, V.M.; Pătrașcu-Pană, D.M.; Petcu, F.L.; Reșceanu, I.C.; Cismaru, Ș.I.; Trășculescu, A.; Bîzdoacă, N. The Impact of COVID on Lower-Limb Exoskeleton Robotic System Patents—A Review. *Appl. Sci.* **2022**, *12*, 5393. [CrossRef]
3. Brinkemper, A.; Grasmücke, D.; Yilmaz, E.; Schildhauer, T.A.; Aach, M. HAL Training in Spinal Cord Injured Patients: A Narrative Review of 10 Years Experience. *Appl. Sci.* **2023**, *13*, 1369. [CrossRef]
4. Copilusi, C.; Dumitru, S.; Geonea, I.; Ciurezu, L.G.; Dumitru, N. Design Approaches of an Exoskeleton for Human Neuromotor Rehabilitation. *Appl. Sci.* **2022**, *12*, 3952. [CrossRef]
5. Deaconescu, A.; Deaconescu, T. Energy-to-Mass Ratio—A Novel Selection Criterion of Pneumatic Motors Used for the Actuation of Wearable Assistive Devices. *Appl. Sci.* **2022**, *12*, 6459. [CrossRef]
6. Pană, C.F.; Manta, L.F.; Vladu, I.C.; Cismaru, Ș.I.; Petcu, F.L.; Cojocaru, D.; Bîzdoacă, N. The Design of a Smart Lower-Limb Prosthesis Supporting People with Transtibial Amputation—A Data Acquisition System. *Appl. Sci.* **2022**, *12*, 6722. [CrossRef]
7. Tarnita, D.; Geonea, I.D.; Pisla, D.; Carbone, G.; Gherman, B.; Tohanean, N.; Tucan, P.; Abrudan, C.; Tarnita, D.N. Analysis of Dynamic Behavior of ParReEx Robot Used in Upper Limb Rehabilitation. *Appl. Sci.* **2022**, *12*, 7907. [CrossRef]
8. Geonea, I.D.; Tarnita, D.; Pisla, D.; Carbone, G.; Bolcu, A.; Tucan, P.; Georgescu, M.; Tarniță, D.N. Dynamic Analysis of a Spherical Parallel Robot Used for Brachial Monoparesis Rehabilitation. *Appl. Sci.* **2021**, *11*, 11849. [CrossRef]
9. Böhme, M.; Köhler, H.-P.; Thiel, R.; Jäkel, J.; Zentner, J.; Witt, M. Preliminary Biomechanical Evaluation of a Novel Exoskeleton Robotic System to Assist Stair Climbing. *Appl. Sci.* **2022**, *12*, 8835. [CrossRef]
10. Andersson, R.; Björsell, N. The Energy Consumption and Robust Case Torque Control of a Rehabilitation Hip Exoskeleton. *Appl. Sci.* **2022**, *12*, 11104. [CrossRef]

Article

Investigating Inter-Day Variations in the Physical Effects of Exoskeletons: Requirements for Long-Term Biomechanical Studies

Julia Riemer [1,*], Thomas Jaitner [2] and Sascha Wischniewski [1]

[1] Federal Institute for Occupational Safety and Health, 44149 Dortmund, Germany; wischniewski.sascha@baua.bund.de
[2] Institute for Sport and Sport Science, TU Dortmund University, 44227 Dortmund, Germany; thomas.jaitner@tu-dortmund.de
* Correspondence: julia.riemer@baua.bund.de

Abstract: Exoskeletons potentially reduce physical strain on workers. However, studies investigating the long-term effects of exoskeletons in the workplace are rare, not least because demonstrating physical long-term impacts faces several challenges, including the collection of reliable biomechanical data with the exoskeleton. By examining the potential impact of using an exoskeleton on inter-day measurements, we can provide valuable insights into the suitability of long-term studies. Therefore, this study aims to investigate the inter-day variation in muscle activity (MA) and kinematics of the trunk and legs during lifting, carrying, walking, and static bending with and without a passive back exoskeleton. The majority of results show no significant differences in inter-day variation. However, we found minor significant unilateral variation in knee and ankle kinematics when using the BSE during the lifting, carrying, and walking tasks, as well as in MA of M. biceps femoris when measuring without the BSE during the lifting tasks. Cohen's d showed small effect sizes, ranging from $-0.0045 \leq d \leq 0.384$ for all significant p-values. While we classify the observed significant differences as minor, it is still crucial to consider day-to-day variations in long-term studies. However, by implementing high levels of standardization in study designs, including precise exoskeleton fitting, consistent assistance level, familiarization with measurement systems, and standardized working tasks, the impact of the exoskeleton on inter-day measurements can be minimized. Additional field studies are necessary to validate our findings in real work conditions.

Keywords: occupational health; reproducibility of results; statistical parametric mapping; muscle activity; kinematics

Citation: Riemer, J.; Jaitner, T.; Wischniewski, S. Investigating Inter-Day Variations in the Physical Effects of Exoskeletons: Requirements for Long-Term Biomechanical Studies. *Appl. Sci.* **2023**, *13*, 6483. https://doi.org/10.3390/app13116483

Academic Editor: Ionuţ Daniel Geonea

Received: 28 April 2023
Revised: 19 May 2023
Accepted: 23 May 2023
Published: 25 May 2023

1. Introduction

Exoskeletons provide physical support and potentially reduce strain while performing tasks in the workplace [1]. In addition to the benefits of exoskeleton use in terms of reduced muscle activity (MA) and improved endurance [2], studies to date also confirm their potential negative effects on kinematics and MA, including a limited range of joint motion and a higher load on muscle groups [3–5]. As such, it is not yet possible to predict the long-term consequences of movement changes and altered MA patterns caused by using exoskeleton in the workplace [6].

Studies addressing the long-term effects of wearing an exoskeleton at work are scarce, not least because demonstrating long-term relief or strain on the musculoskeletal system from wearing exoskeletons faces several prerequisites [7]. The collection of reliable day-to-day biomechanical data is particularly challenging, because the signals can be affected by intrinsic and extrinsic factors, such as sensor position or fluctuating movement patterns of the target group of workers [8,9].

The most important prerequisite for conducting a biomechanical long-term study is the use of measurement parameters and systems, which can provide comparable results at different points in time. In addition to others, the measurement of MA using surface electromyography (sEMG) and the assessment of movement patterns using wearable sensors, such as inertial measurement units (IMUs), are suitable for this purpose [10]. Both measurement systems enable the comparison of workloads and movement patterns between different individuals and days [11,12] and are reported as being reliable during static [9,13,14] and dynamic conditions [9,15,16].

In order to ensure inter-day reliability of measurement with the described systems, a high degree of execution standardization is required, e.g., consistent fitting and support of the exoskeleton, familiarization with the work task, as well as the normalization task of sEMG measurements [17], and accurate placement of sEMG and IMU sensors [18]. However, even with high standardization, we cannot exclude the exoskeleton-user interaction as a factor influencing the measurements reliability. An exoskeleton, as a rigid mechanical system, has the same effect on the user's body at all times during standardized use. However, we cannot rule out the possibility that the interaction between the user and the exoskeleton may vary on inter-day measurement due to intrinsic factors, which would skew the results. Results of biomechanical studies using exoskeletons vary widely even when using similar or the same exoskeletons [19], which can also be explained by individual user characteristics [10].

Several investigations have already demonstrated significant biomechanical inter-day variation in dynamic and static work tasks without an exoskeleton [16,20–22] using methods such as interclass correlation coefficient (ICC), coefficient of variation (CV) and standard error of measurement (SEM). However, these methods do not enable such a detailed analysis, as required for clinical inference, or the determination of reliability at different stages of a work movement, because they are pooling discrete time points (e.g., peaks) or the entire time series (e.g., root mean square or mean frequency). Previous studies applying exoskeletons in inter-day measurements have only used subjective evaluation criteria to assess the exoskeletons effect [23–26]. To date, only one study by Kozinc et al. [27] has addressed inter-day repeatability and found good reliability in the repeated administration of a test battery for functional assessment of a back supporting exoskeleton. However, they did not measure any biomechanical parameters.

Whether the interaction between exoskeletons and user influences the reliability of the measurements of kinematics and MA on consecutive days has not yet been determined in any study. The objective of this study is therefore to investigate the inter-day variation for MA and kinematics of trunk and leg, in lifting, carrying, walking, and static bending, and examine whether the utilization of a passive back exoskeleton may have an impact on day-to-day results comparability. In order to assess potential inter-day differences throughout the entire duration of the motor tasks, we utilized statistical parametric mapping (SPM) to compare the parameters of each task between day one and day two, both with and without the exoskeleton.

2. Materials and Methods

2.1. Passive Back Exoskeletons

To ensure that any results were related to the use of a specific exoskeleton model, we decided to select two different back exoskeletons with similar mechanisms for this study, the V2.5 (Laevo) and the PaexoBack (Otto Bock). Both models are attached to the user's hip and secured to the upper body with a chest pad. The leg shells of both models are attached to the thighs in a similar position. Both exoskeletons transfer forces from the lower back to the chest and leg pads. A feature of both exoskeletons was that the support could be switched on and off manually while wearing them.

2.2. Subjects

Twenty subjects participated in the study, representing an average user population for the occupational sciences. We therefore recruited the participants from a subject list of the Federal Institute for Occupational Safety and Health including individuals of working age, mostly with either recreational or with no previous sport experience. They were, respectively, ten healthy men and women aged 26 to 49 years (age 34 ± 7 years, height: 174.7 ± 8.2 cm, weight: 74.4 ± 11.6 kg, BMI: 24.2 ± 2.4 kg/m^2), mostly with either recreational or no previous sport experience. Subjects were asked to abstain from physical activity the day before and the days of the tests, to avoid the effects of cumulative muscle fatigue. All subjects signed an informed consent form before the test. Ethical approval was obtained from the local institutional ethics committee.

2.3. Procedure

Subjects performed four different tasks (lifting, carrying, walking, and static bending task), with and without wearing an exoskeleton on two consecutive days. All subjects started with the lifting task, followed by the carrying, walking, and static bending tasks. The order of the two conditions (with and without exoskeleton) within the tasks was systematically varied across subjects. The subjects were randomly assigned to one of two exoskeleton groups, PaexoBack and V2.5, by drawing a number. The number 1 represented the PaexoBack group, while the number 2 represented the V2.5 group. This randomization process ensured an unbiased distribution of participants across the two exoskeleton groups. The time interval between the two examinations was 24 h to 48 h.

In order to familiarize the subjects with the exoskeleton and to minimize any training effect during the measurements, the entire study procedure was performed over two additional training days. On each of the two training days, the entire study protocol was conducted. This included both the testing of MVIC normalization contractions and the complete procedure of the work tasks with and without the exoskeleton, as described in Sections 2.3.1–2.3.4.

On the first training day, the exoskeleton was adjusted to the participants' body dimensions according to the manufacturer's instructions, and these same settings were used for all subsequent days.

2.3.1. Task 1—Lifting and Lowering a Box

The first task was a repetitive box-lifting and lowering task with a load of 10 kg for male subjects and 5 kg for female subjects (Figure 1). The subjects stood in front of the box and performed six lifting and lowering cycles using their own technique: The lift started in an upright position. The subjects lifted the box and straightened up into an upright position. After a two-second standing phase with the box in the hand, the lowering phase began. The box was returned to the ground position and the subjects straightened up again. After standing upright without the box, the next cycle started.

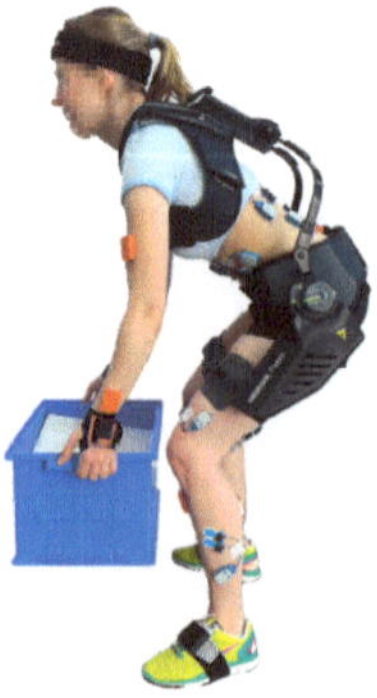

Figure 1. Task 1. lifting and lowering a box with the exoskeleton PaexoBack.

2.3.2. Task 2—Carrying a Box

The subjects were instructed to carry a box with a load of 10 kg for male and 5 kg for female over a distance of 20 m at their own pace (Figure 2). The subjects lifted the box at the marked starting point and straightened up into an upright position to start walking. At the end of the walk, the box was placed back on the ground in a marked position.

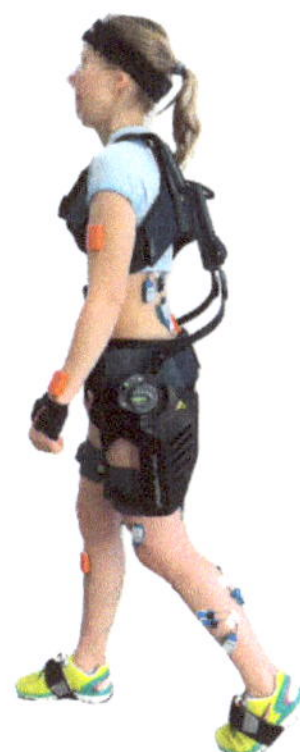

Figure 2. Tasks 2 and 3: carrying a box (**left**) and walking without load (**right**) with the exoskeleton PaexoBack.

2.3.3. Task 3—Walking without Load

The same distance as when carrying the box was completed by the subjects without the box at their own pace (Figure 2). As we wanted the study to represent realistic use scenarios of the exoskeleton and in working conditions, the support function of the exoskeleton was switched off for the walking distance.

2.3.4. Task 4—Static Bending Task

Subjects were instructed to stand at a marked position in front of an engine block (Figure 3). They assumed a forward bending trunk posture (30–40°) and removed spark plugs from the engine block. The total duration of the task was 60 s.

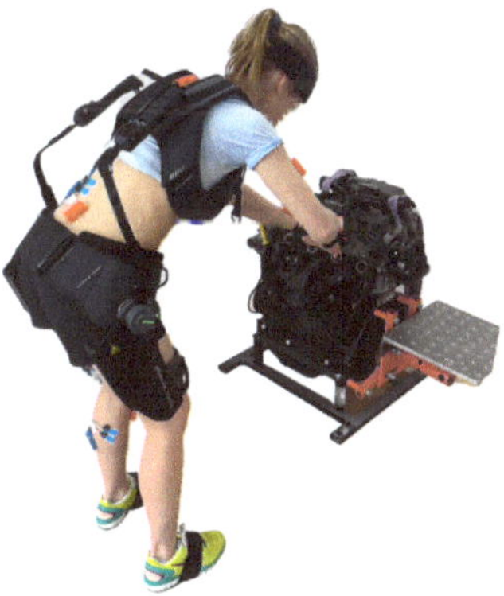

Figure 3. Task 4: Static bending task with the exoskeleton PaexoBack.

2.4. Measurements

2.4.1. Kinematics

The movements were recorded using a full body inertial motion capture system (MTw Awinda, Xsens Technologies, Enschede, Netherlands). The 17MTx sensors were placed directly on the skin using skin adhesive, to achieve the most accurate measurement result possible (Figure 4) and position standardized via procedural instructions. Prior to the experiment, participants' body dimensions and calibration poses were measured according to Xsens calibration protocol with the MVN software (MVN Analyse Pro 2021.0.1), to fit and scale the MVN biomechanical model to the subject. The kinematic data was collected at a sample rate of 60 Hz and exported in an excel format.

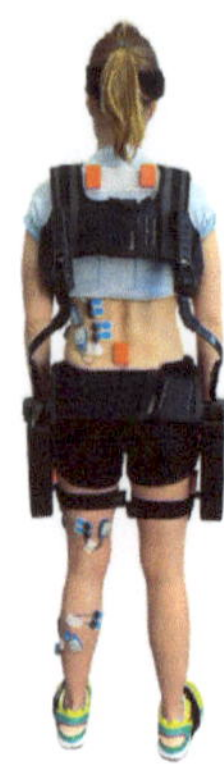

Figure 4. Sensor placement of sEMG and Xsens sensors.

2.4.2. Electromyography

We equipped the subjects with sEMG sensors, connected to a mobile sEMG system (Ultium, Noraxon). Beforehand, skin was prepared to achieve stable electrode contact and high skin conductance by lowering the impedance. For this purpose, hair was removed from the skin positions to be covered with a disposable razor. The skin was cleaned with alcohol and treated with an abrasive gel. This method is suggested for clinical use [28]. Adhesive gel dual disposable electrodes served as sEMG electrodes; we placed these on dominant body side on the M. erector spinae thoracic (EST) and lumbar (ESL), M. obliquus externus abdominis (OE), M. vastus medialis (VM), M. biceps femoris (BF), M. tibialis anterior (TA), and M. gastrocnemius medialis (GM) according to SENIAM guidelines (Figure 4). We took care to place the electrodes on the midline of the abdominal area, perpendicular to the length of the muscle fibers between the muscle tendon junction and the nearest innervation zone. The muscles were chosen and the electrodes were attached, standardized in such a way that they did not come into contact with the exoskeleton.

An MVIC was performed for all muscles prior to the test. Based on the results of a previous study [17], the performance of the MVICs was also trained on the two previous days, in order to achieve the highest possible level of standardization.

The rectified sEMG signal was set to a sampling rate of 2.000 Hz per channel and filtered using a fourth-order Butterworth filter with a bandpass of 20–500 Hz for GM and TA. To remove heart artefacts in OE and RA signals, a hampel filter was applied, which is a useful method to remove sEMG signal outliers [29]. SEMG signal was root mean squared (RMS) with a window size of 250 ms and normalized using MVICs.

2.5. Data Processing

For tasks 2 (carrying a box) and 3 (walking without load), the gait cycles were considered as follows: A step of a gait cycle started with the right foot terminal contact and its subsequent terminal contact (toe off to toe off). The first and last step of a gait cycle were not considered. A total of eight steps per subject were included in the further analysis. The lifting cycles were divided into lifting and lowering, as described in Section 2.3.1. Five lifting and lowering cycles per subject were included in the analysis. For the bending task, the middle 50 s were selected and divided into 2 s segments. The first and last 2 s segments were selected for further analysis.

The joint angles and muscle signals acquired during each measurement were normalized to a standardized time scale of 101 points. Subsequently, all time-normalized data was averaged per measurement day, comprising the eight gait cycles and five lifting/lowering cycles for each subject. This resulted in a single time-normalized dataset per measurement day for each individual, which was utilized for further comparative analysis using SPM. All data processing steps were programmed and executed in R (R Core Team) [30].

Statistical Parametric Mapping

To assess the reliability of the measurements, we employed statistical parametric mapping (SPM) to compare the parameters of each task between day one and day two. SPM offers a comprehensive statistical analysis framework that enables the identification of significant differences across the entire duration of the motor task. All SPM analyses were conducted in MATLAB Version: 9.13.0 (R2022b), using the open-source software package from spm1D 0.4 [31]. Prior to the analysis, all data were tested for normal distribution. Paired t-tests were then used to compare the paired conditions of day one and day two.

The entire time series data representing the lifting, carrying, and walking cycles, as well as the static holding sections, were compared for both kinematic and sEMG data of day one and day two. At each time point, a scalar output statistic called SPM{t} was computed and evaluated by determining the critical threshold at α = 0.05 significance level. If the trajectory of SPM{t} crosses this critical threshold at any time point, they were marked as statistically significant. For significant areas, the effect size Cohen's d (d) was also determined.

3. Results

All subjects completed the training sessions and both measurement days. One subject's gait data could not be used for further analysis because the gait pattern was too abnormal. Therefore, this subject was excluded from the analysis of tasks 2—carrying a box—and 3—walking without load. For the other tasks, all the data collected were used. The results for each task are presented below. The blue plots represent the comparison of day 1 and day 2 with the exoskeleton, and the red plots the comparison without the exoskeleton. Regions of significant difference from the SPM results are indicated with background shading (Figures 5–9).

In combination with the shaded regions indicating significant differences, the SPM {t} value plots directly below each plot can be used as an indicator of practical significance. If a significant difference can be detected in a region, Cohen's d is also shown in the plot next to the corresponding p-value, as representation of the effect size. For the kinematic data, we present the variables that are particularly relevant to the exoskeleton and have been used in previous studies [1]. These include ankle, hip, knee and spine flexion.

The first task we analyzed was the lifting and lowering (task 1). In the kinematic data, there was no significant difference between the measured joint angles on day 1 and day 2 for lifting. However, for lowering the box, a significant difference in knee flexion was measured with the exoskeleton, as seen in Figure 5. The joint angle at the start of weaning is significantly greater on day 1 (p = 0.038, d = 0.384) than on day 2. The difference is also visible on the left side of the body at the same point, but is not significant. This section may be divided by subheadings. It should provide a concise and precise description of the experimental results, their interpretation, as well as the experimental conclusions that can be drawn.

For the sEMG data, there is a significant difference in MA of the BF in the downward movement on day 1 and day 2 for both lifting and lowering in Figures 6 and 7 (lifting: p = 0.001, d = −0.243; lowering: p = 0.040, d = −0.268) without the exoskeleton. For the other muscles there were no measurable significant differences in MA from day 1 to day 2. For the task of carrying a box, we found a significant difference in knee flexion for the exoskeleton measurement in the kinematic data (Figure 8). There was significantly higher knee flexion in the terminal contact area for the right knee on day 2 (p = 0.037, d = −0.14). For the left body side, we could not confirm this effect.

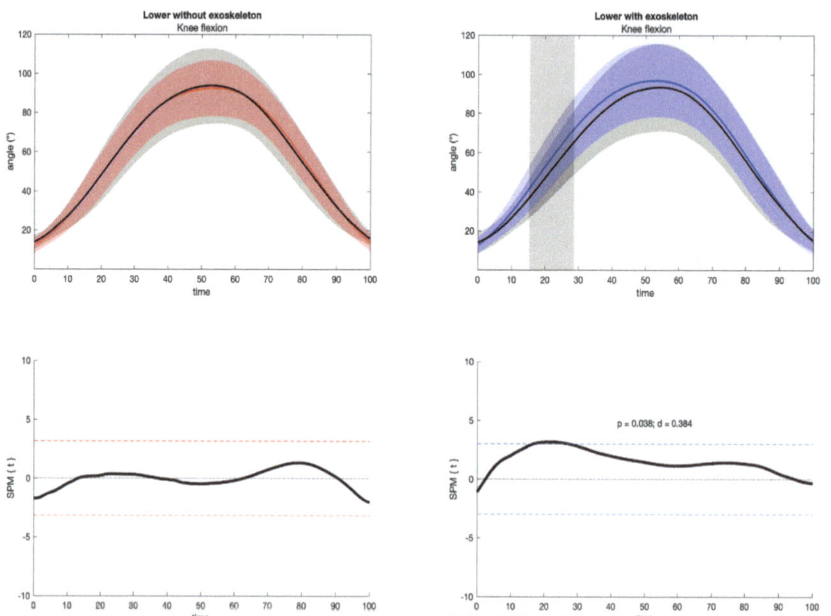

Figure 5. SPM *t*–test results and profile plots of kinematic data in lowering a box, comparing joint angles (°) of the right body side in knee flexion on day 1 (blue resp. red lines) and day 2 (black lines). The blue plot represents the comparison of the measurements with exoskeleton, and the red plot shows the measurements without exoskeleton (n = 20).

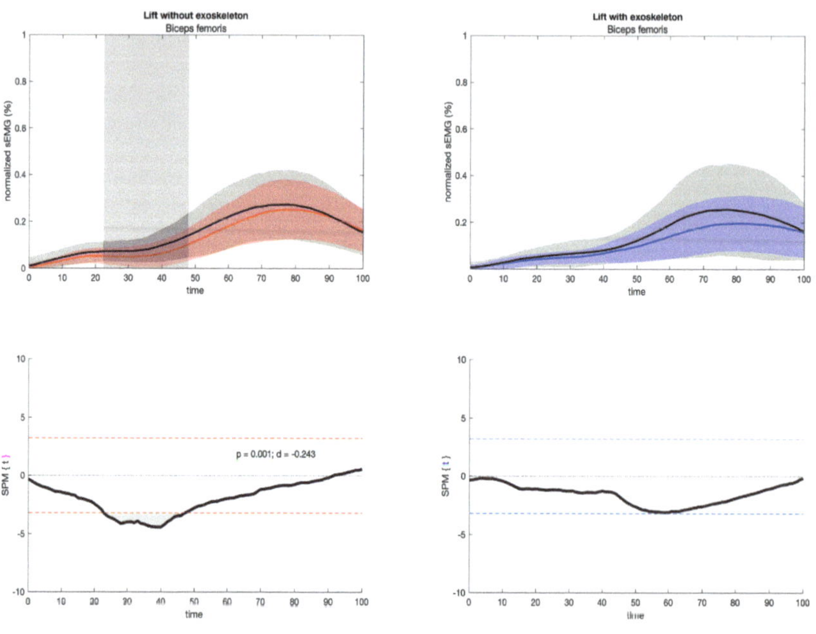

Figure 6. SPM *t*–test results and profile plots of MA in lifting a box, comparing normalized sEMG (%) of leg muscles BF on day 1 (blue resp. red lines) and day 2 (black lines). The blue plot represents the comparison of the measurements with exoskeleton, and the red plot shows the measurements without exoskeleton (n = 20).

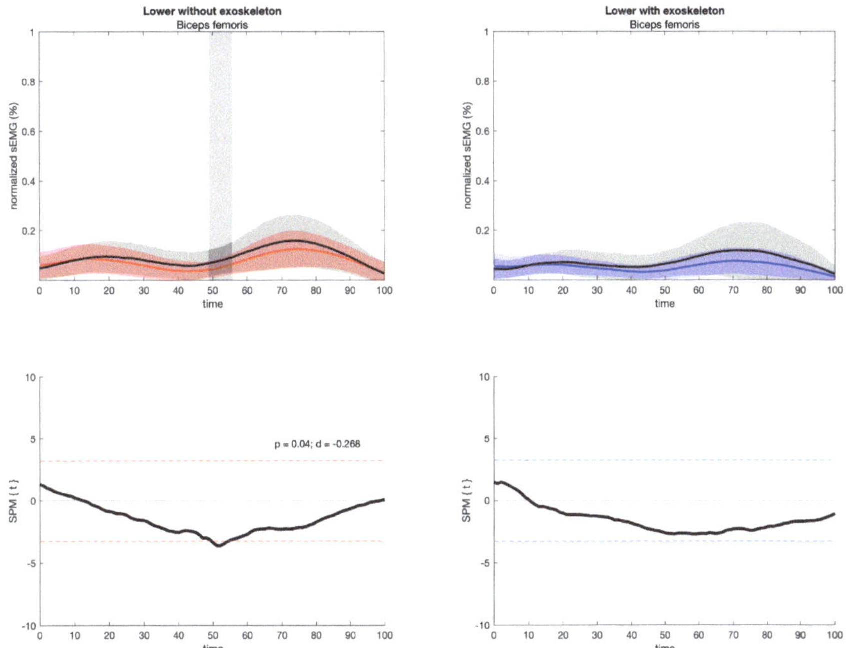

Figure 7. SPM *t*–test results and profile plots of MA in lowering a box, comparing normalized sEMG (%) of leg muscles BF on day 1 (blue resp. red lines) and day 2 (black lines). The blue plot represents the comparison of the measurements with exoskeleton, and the red plot shows the measurements without exoskeleton (n = 20).

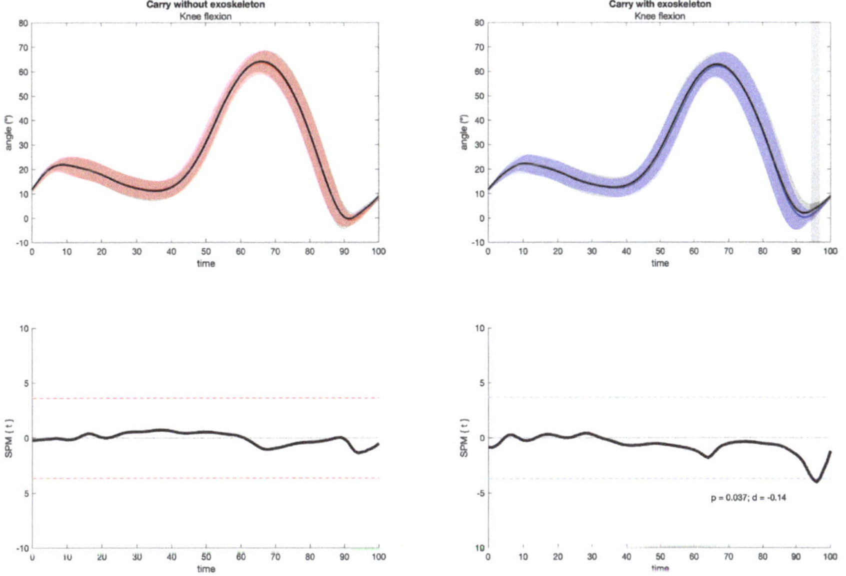

Figure 8. SPM *t*–test results and profile plots of kinematic data in carrying a box, comparing joint angles (°) of the right body side in knee flexion on day 1 (blue resp. red lines) and day 2 (black lines). The blue plot represents the comparison of the measurements with exoskeleton, and the red plot shows the measurements without exoskeleton (n = 19).

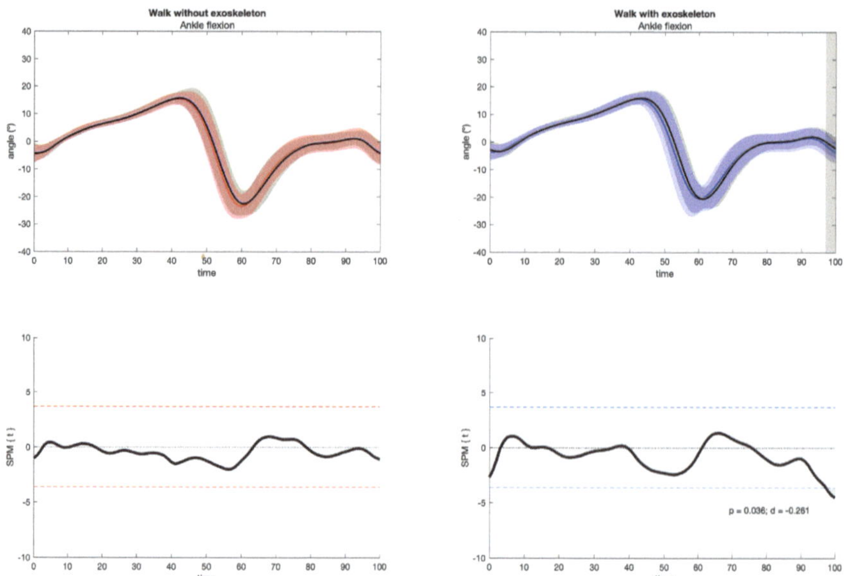

Figure 9. SPM *t*–test results and profile plots of kinematic data in walking without load, comparing joint angles (°) of the right body side in ankle flexion on day 1 (blue resp. red lines) and day 2 (black lines). The blue plot represents the comparison of the measurements with exoskeleton, and the red plot shows the measurements without exoskeleton (n = 19).

We found similar results to those in walking without load. The difference was significant in the kinematic data with the exoskeleton in the area of terminal contact, but this time in the joint angle of right ankle flexion (Figure 9). It was significantly greater on day 2 than on day 1, when the exoskeleton was worn during the measurement ($p = 0.038$, d = -0.261).

In contrast, the measured MA was comparable for all muscles on day 1 and day 2 detected in the carrying and walking tasks.

For the static bending task, no significant differences were found in either the kinematic data or the MA data. This was true for both the baseline (the first 2 s after the start of the measurement) and the end of the measurement (the last 2 s of the measurement). Here, all differences between day 1 and day 2, with and without exoskeleton, were in a non-significant range with $p \geq 0.05$.

4. Discussion

This study was designed as a preliminary study for a long-term biomechanical evaluation of exoskeletons in the workplace. We investigated the inter-day variation for MA and kinematics of trunk and leg, in lifting, carrying, walking, and static bending, with and without a passive back exoskeleton and examined whether the utilization of a passive back exoskeleton may have an impact on the day-to-day results comparability.

When biomechanical methods are obtained on different measurement days in a long-term study, we need to be sure that these effects are not due to fundamental measurement variations caused by confounding factors. We therefore based our study design on previous research and considered several factors that can improve the reliability of inter-day biomechanical measurements [10].

Firstly, we emphasized the importance of standardized setup and use of the exoskeleton so it was adapted to the individual requirements of each subject according to the manufacturer's specifications and the optimal, most comfortable support performance. Over the course of two familiarization days, any discomfort in using the exoskeleton was identified and appropriate adjustments were made. This is important because several studies have reported discomfort with the exoskeleton [23,32], which could affect the exoskeleton–user interaction. We established a high level of standardization in the exoskeleton's assistance performance, which has been a critical issue in previous studies. At the same time, the two training days were designed to counteract any training effects.

In addition to the use of validated measurement systems [9,15], we included standardized sensor placement via procedural instructions [15] as well as sufficient two-day familiarization with the working tasks. As a previous study we conducted suggests that two days of familiarization may provide a high degree of standardization to ensure reliable MVIC normalization for an occupational target group [17], we also included MVIC normalization tasks in the two training days.

The results show minor but significant differences between the measurements of day 1 and day 2. Regarding the IMU measurements, we found differences especially in the day-to-day measurements with the exoskeleton. For example, we measure significant unilateral differences during the lifting tasks in the area of knee flexion ($p = 0.038$), as well as in the carrying and walking task, in the area of initial contact in knee and ankle flexion ($p = 0.037$ & $p = 0.036$). For the muscle activation, we found significant differences in the measurements without exoskeleton, in the M. biceps femoris, during lifting ($p = 0.001$), and also during lowering a box ($p = 0.040$).

In order to quantify our resulting significant differences, we also determined the effect size Cohen's d. It ranged from $-0.0045 \leq d \leq 0.384$ for all significant p-values. After analyzing relevant publications, we assume that our d values < 0.3 can be classified as small effects [33–35], with exception of the measurement of M. biceps femoris when lifting a box ($p = 0.001$), where the effect size is $d = 0.384$. In this context, Button et al. [34] also discuss the influence of sample size on the effect size. They report that a decrease in the effect size is to be expected as the sample size increases. In our case, this means that the measured effect size of M. biceps femoris > 0.3 can also be regarded as low, as our sample is a small number of subjects (n = 20) according to ibid.

The minor yet significant differences, with small effect sizes, affirm that the impact of the exoskeleton on the user's body remains relatively consistent from day to day. However, considering the context of previous studies, it is crucial to take into account the measured differences for long-term investigations. This is particularly important as earlier studies predominantly report unfavorable outcomes when assessing the reliability of biomechanical parameters in inter-day measurements, especially in dynamic tasks. For example, Sood et al. [36] found poor day-to-day reliability of muscle activity in the shoulder muscles in a laboratory-based simulation of overhead work at different working heights. Van Helden et al. [16] performed dynamic voluntary movements of the trunk during functional reaching tasks and measured moderate to poor reliability of the back muscles. Brandt et al. [22] also found moderate results when assessing the muscle activity of the back muscles during box lifts with different weights and lifting tasks. Ghofrani et al. [37] also confirmed these results regarding the muscle activity. In terms of kinematic day-to-day reliability, Graham et al. [38] revealed significant differences when evaluating a repetitive dynamic trunk flexion/extension task in automotive manufacturing. Howarth and Graham [21] also found poor to moderate results in the evaluation of joint angles during a repetitive pipetting task on three different days. In this context, they also highlight the importance of standardized sensor placement. Koumantakis et al. [20] found poor repeatability in the repetition of predefined joint angles during trunk flexion exercises. Certainly, when comparing our findings to previous studies, it is important to consider that we compared the parameters of each task between days one and two across their entire duration. By using statistical parametric mapping (SPM) and analyzing the entire time series data, we were able to avoid potential limitations that may arise from reducing movement phases to discrete points in time or summarizing time series, as done in previous research. Instead of focusing solely on specific points or summary statistics, SPM allows for a comprehensive analysis of the complete temporal profile of the measurements. This approach provides a more detailed and accurate representation of the data, ensuring that potential impairments or biases resulting from data reduction methods are minimized. The ability to detect significant differences in the temporal profile of the movement allows for a more nuanced understanding of the data. However, it also means that drawing generalized conclusions becomes more difficult in terms of comparing our results to those of previous studies.

In addition our chosen analytical method, it should be noted that the high degree of measurement standardization in our study influences the comparability of our results to those of previous studies. By executing two familiarization days, we counteracted habituation affecting the inter-day measurement sEMG and kinematic results. During the familiarization phase, we placed special emphasis on acclimating the participants to the measurement systems with their calibration and the MVIC normalization procedures, which are essential for the standardized execution of measurements. We have endeavored to achieve consistent and highly standardized MVC measures for sEMG normalization as important prerequisite for inter-day reliability, because we expected our subjects to possible lack of experience in producing high muscle forces limits the ability to reproduce a relatively constant MVIC [39]. It is therefore reasonable to assume that our normalized sEMG data additional value to a high level of reliability. Appropriately, Van Helden et al. [16] report in their study, in which such MVIC-familiarization was not performed, that the relative inter-day reliability was higher for absolute sEMG amplitudes compared to normalized sEMG amplitudes. This finding underscores the importance of familiarization in MVIC normalization.

However, other previous studies do not describe any familiarization approaches when measuring inter-day reliability of muscle activity [16,37] or kinematics [20,21]. Ibid counteracted day-to-day changes through habituation effects by performing a sufficient number of repetitive tasks during the measurement, but no separate training was performed prior to the measurements for movement familiarization. Therefore, we assume that the extensive measures for familiarization contribute to a minor inter-day variance in our study, whereas habituation occurring during the day-to-day measurements may possibly explain the moderate to poor results in reliability of previous studies.

The results demonstrate that the interaction between the user and exoskeleton during inter-day measurements can result in significant differences. While we classify the observed significant differences as minor, we conclude that achieving consistent inter-day measurements is possible, despite minor variations in measurements obtained using exoskeletons. Still, our results strongly indicate that high standards of measurement are mandatory in a long-term study with exoskeletons. It is necessary to carry out further studies in this field, optimally under field conditions with a large sample size, to enable a generalizability of the results.

5. Conclusions

We show the feasibility of reliable biomechanical measurements over repeated days with and without a back-supporting exoskeleton. To ensure reliable measurements, a high degree of standardization is required, including standardized placement of sEMG and IMU sensors and sufficient familiarization with the work tasks and the exoskeleton. Despite potential non-significant differences through a relatively large standard deviation in our sEMG results, we offer an approach to achieve a high degree of day-to-day reliability in measuring long-term biomechanical effects.

By using an ergonomically relevant subject population, we are able to transfer the results to further field evaluations. At the same time, by using two different exoskeleton models, we can assume that the interaction of the passive back exoskeletons with the users is independent of the model. Due to the unique characteristics of our study, further studies under field conditions are necessary to confirm our findings.

Author Contributions: Conceptualization, J.R., T.J. and S.W.; validation, J.R., T.J. and S.W.; formal analysis, J.R.; investigation, J.R.; resources, J.R.; data curation, J.R.; writing—original draft preparation, J.R.; writing—review and editing, T.J. and S.W.; visualization, J.R.; supervision, S.W. and T.J.; project administration, J.R. All authors have read and agreed to the published version of the manuscript.

Funding: This research received no external funding.

Institutional Review Board Statement: The study was conducted in accordance with the Declaration of Helsinki, and approved by the Institutional Ethics Committee of Federal Institute for Occupational Safety and Health on 19.04.2021).

Informed Consent Statement: Informed consent was obtained from all subjects involved in the study.

Data Availability Statement: Data not available due to restrictions e.g. privacy or ethical reasons. Data not publicly available due to data protection regulations.

Conflicts of Interest: The authors declare no conflict of interest.

References

1. Bär, M.; Steinhilber, B.; Rieger, M.A.; Luger, T. The influence of using exoskeletons during occupational tasks on acute physical stress and strain compared to no exoskeleton–A systematic review and meta-analysis. *Appl. Ergon.* **2021**, *94*, 103385. [CrossRef] [PubMed]
2. Bosch, T.; van Eck, J.; Knitel, K.; de Looze, M. The effects of a passive exoskeleton on muscle activity, discomfort and endurance time in forward bending work. *Appl. Ergon.* **2016**, *54*, 212–217. [CrossRef] [PubMed]
3. Kermavnar, T.; de Vries, A.W.; de Looze, M.P.; O'Sullivan, L.W. Effects of industrial back-support exoskeletons on body loading and user experience: An updated systematic review. *Ergonomics* **2021**, *64*, 685–711. [CrossRef] [PubMed]
4. Koopman, A.S.; Kingma, I.; de Looze, M.P.; van Dieën, J.H. Effects of a passive back exoskeleton on the mechanical loading of the low-back during symmetric lifting. *J. Biomech.* **2020**, *102*, 109486. [CrossRef]
5. Picchiotti, M.T.; Weston, E.B.; Knapik, G.G.; Dufour, J.S.; Marras, W.S. Impact of two postural assist exoskeletons on biomechanical loading of the lumbar spine. *Appl. Ergon.* **2019**, *75*, 1–7. [CrossRef]
6. Howard, J.; Murashov, V.V.; Lowe, B.D.; Lu, M.L. Industrial exoskeletons: Need for intervention effectiveness research. *Am. J. Ind. Med.* **2020**, *63*, 201–208. [CrossRef]
7. Crea, S.; Beckerle, P.; De Looze, M.; De Pauw, K.; Grazi, L.; Kermavnar, T.; Masood, J.; O'sullivan, L.W.; Pacifico, I.; Rodriguez-Guerrero, C.; et al. Occupational exoskeletons: A roadmap toward large-scale adoption. Methodology and challenges of bringing exoskeletons to workplaces. *Wearable Technol.* **2021**, *2*, e11. [CrossRef]
8. Merletti, R.; Muceli, S. Tutorial. Surface EMG detection in space and time: Best practices. *J. Electromyogr. Kinesiol.* **2019**, *49*, 102363. [CrossRef]
9. Kobsar, D.; Charlton, J.M.; Tse, C.T.F.; Esculier, J.F.; Graffos, A.; Krowchuk, N.M.; Thatcher, D.; Hunt, M.A. Validity and reliability of wearable inertial sensors in healthy adult walking: A systematic review and meta-analysis. *J. Neuroeng. Rehabil.* **2020**, *17*, 62. [CrossRef]
10. Kuber, P.M.; Abdollahi, M.; Alemi, M.M.; Rashedi, E.A. Systematic Review on Evaluation Strategies for Field Assessment of Upper-Body Industrial Exoskeletons: Current Practices and Future Trends. *Ann. Biomed. Eng.* **2022**, *50*, 1203–1231. [CrossRef] [PubMed]
11. Besomi, M.; Hodges, P.W.; Clancy, E.A.; Van Dieën, J.; Hug, F.; Lowery, M.; Merletti, R.; Søgaard, K.; Wrigley, T.; Besier, T.; et al. Consensus for experimental design in electromyography (CEDE) project: Amplitude normalization matrix. *J. Electromyogr. Kinesiol.* **2020**, *53*, 102438. [CrossRef] [PubMed]
12. Mahdavi, N.; Dianat, I.; Heidarimoghadam, R.; Khotanlou, H.; Faradmal, J. A review of work environment risk factors influencing muscle fatigue. *Int. J. Ind. Ergon.* **2020**, *80*, 103028. [CrossRef]
13. Błaszczyk, A.; Ogurkowska, M.B. The use of electromyography and kinematic measurements of the lumbar spine during ergonomic intervention among workers of the production line of a foundry. *PeerJ* **2020**, *10*, e13072. [CrossRef] [PubMed]
14. Hellig, T.; Rick, V.; Mertens, A.; Nitsch, V.; Brandl, C. Investigation of observational methods assessing workload of static working postures based on surface electromyography. *Work* **2019**, *62*, 185–195. [CrossRef] [PubMed]
15. Cudejko, T.; Button, K.; Al-Amri, M. Validity and reliability of accelerations and orientations measured using wearable sensors during functional activities. *Sci. Rep.* **2022**, *12*, 14619. [CrossRef]
16. Van Helden, J.; Martinez-Valdes, E.; Strutton, P.; Falla, D.; Chiou, S.-Y. Reliability of high-density surface electromyography for assessing characteristics of the thoracic erector spinae during static and dynamic tasks. *J. Electromyogr. Kinesiol.* **2022**, *67*, 102703. [CrossRef]
17. Riemer, J.; Jaitner, T.; Wischniewski, S. Effect of familiarization on the reproducibility of maximum isometric normalisation contractions in a worker-specific sample. *Int. J. Ind. Ergon.* **2023**, revised.

18. Beange, K.H.; Chan, A.D.; Beaudette, S.M.; Graham, R.B. Concurrent validity of a wearable IMU for objective assessments of functional movement quality and control of the lumbar spine. *J. Biomech.* **2019**, *97*, 109356. [CrossRef]
19. De Bock, S.; Ghillebert, J.; Govaerts, R.; Tassignon, B.; Rodriguez-Guerrero, C.; Crea, S.; Veneman, J.; Geeroms, J.; Meeusen, R.; De Pauw, K. Benchmarking occupational exoskeletons: An evidence mapping systematic review. *Appl. Ergon.* **2022**, *98*, 103582. [CrossRef]
20. Koumantakis, G.A.; Winstanley, J.; Oldham, J.A. Thoracolumbar proprioception in individuals with and without low back pain: Intratester reliability, clinical applicability, and validity. *J. Orthop. Sport. Phys. Ther.* **2002**, *32*, 327–335. [CrossRef]
21. Howarth, S.J.; Graham, R.B. Sensor positioning and experimental constraints influence estimates of local dynamic stability during repetitive spine movements. *J. Biomech.* **2015**, *48*, 1219–1223. [CrossRef] [PubMed]
22. Brandt, M.; Andersen, L.L.; Samani, A.; Jakobsen, M.D.; Madleine, P. Inter-day reliability of surface electromyography recordings of the lumbar part of erector spinae longissimus and trapezius descendens during box lifting. *BMC Musculoskelet. Disord.* **2017**, *18*, 519. [CrossRef]
23. Kim, S.; Nussbaum, M.A.; Smets, M.; Ranganathan, S. Effects of an arm-support exoskeleton on perceived work intensity and musculoskeletal discomfort: An 18-month field study in automotive assembly. *Am. J. Ind. Med.* **2021**, *64*, 905–914. [CrossRef] [PubMed]
24. Baltrusch, S.J.; Houdijk, H.; Van Dieën, J.H.; Kruif, J.T.C.d. Passive trunk exoskeleton acceptability and effects on self-efficacy in employees with low-back pain: A mixed method approach. *J. Occup. Rehabil.* **2021**, *31*, 129–141. [CrossRef] [PubMed]
25. Iranzo, S.; Piedrabuena, A.; Iordanov, D.; Martinez-Iranzo, U.; Belda-Lois, J.-M. Ergonomics assessment of passive upper-limb exoskeletons in an automotive assembly plant. *Appl. Ergon.* **2020**, *87*, 103120. [CrossRef]
26. Marino, M. Impacts of Using Passive Back Assist and Shoulder Assist Exoskeletons in a Wholesale and Retail Trade Sector Environment. *IISE Trans. Occup. Ergon. Hum. Factors* **2019**, *7*, 281–290. [CrossRef]
27. Kozinc, Ž.; Baltrusch, S.; Houdijk, H.; Šarabon, N. Reliability of a battery of tests for functional evaluation of trunk exoskeletons. *Appl. Ergon.* **2020**, *86*, 103117. [CrossRef]
28. Hermens, H.J.; Freriks, B.; Disselhorst-Klug, C.; Rau, G. Development of recommendations for SEMG sensors and sensor placement procedures. *J. Electromyogr. Kinesiol.* **2020**, *10*, 361–374. [CrossRef]
29. Marateb, H.R.; Rojas-Martínez, M.; Mansourian, M.; Merletti, R.; Villanueva, M.A.M. Outlier detection in high-density surface electromyographic signals. *Med. Biol. Eng. Comput.* **2012**, *50*, 79–89. [CrossRef]
30. R Core Team. *R: A Language and Environment for Statistical Computing*; R Foundation for Statistical Computing: Vienna, Austria, 2021; Available online: https://www.R-project.org/ (accessed on 26 May 2022).
31. Pataky, T.C. One-dimensional statistical parametric mapping in Python. *Comput. Methods Biomech. Biomed. Eng.* **2012**, *15*, 295–301. [CrossRef]
32. Luger, T.; Bär, M.; Seibt, R.; Rieger, M.A.; Steinhilber, B. Using a back exoskeleton during industrial and functional tasks—Effects on muscle activity, posture, performance, usability, and wearer discomfort in a laboratory trial. *Hum. Factors* **2023**, *65*, 5–21. [CrossRef] [PubMed]
33. Knudson, D. Significant and meaningful effects in sports biomechanics research. *Sport. Biomech.* **2009**, *8*, 96–104. [CrossRef] [PubMed]
34. Button, K.S.; Ioannidis, J.P.; Mokrysz, C.; Nosek, B.A.; Flint, J.; Robinson, E.S.; Munafò, M.R. Power failure: Why small sample size undermines the reliability of neuroscience. *Nat. Rev. Neurosci.* **2013**, *14*, 365–376. [CrossRef] [PubMed]
35. Buchanan, T.L.; Lohse, K.R. Researchers' perceptions of statistical significance contribute to bias in health and exercise science. *Meas. Phys. Educ. Exerc. Sci.* **2016**, *20*, 131–139. [CrossRef]
36. Sood, D.; Nussbaum, M.A.; Hager, K. Fatigue during prolonged intermittent overhead work: Reliability of measures and effects of working height. *Ergonomics* **2007**, *50*, 497–513. [CrossRef]
37. Ghofrani, M.; Olyaei, G.; Talebian, S.; Bagheri, H.; Kazemi, P. Reliability of SEMG measurements for trunk muscles during lifting variable loads in healthy subjects. *J. Bodyw. Mov. Ther.* **2017**, *21*, 711–718. [CrossRef]
38. Graham, R.B.; Sheppard, P.S.; Almosnino, S.; Stevenson, J.M. Dynamic spinal stability and kinematic variability across automotive manufacturing work shifts and days. *Int. J. Ind. Ergon.* **2012**, *42*, 428–434. [CrossRef]
39. Frost, L.R.; Gerling, M.E.; Markic, J.L.; Brown, S.H. Exploring the effect of repeated-day familiarization on the ability to generate reliable maximum voluntary muscle activation. *J. Electromyogr. Kinesiol.* **2012**, *22*, 886–892. [CrossRef]

Review

The Impact of COVID on Lower-Limb Exoskeleton Robotic System Patents—A Review

Cristina Floriana Pană, Virginia Maria Rădulescu *, Daniela Maria Pătrașcu-Pană, Florina Luminița Petcu (Besnea) *, Ionuț Cristian Reșceanu, Ștefan Irinel Cismaru, Andrei Trăsculescu and Nicu Bîzdoacă

Faculty of Automation, Computers and Electronics, University of Craiova, Blvd. Decebal nr. 107, RO-200440 Craiova, Romania; cristina.pana@edu.ucv.ro (C.F.P.); daniela.pana@edu.ucv.ro (D.M.P.-P.); ionut.resceanu@edu.ucv.ro (I.C.R.); stefan.cismaru@edu.ucv.ro (Ș.I.C.); andrei.trasculescu@edu.ucv.ro (A.T.); nicu.bizdoaca@edu.ucv.ro (N.B.)
* Correspondence: virginia.radulescu@edu.ucv.ro (V.M.R.); florina.petcu@edu.ucv.ro (F.L.P.)

Abstract: In recent decades, the field of physical rehabilitation, with the help of robotic systems that aid the population of any age with locomotor difficulties, has been evolving rapidly. Several robotic exoskeleton systems of the lower limbs have been proposed in the patent literature and some are even commercially available. Given the above, we are asking ourselves at the end of the COVID-19 pandemic: how much has this pandemic affected both the publication of patents and the application of new ones? How has new patents' publication volume or application in robotic exoskeleton systems changed? We hypothesize that this pandemic has caused a reduction in the volume of new applications and possibly publications. We compare pandemic analysis and the last decade's analysis to answer these questions. In this study, we used a set of statistical tests to see if there were any statistically significant changes. Our results show that the pandemic had at least one effect on applying for new patents based on the information analyzed from the three databases examined.

Keywords: exoskeleton; lower limb; rehabilitation robotics

Citation: Pană, C.F.; Rădulescu, V.M.; Pătrașcu-Pană, D.M.; Petcu, F.L.; Reșceanu, I.C.; Cismaru, Ș.I.; Trăsculescu, A.; Bîzdoacă, N. The Impact of COVID on Lower-Limb Exoskeleton Robotic System Patents—A Review. *Appl. Sci.* **2022**, *12*, 5393. https://doi.org/10.3390/app12115393

Academic Editor: Manuel Armada

Received: 5 April 2022
Accepted: 23 May 2022
Published: 26 May 2022

Publisher's Note: MDPI stays neutral with regard to jurisdictional claims in published maps and institutional affiliations.

1. Introduction

According to the literature, exoskeleton systems can be classified into two broad categories: medical and non-medical systems, depending on the field of applicability. Medical ones include the exoskeleton systems used for rehabilitation, compensation, paraplegics, and amputees. In contrast, the type of non-medical ones consists of those used in the army (help soldiers transport heavy equipment on rough roads, improving mobility), rescue operations (rescuers can use it to transport supplies or firefighters can use it to transport firefighting equipment), construction (workers can use it to transport construction materials), and support (for healthy older adults). Although the study of the exoskeleton of the lower limbs came from the military, in recent years, the medical field has become a priority. In the following, we will focus on the systems of the exoskeletons of the lower limbs mainly, and only the patents of the exoskeletons that treat the whole leg or certain parts of the lower limb are considered. Exoskeletal patents involving the upper body, although vital in themselves, do not significantly contribute to mobility and are therefore omitted from this study.

The oldest exoskeleton patent was probably registered by Nicholas Yagn from Russia in 1890 [1] under the name "Apparatus for facilitating walking, running and jumping" (Figure 1a). As it stands in the original request, this patent aims to increase the efficiency of walking, running, and jumping, thus reducing the inherent fatigue that arises as a result of these actions. Such a "device", which acts parallel with the human body to aid human locomotion, is now called the exoskeleton and can be considered a starting point for modern exoskeleton systems [2,3].

Exoskeleton lower-limb rehabilitation robots constitute a significant class of robotic rehabilitation systems. Rehabilitation of the lower limbs with the help of robotic exoskeletons appeared over 40 years ago as an alternative to conventional manual gait training [4,5]. These exoskeletal systems connect to the human body in a portable way. Compared to traditional therapy, walking rehabilitation using a robotic exoskeletal system can provide highly controlled training (being able to control the movement of all joints in the training and recovery process), repetitive and intensive [6], and can reduce the physical load on the therapist and provide accurate and fast progress of patients [7,8]. Among the latest patents applied for and published in 2021 (according to the Google Patents platform), it is registered in South Korea. It is entitled "Lower extremity exoskeleton robotic device" (Figure 1b). The patent application states that the present invention relates to a robot device with a lower exoskeleton used for rehabilitation exercises. The exoskeleton robotic system is a device mainly used for rehabilitation by supporting the muscular strength of the wearer when the lower extremities are paralyzed by accident or when the muscular strength of the lower extremities is insufficient due to old age. This device can also be used for industrial or military purposes, such as running [9].

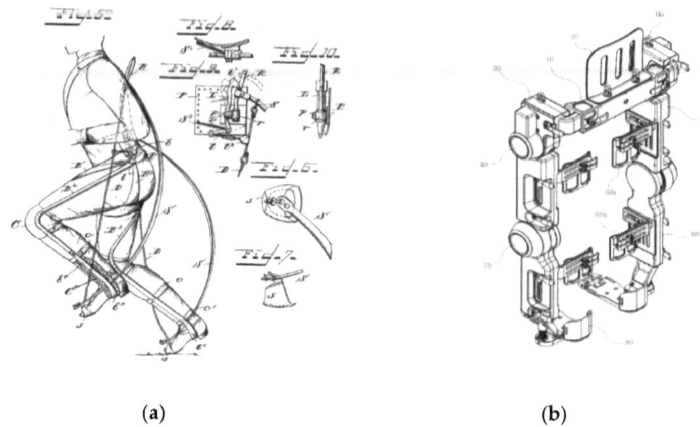

(a)	(b)

Figure 1. (**a**) US420179—Apparatus for facilitating walking, running and jumping (see the reference [1]); (**b**) KR102292983B1—Lower extremity exoskeleton robotic device (see the reference [9]).

Several revisions have emerged over the last decade that address various aspects of lower extremity rehabilitation and support devices. For example, in 2013, Chen et al. [4] focused on robotic lower extremity care exoskeletons used in rehabilitation therapy. In 2015, Yan et al. [10] presented a comprehensive analysis of all lower-limb exoskeletons developed after the 1990s; all devices analyzed were classified based on four criteria: mechanical structure, robot control system, user validation, and sensory devices used. Moreover, in 2015, Meng et al. [11] analyzed the recent evolution of control mechanisms and strategies for robotic lower-limb rehabilitation devices. Then, in 2016, Rupal et al. [12] conducted a review of exoskeleton robots' construction and technological features for existing lower extremities, classifying them based on amplifying and rehabilitating human power into commercial or prototype versions. Moreover, in 2016, Chen et al. [13] described the evolution of the exoskeletons of the lower limbs in terms of their structure, the control algorithms used, the detection technology to determine the wearer's intention to move, and the detection technology, the remaining challenges in this area. The walking rehabilitation exoskeletons presented in this review were classified as treadmill and ground training/assistance exoskeletons. Then, Shi et al. [14] published a comprehensive study in 2019 summarising the current state of exoskeleton robots for lower limbs, focusing on human gait analysis and the design of the drive system and control of these devices. Moreover, in 2019, Sanchez-Villamañan et al. [15] analyzed 52 wearable exoskeletons of the lower limbs in their review, focusing on three main aspects of compliance: drive, structure, and interface attachment components. The authors highlighted the disadvantages and advantages of different solutions and suggested several promising lines of research. In 2021, Koch and Font-Llagunes [16] conducted a review in which the primary purpose was to provide a comprehensive overview of the technological status of exosuits and the clinical results obtained when applied to users with reduced mobility based on 19 studies identified by the authors as relevant. Finally, in March 2022, Kian et al. [17] published a study that provides a comprehensive review of how portable sensor technology has contributed to the activation and control of motorized ankle exoskeletons developed over the past two decades. The authors also investigated the control schemes and operating principles used in the revised motorised ankle exoskeletons and their interaction with the integrated sensor systems. Other reviews have focused on reviewing the technological aspects of exoskeletons from a general perspective [2], while others have concentrated on bipedal locomotion [18,19] or designing specific joints [20]. This review provides an overview of robotic lower-limb exoskeletal systems for rehabilitation that are intended primarily for use by people with gait disorders.

Compared to other studies, we set out to analyze the effect of the COVID-19 pandemic on the publication and application of new patents. According to the World Health Organization, the year 2020 began with the extremely rapid spread of the COVID-19 pandemic and, by 2022, has already reached multiple peaks. This pandemic has affected almost every aspect of life, from industry, economics, and tourism to politics, arts, sports, and in particular health, and yet there has been a global union and effort, especially from the scientific research community, to find ways to deal with it. As such, the patterns of innovative research have also been severely affected by this crisis. Therefore, this systematic review was conducted to answer the following questions: (1) how much has this pandemic affected the publication of patents and the application of new ones? (2) how has new patents' publication volume or application in robotic exoskeleton systems changed? According to our literature analysis, scientometric analysis has not been performed on research involving robotic exoskeleton patents for lower limbs. Thus, our study is the first evaluation analysis of this field of research using this method. We aim to provide an objective picture of the development of science and the productivity of researchers, as well as to evaluate the topicality of this field of research on the following aspects: (1) a historical map of the subject; (2) a ranking of countries in terms of patent application/publication; (3) the distribution between exoskeletons dealing with the whole of the respective lower system, which deals in part with segments; (4) the time elapsed between the filing and publication of the patent; (5) type of exoskeleton structure: rigid or suit, and mode of wearing: portable or non-portable; (6) patents applied/published during the COVID-19 pandemic. We hope that this study could provide a guide for researchers when they want to file a new patent on robotic exoskeleton rehabilitation for lower limbs, and, at the same time, encourage them in the sense that whenever a situation arises—a crisis (even worldwide) or any impediment of any kind—solutions will be found to help continue their research work. Although work information (patents) collected from public platforms are known and can be accessed by the general public, this paper seeks to generate debate focused on improving current systems and also on how to cover the shortcomings of other identifiable solutions.

2. Review Methodology

2.1. Search Strategy

2.1.1. Inclusion Criteria

All patterns must be:

Filter 1: published in or translated to English and directly related to rehabilitation of lower-limb exoskeletons robot systems.

Filter 2: related to the International Patent Classification (IPC) A61H3/00—appliances for aiding patients or disabled persons to walk about; A61F 5/0102—orthopedic devices for correcting deformities of, or supporting, limbs; A61F 5/0106—for the knees, A61F 5/0111—for the feet or ankles.

Filter 3: describing the design, manufacturing method, and control of an exoskeleton.

Filter 4: able to be registered in any patent office in any country.

Filter 5: granted.

Filter 6: in the COVID-19 pandemic period 12 March 2020–12 March 2022.

2.1.2. Exclusion Criteria

All patterns must not be:

Filter 7: patent application or limited patent.

Filter 8: inactive, discontinued or pending legal status.

Filter 9: related to upper rehabilitation limb (hand or arm) exoskeletons robot systems.

Filter 10: published in a non-English language and those whose translations to English were very inadequate.

Filter 11: with an application date before 2012.

2.1.3. Information Databases and Search Methodology

We considered that portable exoskeletons are all those that have a rigid external structure, but also soft exoskeletons or exosuits were included in the present study. Exoskeletons that have used bodyweight support or a treadmill have been excluded to focus only on patents with the effect of wearable technology.

We searched for pattern publications in three online databases: Google Patents, PatentScope, and Lens, from 1 March 2012 until 31 March 2022, using the following search terms: *(exoskeleton) AND (robot) AND (walk OR gait) AND ((leg OR lower) AND (limb OR extremity)) AND (rehabilitation)*.

With the above keywords and filters: 1, 2, 4, and 9, we initially retrieved 1021 patents on Google Patents, 128 patents on PatentScope and 161 patents on Lens. After applying filters 5, 7, and 8, we get the following: 281 patents on Google Patents, 128 patents on PatentScope, and 74 patents on Lens.

An example of the application of the search methodology presented above can be seen in Figure 2 for the Lens platform. Initially, we obtained 161 patents that fulfilled the respective filters: of which 84 appear as patent applications, and the remaining 77 are published patents. Moreover, out of the 161 patents, 120 complied with the requirement of filter 8. After applying filters 5 and 7, we obtained a number of 74 patents. Further, using filters 6 and 12, we get 74 patents. The numerical distribution of these patents by year of publication and by inventors can be seen in Figure 2.

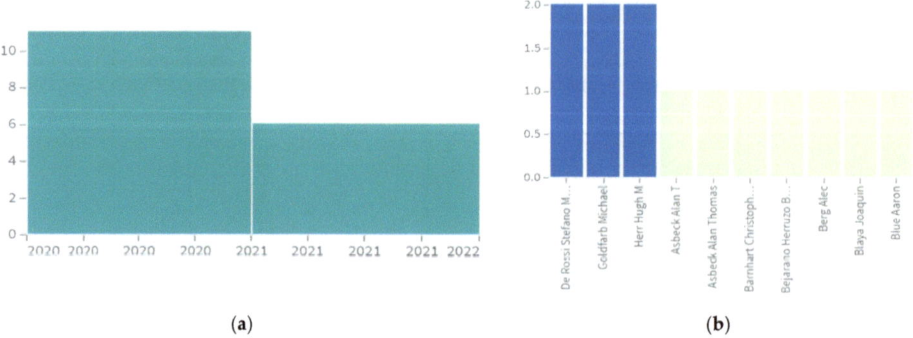

(a)   (b)

Figure 2. (a) Publications in COVID-19 pandemic period—highlight a selection to filter 6; (b) Inventors—patents count.

2.2. Data Collection

The data obtained from the queries of the three patent platforms are presented in the tables in Appendix A (Tables A1–A3). Moreover, in the tables in Appendix A, a classification of the exoskeletons was made according to the area of the lower limb that treats it (in whole or part), the type of rigid or suit structure and the portability mode: portable, portable with wheels (which also contain a set of auxiliary wheels for locomotion) and non-portable (which are fixed and can only be used in the spaces where they were placed).

2.3. Analysis Method

We used both a short-term analysis technique, focusing on the pandemic period 13 March 2020–13 March 2022, and an analysis technique in the longer term by which we compared the publication patterns of the last decade (2012–2022). Our study used a set of statistical tests to find statistically significant changes. These tests were performed both short-term and longitudinally on each of the three platforms, Google Patents, PatentScope, and Lens. At the short-term level, these tests indicate whether there are statistically significant differences when comparing the number of patents applied with those published. At the longitudinal level, these tests show whether there are statistically significant differences when comparing patents published before the pandemic with those published during the pandemic.

3. Results

To answer the two questions that were the main objectives (Section 1), we used the method of analysis presented in Section 2.3. Following these tests, we obtained the following results presented in the subsections below.

3.1. Google Patents Platform

In the first analysis of the data collected, one can see a particular concern for the development of exoskeletons of Asian researchers. We studied the applications for granting patents worldwide aimed at exoskeletons. We observed that out of 281 published patents (for the period 2012–2022), researchers from Asia applied for a percentage of 70% of them. We can see these data in Table 1 and Figure 3, respectively.

Table 1. The number of application patents by region.

Regions	Number of Application Patents
Europe	27
Asia	197
SUA+Canada	57
Total	281

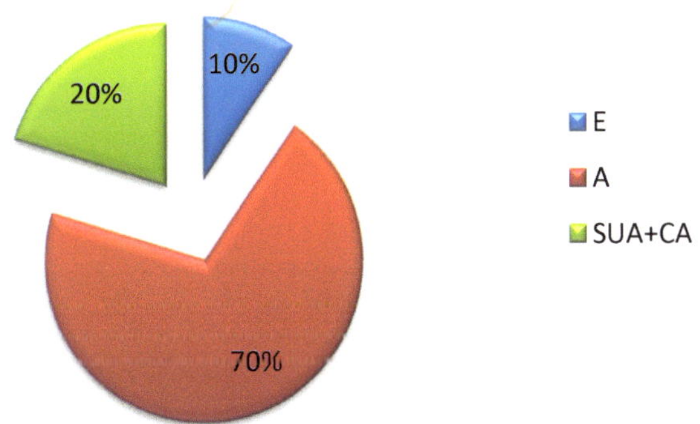

Figure 3. Percentage distribution by region after applying filters 5, 7 and 8.

The distribution of these requests, divided by country, reveals, and reinforces the fact that there is a concern in finding the best and most refined solutions to the problems that exoskeletons can solve. Furthermore, Chinese researchers are very active, and an extraordinary number of applications support this. Our statements are reinforced by the data in Table 2 and the graphs in Figure 3.

Table 2. The number of application patents by country.

Country Code	Number of Application Patents
CA	3
CN	129
DE	1
ES	16
EP	10
JP	13
KR	50
RU	4
TW	1
US	54

Once we have established the areas of interest, we can move on to data analysis. An important aspect is the date on which the patent applications were filed. The period analyzed in this study is from the year beginning 2005. It seems a timid beginning for what this field can offer today. Still, the researchers of a US team started with gait rehabilitation methods and apparatuses, the essential operation for the development of new techniques and technologies. It was a timid start because the patent was validated after nine years. From Table 3, it can be easily seen that when new roads are opened, a period of adaptation and understanding of the proposed new techniques and technologies is needed. From here, a series of waiting for the results appears. From our study, the period in which nothing was published is quite long (applications submitted in the period 2005–2011 began to be approved only in 2012), strengthening the abovementioned beliefs.

Table 3. The number of patent applications and publications distributed per year.

Year	Application Date	Publication Date
2005	1	0
2006	1	0
2007	1	0
2008	2	0
2009	9	0
2010	2	0
2011	2	0
2012	6	3
2013	17	5
2014	16	10
2015	23	9
2016	59	9
2017	42	22
2018	51	40
2019	34	48
2020	12	65
2021	4	58
2022	0	13

What we found earlier entitles us to refine our study further, and we will take for analysis the last decade, i.e., the period 2012–2022. In 2012, there were more patent applications in North America and Asia.

We now consider only patent applications. At the beginning of the analyzed period, a typical trajectory of things can be observed in Table 3. We can say that the ascending trend was set in terms of research in exoskeletons and filing patent applications. However, we see that starting in 2020, the presence of the pandemic is starting to make itself felt. For example, if in 2018, 51 applications were submitted, and in 2019 they decreased to 34 (i.e., the applications decreased by more than 30%), then in 2020 and 2021 they were likely to be drastically reduced. One of the ideas that emerges from the analysis of the above data would be the one that affected entire areas during the pandemic, namely the limitation of the interaction between the members of the research teams.

The analyzed phenomenon has two components (the patent filing component and the one for their approval); the second one must be investigated. If we examine the above, we can conclude that the pandemic was a disaster on all levels, but there is still a positive aspect. Analyzing Table 3 and the data collected on granting patents, we can observe an increasing trend even during the pandemic. We can say that we tried to help put into practice as many scientific ideas as possible used by all humanity during this period.

Another interesting issue is the problem that these exoskeletons deal with of exoskeletons. Here are two approaches:

— Treatment of the entire lower system
— Partial treatment of the lower system (its different subsystems).

During the analysis period (2012–2022), out of the 264 patents applied in the Google Patents platform, 82% treated the entire lower system, and another part focused on different subsystems. This aspect can also be seen in Figure 4.

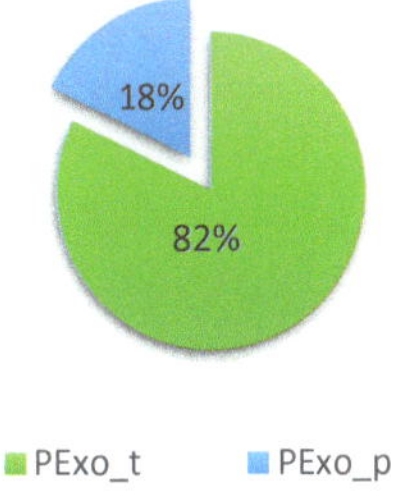

Figure 4. Distribution of patents by design (Complete Exoskeleton versus Exoskeleton Subsystems).

Another aspect can be observed by analyzing Tables 4 and 5, namely that the pandemic, in a way, prevented the development of things. Until 2018, we follow an increasing trend in the analysis and implementation of complete exoskeletons and the performance and improvement of research in subsystems. From 2019 we can see, unfortunately, a downward trend. This trend is seen to have affected subsystem research more strongly. From the analyzed patents, the research teams focused on updating entire exoskeleton structures to help patients move in conditions where they could no longer move to specialized centers.

Table 4. Distribution by year of the number of patents according to the time elapsed between application and publication conforms with the Google Patents platform.

Application Date	The Time Elapsed between the Application and Publication of the Patent		Number of PExo_t	Number of PExo_p
	Minimum Period	Maximum Period		
2012	1 year	5 years	5	1
2013	1 year	7 years	15	2
2014	1 year	7 years	11	5
2015	2 years	7 years	20	3
2016	1 year	5 years	50	9
2017	1 year	5 years	32	10
2018	1 year	4 years	41	8
2019	1 year	3 years	29	4
2020	4 months	2 years	9	5
2021	7 months	1 year	4	1

Table 5. Distribution of patents by year of publication.

Year	PExo_c.t	PExo_c.p
2012	3	0
2013	2	3
2014	9	1
2015	9	0
2016	5	4
2017	19	2
2018	34	6
2019	37	10
2020	55	11
2021	48	11
2022	11	2

Regarding publishing patents, we can say that an increasing trend is observed, as shown in Table 4. The data analysis makes us say that this trend was ensured by the conditions imposed by the appearance of the COVID-19 pandemic.

Furthermore, we will resume a previously mentioned aspect, namely that the research, during the pandemic, was directed towards ensuring the movement of patients and, therefore, this research was focused on the entire lower system and not only on its subsystems. It is also easy to see that more patents deal with the total lower exoskeleton than those dealing with the exoskeleton that treats the same leg parts.

The subject of interest is the time interval between the date of application of a patent and the date of its publication. The patents analyzed by us in this platform reveal a fascinating fact. During the pandemic, the time elapsed between the moment of application for a patent and the date when it was published was reduced. Therefore, we can say that more attention has been paid to these patents to implement the obtained results faster.

Further, in Figure 5, you can see the numerical distribution of patent types based on the year of publication. Figure 5 shows two peaks of a substantial increase in the number of patents published on PExoRP in 2018 (32 patents) and 2020 (48 patents). The number of patents published on PExoRNP follows a growth curve with the maximum growth peak in 2021 (12 patents). Patents published on PExoPR have gained more interest since 2019. There has also been an interest in PEoxS since 2013 (one patent) continuing to grow timidly in 2021 (three patents). We can say that patents published on PEoxRP, compared to other types, predominate throughout the period.

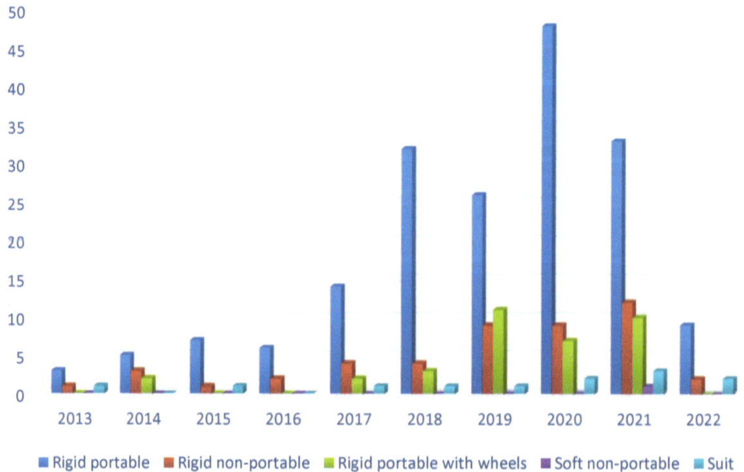

Figure 5. Distribution by year of the number of published patents according to the type of structure and portable mode.

3.2. PatentScope Platform

As we did in the previous case, we analyzed the number of applications and the number of patent publications on this platform. The distribution by region according to the year of application of the patent after the use of the filters 5 and 7 can be seen in Figure 6. The highest percentage of patent applications is registered in South America.

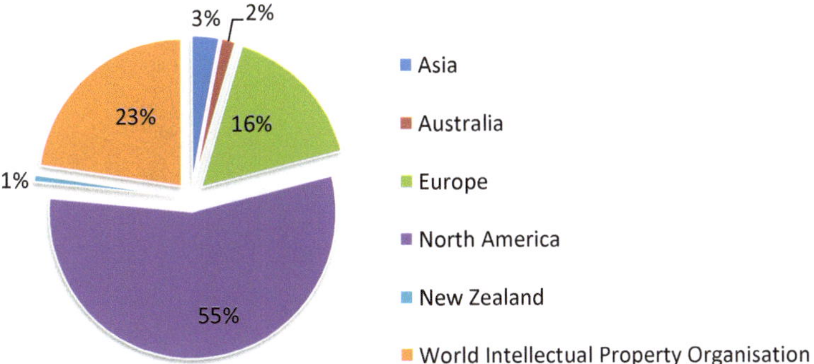

Figure 6. Distribution of patent applications by region after applying filters 5 and 7.

As can be seen from Table 6, the period in which we find information is longer. It includes requests since the early 2000s. As in the case of the analysis in the previous subsection, due to the minimal number of applications/publications, we can conclude that the valuable research period can be considered as starting with the year 2012. Hence, in this case, the data distribution confirms that the period in which we did the analysis is correct and fair.

Table 6. The number of patent applications and publications distributed per year.

Year	Application Date	Publication Date
2000	3	0
2001	0	2
2002	0	0
2003	1	1
2004	0	1
2005	1	0
2006	2	3
2007	0	0
2008	6	3
2009	4	1
2010	0	3
2011	0	2
2012	4	1
2013	16	7
2014	12	6
2015	15	13
2016	14	21
2017	20	10
2018	10	15
2019	7	14
2020	5	9
2021	6	16
2022	0	0

Going further with the data analysis, for the period considered in our study to be the reference period, we find some ideas previously stated in Section 3.1. In Figure 7, we can see the direction of declining patent applications as the pandemic begins. In terms of patent approval, an upward trend can be seen. Following these findings, we maintain the idea stated at the beginning of the study, namely that the pandemic negatively affected the research teams and their results. However, we can also say that the publishing process has shown an upward trend.

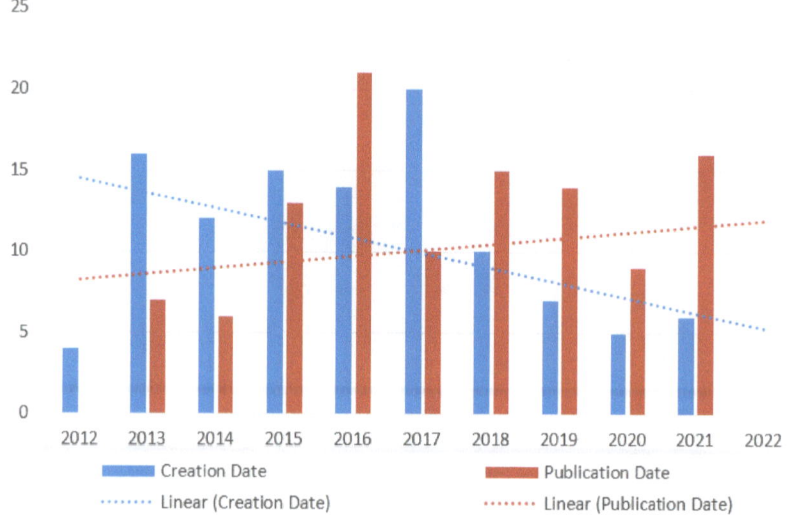

Figure 7. Distribution of patents in the period 2012–2022.

From 2012 to 2022, 74% treated the total lower system, and 26% focused on different subsystems of the lower limb. This aspect can also be seen in Figure 8.

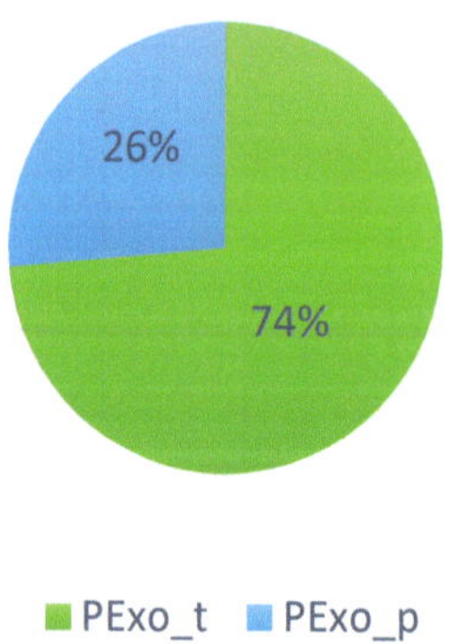

Figure 8. Distribution of patents in the period 2012–2022.

Next, for a more in-depth study, we looked at the problem that these exoskeletons treat. We found that most research that debates and applies issues of the entire lower patient system has been preserved, as shown in Figures 9 and 10.

As we did in the case of the analysis of the Google Patents Platform, we also determined for this platform the time intervals between the date of filing the patent and the date of its publication. The results obtained can be seen in Table 7 It is also observed that during the COVID-19 pandemic, the maximum interval decreased.

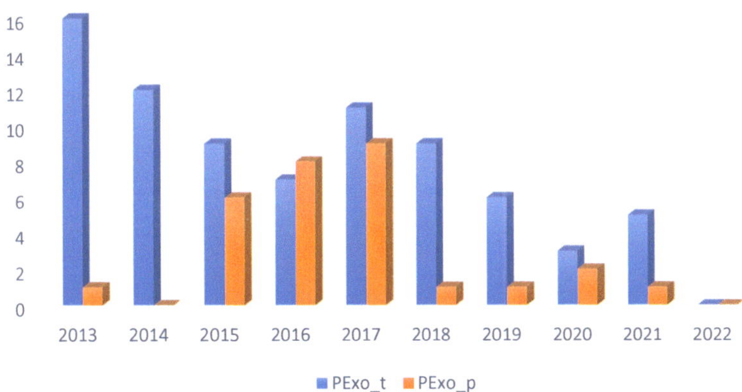

Figure 9. Distribution of patents by year of application and the problem that these exoskeletons deal with.

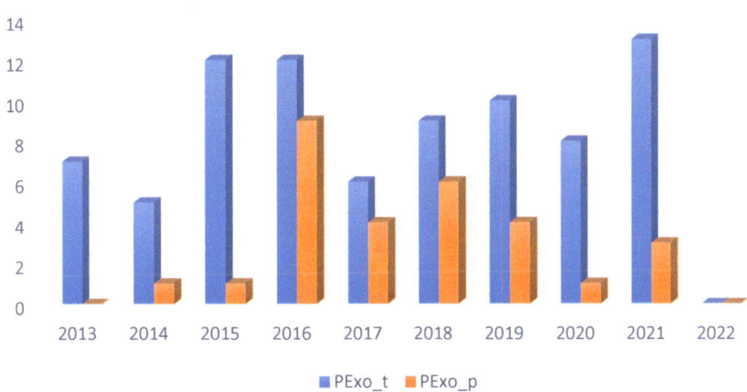

Figure 10. Distribution of patents by year of publication and the problem that these exoskeletons deal with.

Table 7. Distribution by year of the number of patents according to the time elapsed between application and publication conforms with the PatentScope platform.

Application Date	The Time Elapsed between the Application and Publication of the Patent		Number of PExo_t	Number of PExo_p
	Minimum Period	Maximum Period		
2012	1 year	2 years	4	0
2013	7 months	8 years	15	1
2014	7 months	7 years	12	0
2015	8 months	3 years	9	5
2016	6 months	5 years	7	8
2017	6 months	4 years	11	9
2018	4 months	2 years	9	1
2019	6 months	2 years	6	1
2020	6 months	1 year	3	2
2021	7 months	10 months	5	1

Figure 11 shows the following: a fluctuation in the number of patents published on PExoRP, in the sense that their number increased significantly in 2016, decreased in 2017, increased in 2018, and then fell and increased again in 2021. As for the number of patents published on PExoRNP, it seems to keep the same trend of increase and decrease every 2 years $(+/-1)$ until 2018, and then there will be a more considerable difference in the period 2019–2020 (+3), and 2021–2022 (−2). Patents published on PExoPR gain interest only in 2019 and then decrease. There is also an interest in PEoxS since 2015, continuing with growth peaks in 2019 and 2021. Overall, we can say that patents published on PEoxRP predominate throughout the period compared to other types.

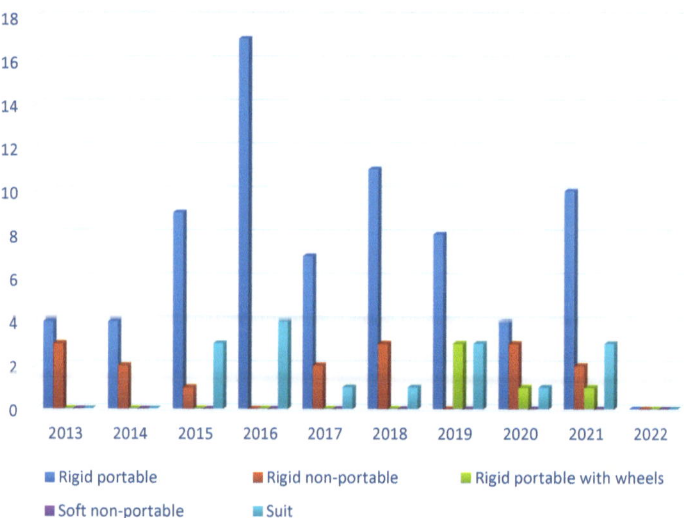

Figure 11. Distribution by year of the number of published patents according to the type of structure and portable mode.

3.3. Lens Platform

Our analysis on this platform is presented in Section 2.1.3 and presented in Figure 12. However, due to the small number of patents obtained because of the consequences of filter use, we consider it unnecessary to detail further. Here, it should be mentioned that no patents have been found on this platform with the application date being 2020–2022 and complying with the filtering criteria considered.

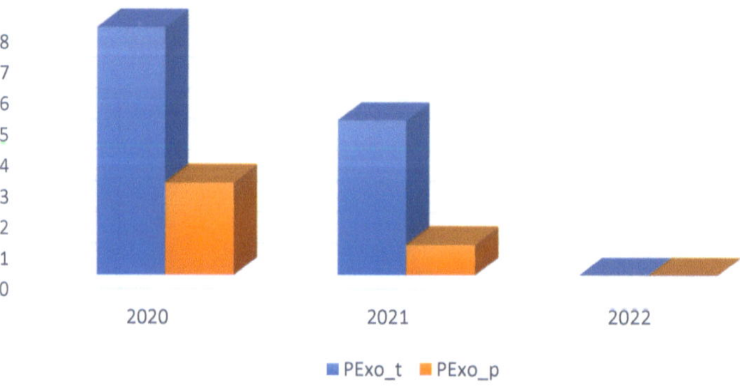

Figure 12. Distribution of patents by year of publication and the problem that these exoskeletons deal with.

According to Figure 12, most published patents dealt with the total foot.

In Figure 13, you can see the numerical distribution of patent types based on the year of publication. For example, according to Figure 13, there is a decrease in published patents. Their distribution by structure and applicability is as follows: in 2020, eight PExoRP, two PExoRP and one PExoS; in 2021, five PExoRP and one PExoRP, and in 2022 it is completely absent.

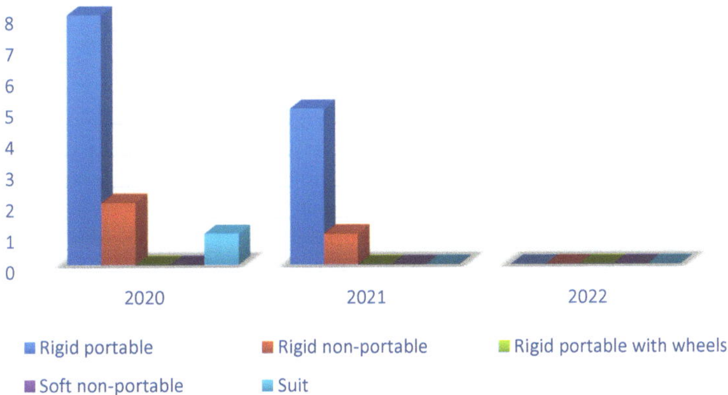

Figure 13. Distribution of patents by year of publication, the type of structure, and portable mode.

4. Discussion

Lower-limb exoskeletal robotic systems integrate advanced mechanical structures, materials, electronics, bionics, control, and even artificial intelligence. In the last decade, the progress in their development has been remarkable. Significant improvements have been made in performance and design. This thing is also clear from the reviews that have been written on this topic. Over time, several studies have looked at the exoskeletons of the lower limbs for rehabilitation. Some of these reviews have focused on reviewing general aspects of exoskeleton technology [2,20–23]; others have focused on more specific issues, such as control strategies [19,24] or joint design [25]. In [20], Meda-Gutiérrez and the team aim to identify state of the art medical device designs, based on an analysis of patents and literature. Although they encountered some difficulties in processing records due to a lack of filters and standardization of names (discrepancies appearing between search engines), the conclusion obtained from the study reflects a tendency to use the mechanical design of exoskeletons based on rigid structures, joints, and elements that provide strength for the movement of the system.

Based on our research in the literature at the time of this study, there have been no reviews of the COVID-19 pandemic patent study results. The only review [26] found in the literature studied the influence of the COVID-19 pandemic on publishing research articles. In this paper, Aviv-Reuven and Rosenfeld analyzed changes in biomedical publication patterns due to the pandemic.

4.1. The Context of the Main Objective Analysis

The COVID-19 pandemic posed an unprecedented challenge to humanity and science. As a result of the new outbreaks of Coronavirus, a state of emergency was declared in almost all countries on 13 March 2020. At the same time, all national institutions closed their activities with the public, with employees being forced to work from home through video conferencing, telephone, or online platforms. This has also happened with national and international patent or trademark registration offices. Moreover, starting on 13 March, these offices announced some different forms of exemption available to patent and trademark applicants. These exemption forms can be divided into two broad categories: term extensions and tax exemptions. For example, in the US, extensions were not granted automatically but were conditional on the actual existence of the COVID-19 outbreak. In addition, for an extension to be valid, the party requesting the exemption had to submit a statement that the delay in filing or payment was due to the COVID-19 outbreak. The applicant also had to be a micro or small institution that was in the process of processing or paying pre-examination maintenance fees.

4.2. General Discussion on the Results of the Analysis of the Patent Platforms

A pertinent observation, we would say, would be that the pre-pandemic period was a perfect time for researchers/research teams. They have been able to develop different relationships and have been able to carry out their activity without limitations. Therefore, the results obtained have led to many patent applications. However, this large volume of applications probably hindered the approval process and hence the time differences between the application and approval periods (a statement from our study). With the onset of the COVID-19 pandemic, research teams have been limited in their interactions physically (at least) and hence there have been fewer patent applications. However, this was a "plus" (if we can say so) for the patent application evaluation teams having, on the one hand, a small number of applications to analyze and sufficiently thorough time. Another aspect that we noticed was that during the pandemic, the difference between the date of the filing of the patent application and its publication decreased a lot. Despite introducing tax exemptions, the number of patents granted is not high due to the introduction of term extensions for regions with a COVID-19 outbreak.

If we analyze the data in Table 7 versus the data in Table 6 that deal with the same issue, we can see the effects of processing a large volume of data. The publication period on the PatentScope platform is significantly shorter in most cases; this may be due to prioritizing patent applications.

Another predominant general idea that emerged from the analysis of the mentioned platforms is that the number of patents published during the entire analyzed period (as well as during the pandemic) on rigid and portable structured exoskeletons is much higher than the rest of the types. This fact is essential and beneficial in gaining the independence of users from hospital spaces in particular and the physical presence of specialists specialized in recovery, especially during the pandemic.

Another interesting aspect that emerges from the Google Patents platform is that although the COVID-19 pandemic had as its starting point in China, the most patent applications have been filed in this country. On the other hand, according to data extracted from the PatentScope platform, the country with the most patent applications during the pandemic was the US. According to Lens platform, no patent application has been filed during this period.

Over the last decade, research on real-life exoskeletons has grown significantly; this can be estimated from the number of patents reported in Section 3. The main reason would be that the lifestyle of modern society is constantly growing and developing rapidly, and people always want to stay active, independent, and live a quality life. Portable exoskeleton technology (whether rigid or suitable) is the key to providing individualized mobility recovery solutions for millions of people immobilized from various causes (injuries, trauma, disabilities, or ageing) to continue their desired activities. Despite the above, the technology is in its infancy. There are still factors that are not/are less addressed about obtaining a fully efficient exoskeleton, in terms of performance and cost.

To see the chronological evolution of the development of this technology, we arbitrarily analyzed some of the patents on the PatentScope platform, selecting the patents published at the beginning, middle, and end of the analyzed period.

The first patents published since 2013, as presented, are designed to reinforce joints. After that, the patents designed for this specific domain gradually became more complex and complementary. From this point forward, several improvements regarding gait rehabilitation have been made, focusing on enhancing previous studies and patents. For example, suppose the first patent focused on knee joint reinforcement. In that case, the following patent is designed to adjust pelvic movement with the same objective: gait training to help people have a more natural walk movement.

The first patent focuses on knee joint reinforcement, followed by concepts designed to adjust pelvic movement with the same objective: gait training to help people have a more natural walk movement.

The Active Knee Rehabilitation Orthotic System (ANdROS) [27] is intended to assist everyday tasks rather than gait retraining. The invention describes wearable and portable assistive equipment for gait neuro-rehabilitation that addresses primary gait abnormalities by reinforcing around the knee joint a preferred gait pattern via correcting torque fields applied. The patent is a wearable assistive device for gait rehabilitation in patients who have lost motor control due to a neurological condition, such as a stroke.

A motorized footplate with force feedback and an exoskeleton to regulate pelvic movement during gait training [28] makes this innovation perfect for implementing ecologically sound procedures that accurately simulate real-life settings and maximize motor improvements in stroke patients. In addition, to optimize rehabilitation effects, an active pelvic prosthesis with a non-treadmill pedal system is necessary.

The following published patents were born from ideas intended to improve mobility further. Therefore, exosuit-based systems with actuators were developed to provide active support for the ankle. Fixed in three points on the leg and created from a lightweight material, it shows comfort for the user and improves torque.

This patent [29] discusses three-point contact with the leg and a series of elastic elements for improved torque control. Apart from their small size (0.88 kg), these devices can monitor joint angle and torque. A combination of proportional feedback and suspension injection is employed during walking trials to modulate torque. The closed-loop torque control of the exoskeleton devices was evaluated by controlling 50 N-m and 20 N-m linear chirps in intended torque while using exoskeletons and measuring bandwidths more significant than 16 Hz and 21 Hz, respectively. There was a maximum torque of 120 N-m and a tracking error of 2.0 N-m. These testbeds demonstrate how exoskeletons may be employed to study a variety of control and support paradigms rapidly. The document provides an exoskeleton system that comprises a cable, a lever attached to the cable, a frame having a strut directing the cable to the lever, and a motor coupled to the cable and adapted to cause the cable to deliver torque around the rotating joint.

In addition to previous innovations, it was considered the proper time to introduce smart materials that will block or even actuate joints to expand the possibilities of implementation for wearable robotic devices. However, smart material-based exoskeleton publications are found in a minimal number and provide a substantial implementation complexity; therefore, such a type of patent is to be considered.

The work [30] aims to offer an exoskeletal device with semi-active joints that can lower articular tensions caused by limb weight, reduce physical labor load, and promote physical workouts for motor ability rehabilitation following cerebral or orthopedic accidents. In addition, other objectives are realized via a mechanism, namely, an exoskeleton with at minimum two links joined by an electro- or magneto-rheological fluid joint.

Further improvements of the exoskeletons are being developed concerning the user's biological skeleton and flexible materials to deliver improved customized mobility according to the user's needs.

With a soft exosuit system and an actuator system, the concept [31] provides active support for natural actions, such as normal leg movement. The soft exosuit uses flexible materials and actuators and depends on the user's biological skeleton to aid in applying forces and transferring loads, unlike previous art-rigid exoskeletons.

It has lower mechanical impedance and kinematic constraints than rigid exoskeletons and does not considerably limit or restrict the user's flexibility. Furthermore, by adding regulated energy impulses rather than the direct command of limb position(s), this system can help improve locomotion and minimize the metabolic cost of movement without limiting mobility.

Other powered devices and control methods [32] can greatly aid ambulation, especially in youngsters with cerebral palsy (CP). Less complex knee extension help was tested in the stance and late swing phase. Compared to the baseline condition, the tested gadget reduced crouching. Moreover, it considerably changed lower extremity kinematics, increasing maximal knee extension in both the left and right legs during the stance phase.

4.3. Limitations of This Study

The main limitation of this study is that these platforms do not declare patent applications unless they have been approved and published. Thus, for the period of the COVID-19 pandemic, the analysis of the number of patent applications was performed only on the data extracted from these platforms. However, the primary purpose of this review was to provide a comprehensive overview of the effort to innovate the lower-limb exoskeletons even during this pandemic.

5. Conclusions

This study looked at how the COVID-19 pandemic affected the publication and application of new patents based on data extracted from three platforms: Google Patents, PatentScope and Lens. We used two types of analysis to address our research questions: short-term analysis (pandemic period 2020–2022) and longer-term analysis (2012–2022). Following the analysis made, the hypothesis issued by us at the beginning of the study that the pandemic has caused a reduction in the volume of new applications and, possibly, the publications, is valid. However, only on the data extracted from these platforms did the number of new patent applications drop dramatically. The number of published patents has been much higher than the number of applications. Therefore, we can say that the COVID-19 pandemic has negatively affected the number of applications for new patents in the field of exoskeletal robotic systems of the lower limbs and in a positive sense (at least kept the trend of recent years) affected the number of patents published (granted).

The main challenge of the paper was to conduct an in-depth analysis of the systems and current patents to summarize their need for development in terms of the benefits of exoskeletal rehabilitation technology. In addition, through its scientometric analysis of patents, this paper wanted to be state of the art and offer/act as a reference point for scientists and researchers. Who wants to develop systems to meet the needs of millions of people with locomotor problems?

In the future, we want to carry out new analysis in which we will use delimiters related to the materials of the components and the actuation and control systems of the exoskeleton robots. Moreover, following the research conducted in the literature, there is a need to perform a qualitative comparative study on the ease of using these robotic exoskeleton systems by patients. The lack of such a study could lead to some biases, especially related to the clinical efficacy of lower-limb exoskeletons for rehabilitation.

Another interesting study that we would like to do is analysis, such as the one made in this review, but post-pandemic.

Author Contributions: Conceptualization, C.F.P., V.M.R. and D.M.P.-P.; methodology, C.F.P. and V.M.R.; validation, C.F.P., F.L.P. and D.M.P.-P.; formal analysis, C.F.P. and V.M.R.; investigation, C.F.P., F.L.P., D.M.P.-P. and I.C.R.; resources, F.L.P., Ș.I.C. and A.T.; data curation, F.L.P., I.C.R., Ș.I.C. and A.T.; writing—original draft preparation, C.F.P., V.M.R. and F.L.P.; writing—review and editing, C.F.P., V.M.R. and F.L.P.; supervision, C.F.P. and N.B.; project administration, C.F.P. and V.M.R.; funding acquisition, N.B. All authors have read and agreed to the published version of the manuscript.

Funding: This research was funded by European Social Fund within the Sectorial Operational Program Human Capital 2014–2020, grant number POCU380/6/13/123990, National Council for the Financing of Higher Education, grant number CNFIS-FDI-2022-0468 and POC-Competitiveness Operational Program.

Institutional Review Board Statement: Not applicable, because this review did not involve humans or animals.

Informed Consent Statement: Not applicable, because this review did not involve humans or animals.

Data Availability Statement: Not applicable.

Acknowledgments: This work was supported by the grant POCU380/6/13/123990, "Entrepreneurial University—higher education and training system for the Romanian labor market by awarding scholarships for doctoral students and postdoctoral researchers and implementing innovative entrepreneurship training programs", co-financed by the European Social Fund within the Sectorial Operational Program Human Capital 2014–2020 and grant "Increasing the quality of the educational act by modernizing the research infrastructure of the University of Craiova", cod CNFIS-FDI-2022-0468 and HUB-UCv-Support Center for International RD Projects for the Oltenia region-cod SMIS 107885.

Conflicts of Interest: The authors declare no conflict of interest.

Abbreviations

The following abbreviations used by us in this manuscript are:

PExo_t	*Patent of the total (treats the whole leg) exoskeleton robot system of the lower limb*
PExo_p	*Patent of the partial (treats a specific part of the foot) exoskeleton robot system of the lower limb*
PExoRP	*Patent involving exoskeleton with rigid structure and is portable*
PExoRPW	*Patent involving rigid structure exoskeleton, is portable and has a pair/pairs of auxiliary wheels*
PExoRNP	*Patent involving exoskeleton with rigid structure and is non-portable*
PexoS	*Patent involving exoskeleton suit*

Appendix A

Table A1. Google Patents platform.

Cite	Country Code	Title	Type of Treatment	Type of Structure	Application Date	Publication Date
[33]	US	Gait device with a crutch	*PExo_t*	Rigid portable	17 January 2013	27 November 2018
[34]	US	Patient aid devices, particularly for mobile upper extremity support in railed devices such as parallel bars and treadmills	*PExo_t*	Rigid portable	23 October 2017	14 April 2020
[35]	US	Recognition method of human walking speed intention from surface electromyogram signals of plantar flexor and walking speed control method of a lower-limb exoskeleton robot	*PExo_t*	Rigid non-portable	20 January 2016	08 October 2019
[36]	KR	Walk assist apparatus	*PExo_t*	Rigid portable	02 April 2012	03 March 2014
[37]	KR	Wearable crutch with lap joints	*PExo_t*	Rigid portable	20 August 2013	23 June 2015
[38]	US	Interactive exoskeleton robotic knee system	*PExo_p*	Rigid portable	21 June 2015	27 August 2019
[39]	KR	Active type step assistance apparatus	*PExo_t*	Rigid portable	09 July2013	12 December 2014
[40]	US	Low-profile exoskeleton	*PExo_t*	Rigid portable	05 November 2015	18 February 2020
[41]	US	Multi-function lower-limb ambulation rehabilitation and walking assist device	*PExo_t*	Rigid non-portable	29 September 2016	17 October 2017
[42]	CN	For aiding in the soft machine armor of human motion	*PExo_t*	Rigid portable	30 May 2014	04 May 2018
[43]	CN	Soft exterior protector for aiding in human motion	*PExo_t*	Suit	17 September 2013	09 June 2017
[44]	KR	Gait assistant robot	*PExo_t*	Rigid portable with wheels	08 November 2013	13 September 2016
[45]	KR	Robot for assisting user to walk	*PExo_t*	Rigid portable with wheels	10 May 2017	20 May 2019

Table A1. *Cont.*

Cite	Country Code	Title	Type of Treatment	Type of Structure	Application Date	Publication Date
[46]	ES	System to assist walk	*PExo_t*	Rigid portable	19 November 2015	06 April 2018
[47]	KR	Robot for assisting user to walk with lower body exoskeleton	*PExo_t*	Rigid portable with wheels	25 March 2016	21 June 2018
[48]	CN	Rehabilitation type lower-limb exoskeleton	*PExo_t*	Rigid portable	07 August 2018	13 March 2020
[49]	US	Exoskeleton ankle robot	*PExo_p*	Rigid portable	21 June 2015	01 October 2019
[50]	US	Mobility system including an exoskeleton assembly releasably supported on a wheeled base	*PExo_t*	Rigid portable with wheels	01 November 2013	26 December 2017
[51]	KR	Rehabilitation robot of legs, boarding and driving method thereof	*PExo_t*	Rigid portable with wheels	29 June 2018	12 March 2019
[52]	ES	Support structure	*PExo_t*	Rigid portable with wheels	18 February 2016	08 May 2020
[53]	CN	Wearable lower-limb exoskeleton driven by lasso artificial muscles	*PExo_t*	Rigid portable	26 July 2017	03 November 2020
[54]	ES	Exoskeleton for human movement assistance	*PExo_t*	Rigid portable	27 November 2014	06 April 2017
[55]	CN	A kind of lower-limb rehabilitation exoskeleton system and its walking control method	*PExo_t*	Rigid portable	15 January 2018	09 July 2019
[56]	KR	Wearable apparatus for support holding posture	*PExo_t*	Rigid portable	22 April 2016	04 September 2017
[57]	US	Pneumatic lower extremity gait rehabilitation training system	*PExo_p*	Rigid non-portable	12 September 2016	21 May 2019
[58]	US	Systems, methods, and devices for assisting walking for developmentally delayed toddlers	*PExo_t*	Rigid portable	05 February 2015	07 May 2019

Table A1. *Cont.*

Cite	Country Code	Title	Type of Treatment	Type of Structure	Application Date	Publication Date
[59]	CN	Variable-rigidity flexible driver for exoskeleton type lower-limb rehabilitation robot	*PExo_t*	Rigid non-portable	02 February 2018	06 December 2019
[60]	US	Actuation system for a joint	*PExo_p*	Rigid portable	13 March 2014	22 November 2016
[61]	CN	Lower-limb rehabilitation training exoskeleton system and its walking control method and hip joint structure	*PExo_t*	Rigid portable	28 January 2018	20 September 2019
[62]	KR	Weight bearing brace	*PExo_t*	Rigid non-portable	12 December 2017	08 July 2019
[63]	KR	Training system for leg rehabilitation having separated treadmill	*PExo_t*	Rigid non-portable	11 October 2011	17 June 2013
[64]	CN	Wearable lower-limb exoskeleton rehabilitation robot	*PExo_t*	Rigid portable	16 November 2017	21 February 2020
[65]	US	Lower extremity exoskeleton for gait retraining	*PExo_t*	Rigid portable	28 September 2012	01 December 2015
[66]	CN	A kind of bionical lower-limb exoskeleton robot driven based on rope	*PExo_t*	Rigid portable	06 March 2017	02 August 2019
[67]	CN	Movable parallel flexible cable driven lower-limb rehabilitation robot and implementation method thereof	*PExo_t*	Rigid portable with wheels	02 May 2018	18 February 2020
[68]	CN	Wearable lower-limb exoskeleton robot	*PExo_t*	Rigid portable	31 August 2016	14 August 2018
[69]	EP	Robotic device for assistance and rehabilitation of lower limbs	*PExo_t*	Rigid portable	07 October2013	21 December 2016
[70]	US	Admittance shaping controller for exoskeleton assistance of the lower extremities	*PExo_t*	Rigid portable	25 June 2015	06 March 2018
[71]	CN	Exoskeleton-type moves walking rehabilitation training device and method	*PExo_t*	Rigid portable with wheels	28 December 2015	01 February 2019
[72]	CN	Based on rope-pulley mechanism drive lacking lower-limb assistance exoskeleton robot	*PExo_t*	Rigid non-portable	03 July2017	17 May 2019

Table A1. *Cont.*

Cite	Country Code	Title	Type of Treatment	Type of Structure	Application Date	Publication Date
[73]	EP	Method for estimating posture of robotic walking aid	*PExo_t*	Rigid portable	21 July2016	17 March 2021
[74]	US	Hybrid terrain-adaptive lower-extremity systems	*PExo_p*	Rigid portable	01 September 2009	16 April 2013
[75]	CN	A kind of dedicated power-assisted healing robot of single lower-limb individuals with disabilities	*PExo_t*	Rigid non-portable	24 July 2017	27 September 2019
[76]	JP	Actuator device, power assist robot and humanoid robot	*PExo_t*	Rigid portable	12 September 2014	20 April 2016
[77]	KR	Monitoring system of walking balance for lower-limb rehabilitation	*PExo_t*	Rigid non-portable	29 December 2017	26 August 2019
[78]	US	Powered lower-limb devices and methods of control thereof	*PExo_t*	Rigid portable	03 November 2017	07 December 2021
[79]	CN	A kind of artificial intelligence motion's auxiliary equipment	*PExo_t*	Rigid non-portable	29October 2015	29 March 2017
[80]	JP	Actuator device, humanoid robot and power assist device	*PExo_t*	Rigid portable	26 December 2014	30 November 2016
[81]	US	Robotic system for simulating a wearable device and method of use	*PExo_t*	Rigid non-portable	20 December 2012	22 November 2016
[82]	US	Passive swing assist leg exoskeleton	*PExo_t*	Rigid non-portable	04 April 2008	02 December 2014
[83]	EP	Forward or rearward oriented exoskeleton	*PExo_t*	Rigid portable	06 May 2015	19 January 2022
[84]	US	Apparatus and system for limb rehabilitation	*PExo_t*	Rigid portable with wheels	29 November 2018	03 December 2019
[85]	ES	Movement assist device	*PExo_t*	Rigid portable	17 June 2013	15 July 2020
[86]	CN	Sitting type walking rehabilitation robot	*PExo_t*	Rigid non-portable	28 October 2015	12 January 2021
[87]	EP	Human movement research, therapeutic, and diagnostic devices, methods, and systems	*PExo_t*	Rigid non-portable	21 April 2015	03 July 2019

Table A1. *Cont.*

Cite	Country Code	Title	Type of Treatment	Type of Structure	Application Date	Publication Date
[88]	CN	Hip joint rehabilitation exoskeleton based on multifunctional driver and motion control method thereof	*PExo_t*	Rigid portable	16 October 2018	23 February 2021
[89]	US	Wearable robot and control method thereof	*PExo_t*	Rigid portable	02 December 2014	12 February 2019
[90]	US	Wearable robot and method for controlling the same	*PExo_t*	Rigid portable	03 September 2014	11 February 2020
[91]	CN	A kind of link-type lower-limb exoskeleton rehabilitation robot	*PExo_t*	Rigid portable	22 May 2017	26 June 2018
[92]	JP	Exoskeleton robot	*PExo_t*	Rigid portable	13 April 2018	29 July 2020
[93]	CN	Interface for the movement by externally applied force motivation of adjustment orthopedic appliance	*PExo_t*	Rigid portable	15 January 2014	09 October 2018
[94]	US	Method and apparatus for providing deficit-adjusted adaptive assistance during movement phases of an impaired joint	*PExo_p*	Rigid portable	20 November 2014	17 April 2018
[95]	CN	Walking aid	*PExo_t*	Rigid portable	29 November 2013	11 September 2020
[96]	CN	A kind of quasi-passive knee ankle-joint coupling lower-limb exoskeleton and its control method	*PExo_p*	Rigid portable	31 March 2017	18 June 2019
[97]	CA	Method and apparatus for providing economical, portable deficit-adjusted adaptive assistance during movement phases of an impaired ankle	*PExo_p*	Rigid portable	20 June 2016	24 March 2020
[98]	EP	Ball screw and tensile member exoskeleton joint actuation device	*PExo_t*	Rigid portable	04 May 2017	16 February 2022
[99]	KR	Robot for assisting user to walk	*PExo_t*	Rigid portable with wheels	12 June 2017	12 June 2019
[100]	CN	Reconfigurable ectoskeleton	*PExo_t*	Rigid portable	11 December 2013	29 September 2017
[101]	US	Gait rehabilitation methods and apparatuses	*PExo_t*	Rigid non-portable	04 February 2005	18 November 2014

Table A1. *Cont.*

Cite	Country Code	Title	Type of Treatment	Type of Structure	Application Date	Publication Date
[102]	US	Methods of operating an exoskeleton for gait assistance and rehabilitation	*PExo_t*	Rigid portable	30 July 2012	17 January 2017
[103]	US	Powered orthosis	*PExo_t*	Rigid non-portable	04 April 2008	03 April 2012
[104]	KR	A treatment device for hemiplegia	*PExo_t*	Rigid portable	25 July 2013	10 April 2015
[105]	US	Wearable robotic device	*PExo_t*	Rigid portable	21 March 2019	12 October 2021
[106]	EP	An exoskeleton and method for controlling a swing leg of the exoskeleton	*PExo_t*	Rigid portable	23 July 2009	04 September 2019
[107]	KR	Wearing tool for measuring biological signal, and wearing-type motion assisting device	*PExo_t*	Suit	10 September 2009	24 January 2013
[108]	JP	Wheelchair walking assist robot	*PExo_t*	Rigid portable with wheels	09 October 2009	29 January 2014
[109]	CN	Reduction exoskeleton joint and exoskeleton power assisting device thereof	*PExo_t*	Rigid portable	03 May 2016	14 April 2020
[110]	US	Cable driven joint actuator and method	*PExo_t*	Rigid portable	15 January 2015	21 March 2017
[111]	US	Orthopedic device including protruding members	*PExo_p*	Rigid portable	10 April 2015	15 December 2020
[112]	CN	A kind of passive exoskeleton device of hip joint based on energy timesharing regulation	*PExo_t*	Rigid portable	05 December 2017	22 November 2019
[113]	US	Methods of enhancing the rehabilitation or training of an exoskeleton wearer	*PExo_t*	Rigid portable	11 November 2015	03 March 2020
[114]	KR	Robot for assisting user to walk	*PExo_t*	Rigid portable with wheels	10 May 2017	20 May 2019
[115]	EP	Leg support device	*PExo_t*	Rigid portable	21 June 2010	18 February 2015

Table A1. *Cont.*

Cite	Country Code	Title	Type of Treatment	Type of Structure	Application Date	Publication Date
[116]	US	Systems and methods for assistive exosuit system	*PExo_t*	Suit	23 August 2017	23 February 2021
[117]	ES	Device and method for reducing a person's oxygen consumption during a regular walk by using a load-bearing exoskeleton	*PExo_t*	Suit	19 May 2009	22 October 2015
[118]	CN	Device and method for decreasing energy consumption of a person by use of a lower extremity exoskeleton	*PExo_t*	Rigid portable	19 May 2009	18 March 2015
[119]	JP	Walking robot system that regenerates energy	*PExo_t*	Rigid portable	27 March 2018	01 September 2021
[120]	CN	It is single to drive bionical gait rehabilitation training robot system	*PExo_t*	Rigid portable	06 October 2016	30 November 2018
[121]	US	Control logic for biomimetic joint actuators	*PExo_t*	Rigid non-portable	28 August 2009	20 May 2014
[122]	KR	Legged robotic device utilizing modifiable linkage mechanism	*PExo_t*	Rigid portable	06 May 2015	13 September 2017
[123]	CN	Lower-limb exoskeleton robot system based on man–machine terminal interaction	*PExo_t*	Rigid portable	23 October 2018	13 October 2020
[124]	US	Systems for neural bridging of the nervous system	*PExo_t*	Rigid non-portable	02 June 2016	08 December 2020
[125]	CN	Use the lower-limb exoskeleton robot control method of air bag sensor	*PExo_t*	Rigid portable	23 February 2016	06 April 2018
[126]	CA	Control system and device for patient assist	*PExo_t*	Rigid non-portable	28 March 2013	09 October 2018
[127]	ES	Semi-motorized exoskeleton of the lower extremities	*PExo_t*	Rigid portable	13 April 2006	05 September 2014
[128]	CN	Contact displacement actuator system	*PExo_t*	Rigid non-portable	17 July 2007	28 January 2015
[129]	EP	System for controlling a robotic device during walking, in particular for rehabilitation purposes, and corresponding robotic device	*PExo_t*	Rigid portable	09 March 2011	22 October 2014
[130]	KR	Robot for assisting user to walk	*PExo_t*	Rigid portable with wheels	10 May 2017	03 December 2021

Table A1. *Cont.*

Cite	Country Code	Title	Type of Treatment	Type of Structure	Application Date	Publication Date
[131]	CN	Lower-limb rehabilitation walking-aid robot supporting omnidirectional movement and control method	*PExo_t*	Rigid portable with wheels	04 March 2016	24 March 2020
[132]	US	Apparatus and method for reduced-gravity simulation	*PExo_t*	Rigid non-portable	18 June 2009	10 April 2012
[133]	ES	Powered orthopedic system for cooperative above-ground rehabilitation	*PExo_t*	Rigid portable	13 March 2014	16 December 2021
[134]	CN	Walking stick type autonomous falling protection rehabilitation walking-aid robot	*PExo_t*	Rigid portable with wheels	02 August 2017	14 July 2020
[135]	CN	The ectoskeleton wheelchair integrated mobile auxiliary robot of telescopic	*PExo_t*	Rigid portable with wheels	29 November 2016	20 April 2018
[136]	CN	A kind of link joint integrated hydraulic driving ectoskeleton	*PExo_t*	Rigid portable	01 March 2016	20 November 2018
[137]	CN	Collapsible mobile lower-limb exoskeleton	*PExo_t*	Rigid portable	08 January 2016	11 July 2017
[138]	CN	Lower-limb rehabilitation robot based on bidirectional neural interface	*PExo_t*	Rigid non-portable	26 July 2018	30 July 2021
[139]	US	Torque control methods for an exoskeleton device	*PExo_p*	Rigid portable	25 May 2017	11 February 2020
[140]	KR	Sensor system for a user's intention following and walk supporting robot	*PExo_t*	Rigid portable	29 December 2009	06 August 2012
[141]	US	Walking assist robot and control method thereof	*PExo_t*	Rigid portable	03 December 2014	23 May 2017
[142]	CN	Lower-limb exoskeleton system with actively adjustable leg rod length and control method thereof	*PExo_t*	Rigid portable	25 May 2020	14 September 2021
[143]	CN	Variable-rigidity lower-limb exoskeleton power-assisted robot	*PExo_t*	Rigid portable	20 August 2019	17 December 2021

Table A1. *Cont.*

Cite	Country Code	Title	Type of Treatment	Type of Structure	Application Date	Publication Date
[144]	CN	Wearable lower-limb healing robot based on exoskeleton	*PExo_t*	Rigid portable	07 May 2016	23 January 2018
[145]	CN	A kind of passive energy storage foot mechanism for lower-limb assistance exoskeleton	*PExo_t*	Rigid portable	23 February 2016	20 October 2017
[146]	CN	Lower-limb rehabilitation robot movement intention reasoning method	*PExo_t*	Rigid non-portable	22 December 2017	22 October 2021
[147]	CN	Portable ankle joint rehabilitation robot based on active intention control	*PExo_p*	Rigid portable	29 November 2017	01 September 2020
[148]	CN	A kind of passive exoskeleton device of hip knee double jointed based on clutch timesharing regulation	*PExo_p*	Rigid portable	14 September 2018	08 October 2019
[149]	CN	Exoskeleton hybrid control system and method for lower-limb walking aid machine	*PExo_t*	Rigid portable with wheels	11 August 2016	22 May 2020
[150]	US	Reconfigurable ankle exoskeleton device	*PExo_p*	Rigid portable	24 June 2010	05 February 2013
[151]	ES	Active orthosis for the neurological rehabilitation of the movement of the lower limbs, a system comprising said orthosis and a process to put said system into operation	*PExo_t*	Rigid portable	20 February 2013	04 May 2018
[152]	CN	Lower-limb exoskeleton control method and device	*PExo_t*	Rigid non-portable	16 October 2019	04 January 2022
[153]	CN	A kind of wearable flexible lower-limb exoskeleton based on negative pressure rotary pneumatic artificial-muscle	*PExo_t*	Rigid portable	25 April 2018	26 July 2019
[154]	CN	A kind of portable waist hunting gear	*PExo_t*	Rigid portable with wheels	12 October 2015	11 July 2017
[155]	CN	A kind of Unweighting walking rehabilitation training robot	*PExo_t*	Rigid portable with wheels	22 April 2016	16 October 2018
[156]	CN	A kind of wearable flexible knee joint robotic exoskeleton equipment based on gait	*PExo_t*	Rigid portable	01 December 2016	09 April 2019

Table A1. *Cont.*

Cite	Country Code	Title	Type of Treatment	Type of Structure	Application Date	Publication Date
[157]	CN	Semi-automatic bone installations of pulling together	*PExo_p*	Rigid non-portable	21 June 2016	09 March 2018
[158]	RU	Actuation system for hip joint orthosis	*PExo_p*	Rigid portable	08 February 2016	04 December 2019
[159]	CN	Lower-limb exoskeleton heterogeneous knee joint based on parallel elastomers	*PExo_p*	Rigid portable	23 October 2018	15 June 2021
[160]	CN	Flexible body harness	*PExo_t*	Rigid portable	25 February 2016	10 April 2020
[161]	US	Lower-limb training rehabilitation apparatus	*PExo_t*	Rigid non-portable	12 March 2018	09 November 2021
[162]	CN	Motion control method suitable for exoskeleton robot	*PExo_t*	Rigid portable	31 May 2020	25 May 2021
[163]	CN	Anti-falling system based on lower-limb exoskeleton robot	*PExo_t*	Rigid non-portable	19 July 2018	29 December 2020
[164]	CN	A kind of lower-limb exoskeleton robot	*PExo_t*	Rigid portable	17 August 2016	05 September 2017
[165]	CN	A kind of unpowered wearable auxiliary walking servomechanism	*PExo_t*	Rigid portable	01 December 2016	25 December 2018
[166]	CN	A kind of sufficient isomorphism deformation type wheelchair exoskeleton robot of wheel	*PExo_t*	Rigid portable with wheels	22 February 2018	03 December 2019
[167]	CN	Overload slipping mechanism of lower-limb exoskeleton robot	*PExo_t*	Rigid portable	26 April 2019	08 October 2021
[168]	CN	The external bone robot of hemiparalysis recovery type	*PExo_t*	Rigid portable	15 July 2016	12 October 2018
[169]	CN	A kind of exoskeleton robot	*PExo_t*	Rigid portable	21 December 2016	28 December 2018
[170]	JP	Life activity detection device and life activity detection system	*PExo_t*	Suit	24 August 2016	19 August 2020
[171]	KR	Walking assistance apparatus and operation method of the same	*PExo_t*	Rigid portable with wheels	06 December 2017	20 June 2019

Table A1. *Cont.*

Cite	Country Code	Title	Type of Treatment	Type of Structure	Application Date	Publication Date
[172]	CN	Human lower-limb assisting device	*PExo_t*	Rigid portable	30 November 2018	18 September 2020
[173]	JP	Control design framework for resistant exoskeleton	*PExo_t*	Rigid portable	15 July 2016	16 December 2020
[174]	CN	Auxiliary exercise system and lower-limb exoskeleton control method	*PExo_t*	Rigid portable	06 June 2019	18 May 2021
[175]	CN	A kind of the lower-limb exoskeleton training method and system of the triggering of Mental imagery pattern brain–computer interface	*PExo_t*	Rigid non-portable	25 January 2016	03 November 2017
[176]	CN	Human motion intention recognition control device and control method	*PExo_t*	Rigid portable	23 November 2018	25 December 2020
[177]	KR	Ankle module for gait rehabilitation robot	*PExo_p*	Rigid portable	04 November 2016	28 December 2018
[178]	KR	Soft Exosuit for Fall Prevention and Gait Assistance	*PExo_t*	Suit	29 November 2018	30 September 2019
[179]	CN	The ectoskeleton walk help system driven with functional muscle electric stimulation	*PExo_t*	Rigid portable	19 September 2016	08 January 2019
[180]	US	Integrated platform to monitor and analyze individual progress in physical and cognitive tasks	*PExo_t*	Rigid portable	18 July 2016	14 January 2020
[181]	CN	Human motion trend detection device and detection method based on force sensor	*PExo_t*	Rigid non-portable	25 March 2016	28 July 2020
[182]	CN	Exoskeleton robot leg exercise system	*PExo_t*	Rigid portable	31 August 2016	07 September 2018
[183]	CN	Online step generation control system for exoskeleton robot contralateral training	*PExo_t*	Rigid non-portable	09 December 2020	16 April 2021
[184]	CN	A kind of detachable recovery set for lower limbs and control method	*PExo_t*	Rigid portable with wheels	02 December 2016	30 April 2019

Table A1. *Cont.*

Cite	Country Code	Title	Type of Treatment	Type of Structure	Application Date	Publication Date
[185]	EP	Adaptive assistive and/or rehabilitative device and system	*PExo_t*	Rigid portable with wheels	12 February 2019	08 September 2021
[186]	US	Wearable assistive device that efficiently delivers assistive force	*PExo_t*	Rigid portable	22 February 2019	06 October 2020
[187]	JP	Leg straightening device and straightening device	*PExo_t*	Rigid portable	24 June 2015	14 February 2020
[188]	CN	The lower-limb rehabilitation ectoskeleton control system and method that subject dominates	*PExo_t*	Rigid portable	11 August 2016	24 May 2019
[189]	TW	Walking rehabilitation robot system	*PExo_t*	Rigid portable	07 September 2018	01 September 2020
[190]	KR	Ankle assist apparatus	*PExo_p*	Rigid portable	17 November 2017	05 September 2019
[191]	CN	A kind of single rope towards gait and balance rehabilitation training suspends active loss of weight system in midair	*PExo_t*	Rigid non-portable	18 September 2016	10 July 2018
[192]	CN	Convalescence device walking trigger control method based on trunk center of gravity shift	*PExo_t*	Rigid portable	25 November 2015	13 April 2018
[193]	CN	Brain-myoelectricity fusion small-world neural network prediction method for human lower-limb movement	*PExo_t*	Soft non-portable	15 July 2020	07 September 2021
[194]	JP	Assisted rehabilitation training robot	*PExo_t*	Rigid non-portable	28 February 2017	07 August 2019
[195]	KR	Wearable soft exoskeleton apparatus	*PExo_t*	Rigid portable	25 July 2016	04 January 2018
[196]	CN	Walking aid for hemiplegia patients	*PExo_t*	Rigid portable	06 June 2018	30 June 2020
[197]	ES	Drive device for motorized orthosis	*PExo_t*	Rigid portable	12 April 2018	30 September 2021
[198]	KR	Wearable robot and control method for the same	*PExo_t*	Rigid portable	31October 2013	20 August 2020

Appl. Sci. **2022**, *12*, 5393

Table A1. *Cont.*

Cite	Country Code	Title	Type of Treatment	Type of Structure	Application Date	Publication Date
[199]	US	System and device for guiding and detecting motions of 3-DOF rotational target joint	*PExo_p*	Rigid portable	16 January 2017	05 October 2021
[200]	CN	Ankle treatment and exoskeleton measurement device not making contact with ground, capable of being worn and capable of being reconstructed	*PExo_p*	Rigid portable	15 April 2013	25 June 2014
[201]	CN	A kind of unpowered walking power-assisted flexible exoskeleton device	*PExo_t*	Rigid portable	27 December 2017	20 September 2019
[202]	CN	Lower-limb exoskeleton driver	*PExo_t*	Rigid portable	29 August 2019	29 September 2020
[203]	EP	Walking training apparatus and state determination method	*PExo_t*	Rigid non-portable	15 March 2017	30 September 2020
[204]	KR	Shoe module for detecting walking phase, method, gait analysis system and active walking assist device using the same	*PExo_p*	Rigid non-portable	10 November 2014	16 August 2016
[205]	KR	Muscle rehabilitation training method using walking-assistive robot	*PExo_t*	Rigid portable	28 January 2014	01 October 2015
[206]	CN	Method for controlling man-machine interactive motion of lower-limb exoskeleton based on joint stress	*PExo_t*	Rigid portable	26 December 2018	11 February 2022
[207]	CN	Pelvic auxiliary walking rehabilitation training robot	*PExo_p*	Rigid non-portable	21 November 2019	09 November 2021
[208]	CN	Auxiliary standing device and auxiliary standing mechanism	*PExo_t*	Rigid portable with wheels	19 July 2019	28 December 2021
[209]	US	Tendon device for suit type robot for assisting human with physical strength	*PExo_t*	Suit	27 October 2017	13 October 2020
[210]	CN	Pneumatic waist assistance exoskeleton robot	*PExo_p*	Rigid portable	20 April 2018	08 September 2020
[211]	CN	Exoskeleton robot line winding driving hip joint	*PExo_t*	Rigid portable	31 August 2016	14 August 2018
[212]	US	Method and system for control and operation of motorized orthotic exoskeleton joints	*PExo_t*	Rigid portable	01 April 2015	07 May 2019

Table A1. *Cont.*

Cite	Country Code	Title	Type of Treatment	Type of Structure	Application Date	Publication Date
[213]	US	System and method for the regeneration of at least one severed nerve conduit	*PExo_t*	Rigid non-portable	20 July 2018	02 February 2021
[214]	KR	Robot for Assistance Exoskeletal Power	*PExo_t*	Rigid portable with wheels	07 March 2012	18 February 2014
[215]	CN	Sit and stand and go multi-functional motion auxiliary robot	*PExo_t*	Rigid portable with wheels	22 August 2018	01 May 2020
[216]	EP	Wearable assistive device performing protection operation for drive system	*PExo_t*	Rigid portable with wheels	26 February 2019	10 November 2021
[217]	CN	A kind of exoskeleton robot follow-up control device	*PExo_t*	Rigid portable	12 June 2015	31 May 2017
[218]	US	Walking assistance method and apparatuses	*PExo_p*	Rigid portable	04 January 2018	07 September 2021
[219]	US	Methods of exoskeleton communication and control	*PExo_t*	Rigid portable	14 April 2016	30 June 2020
[220]	EP	Foot for a robotic exoskeleton for assisted walking of persons suffering from locomotor disorders	*PExo_p*	Rigid portable	14 July 2016	17 July 2019
[221]	CN	External structure holder device	*PExo_t*	Rigid portable	08 July 2014	09 January 2018
[222]	US	Magneto-rheological series elastic actuator	*PExo_t*	Rigid portable	30 March 2018	21 April 2020
[223]	KR	A knee-orthosis to assist the gait by support the knee-joint	*PExo_p*	Rigid portable	01 February 2012	15 November 2013
[224]	CN	A kind of plantar pressure measuring device and method for ectoskeleton control	*PExo_p*	Rigid portable	25 April 2016	12 October 2018
[225]	US	Exoskeleton device	*PExo_t*	Rigid portable	10 May 2019	15 June 2021

Table A1. *Cont.*

Cite	Country Code	Title	Type of Treatment	Type of Structure	Application Date	Publication Date
[226]	US	Exoskeleton device	*PExo_t*	Rigid portable	10 May 2019	17 August 2021
[227]	CN	Power-source-free knee joint mechanism	*PExo_p*	Rigid portable	12 June 2020	29 September 2020
[228]	CN	A kind of ectoskeleton walk help system driven with functional muscle electric stimulation	*PExo_t*	Rigid portable	19 September 2016	05 April 2019
[229]	KR	Robot for lower limb with multi-link type knee joint and method for controlling the same	*PExo_t*	Rigid portable	25 March 2016	13 February 2018
[230]	CN	A kind of hip joint structure of wearable exoskeleton robot	*PExo_p*	Rigid portable	23 May 2016	21 September 2018
[231]	CN	Variable-rigidity lower-limb exoskeleton robot based on shape–memory alloy	*PExo_t*	Rigid portable	03 August 2020	04 January 2022
[232]	CN	Lower-limb exoskeleton inverse motion analysis method under random road surface condition	*PExo_t*	Rigid portable	11 November 2016	19 May 2020
[233]	CN	A kind of light-duty ankle-joint ectoskeleton	*PExo_p*	Rigid portable	06 January 2017	05 February 2019
[234]	CN	Wearable metatarsophalangeal joint walking power assisting device	*PExo_p*	Rigid portable	22 June 2020	27 August 2021
[235]	CN	Flexible exoskeleton robot assisting movement of hip joint and knee joint	*PExo_p*	Rigid portable	14 June 2018	26 June 2020
[236]	KR	Walking Pattern Training and Intension Analysis System Through Complex Stimulus, and Method thereof	*PExo_t*	Rigid portable	27 June 2017	28 December 2018
[237]	US	Belt for effective wearing and wearable assistive device having the same	*PExo_t*	Rigid portable	14 March 2019	31 March 2020
[238]	RU	Femal link of an active foot orthosis	*PExo_t*	Rigid portable	11 July 2016	25 May 2017
[239]	KR	Gait rehabilitation apparatus	*PExo_t*	Rigid non-portable	30 May 2018	06 July 2020

Table A1. *Cont.*

Cite	County Code	Title	Type of Treatment	Type of Structure	Application Date	Publication Date
[240]	CN	The walking trigger control method of convalescence device based on foot pressure sensor	*PExo_t*	Rigid non-portable	25 November 2015	15 May 2018
[241]	KR	Walking assistance apparatus and operation method of the same	*PExo_t*	Rigid portable with wheels	06 December 2017	28 August 2019
[242]	JP	Joint motion assist device	*PExo_p*	Rigid portable	10 January 2014	29 March 2017
[243]	EP	Walking training apparatus and method of controlling the same	*PExo_t*	Rigid non-portable	15 December 2017	23 September 2020
[244]	CN	Control method of lower-limb exoskeleton robot	*PExo_t*	Rigid portable	28 November 2018	22 December 2020
[245]	CN	Sole human–computer interaction measuring device based on multi-source information fusion	*PExo_p*	Rigid portable	23 October 2018	06 July 2021
[246]	CN	Handrail type intelligent tumble protection walking aid rehabilitation robot	*PExo_t*	Rigid portable with wheels	24 July 2018	04 August 2020
[247]	CN	Support member and the self-adapting seat device with the support member	*PExo_t*	Rigid portable	29 September 2015	31 October 2017
[248]	CN	Lower-limb power assisting device	*PExo_t*	Rigid portable	13 May 2019	08 February 2022
[249]	KR	Walking assistance apparatus and operation method of the same	*PExo_t*	Rigid portable with wheels	06 December 2017	28 August 2019
[250]	CN	A kind of convalescence device speed of travel control method rocked based on trunk	*PExo_t*	Rigid portable	13 November 2015	05 January 2018
[251]	CN	A kind of adaptive ectoskeleton knee joint support plate unlocked	*PExo_t*	Rigid portable	17 August 2016	15 June 2018
[252]	CN	Standing mode control method of exoskeleton mechanical leg rehabilitation system	*PExo_t*	Rigid portable	24 February 2017	22 September 2020

Table A1. *Cont.*

Cite	Country Code	Title	Type of Treatment	Type of Structure	Application Date	Publication Date
[253]	CN	Electromyographic signal collection position choosing method based on complex network	*PExo_t*	Suit	23 March 2016	29 May 2018
[254]	CN	The bionical dynamic knee joint system in the wearable list source of one kind and its control method	*PExo_p*	Rigid portable	29 December 2017	30 August 2019
[255]	EP	Exoskeleton and mounting arrangement	*PExo_t*	Rigid portable	18 December 2018	25 November 2020
[256]	US	Active arm passive leg exercise machine with guided leg movement	*PExo_t*	Rigid non-portable	02 April 2019	10 August 2021
[257]	CA	Self-supported device for guiding motions of a target joint	*PExo_t*	Rigid non-portable	13 May 2019	08 June 2021
[258]	CN	A kind of wearable leg power brace of self-regulation	*PExo_t*	Rigid portable	18 July 2016	12 February 2019
[259]	CN	A wheeled drive self-balancing power ectoskeleton of sole for spinal cord injury patient	*PExo_t*	Rigid portable	12 April 2017	20 March 2020
[260]	EP	Controlling position of wearable assistive device depending on operation mode	*PExo_t*	Rigid portable with wheels	26 February 2019	17 November 2021
[261]	RU	Modular orthopedic apparatus	*PExo_p*	Rigid portable	04 September 2017	28 June 2018
[262]	US	Methods and systems for an exoskeleton to reduce a runners metabolic rate	*PExo_p*	Rigid portable	15 April 2019	04 February 2020
[263]	CN	Telescopic structure and exoskeleton robot with same	*PExo_t*	Rigid portable	14 November 2018	02 February 2021
[264]	EP	Advanced gait control system and methods enabling continuous walking motion of a powered exoskeleton device	*PExo_t*	Rigid portable	23 April 2018	05 January 2022
[265]	US	Exoskeleton system	*PExo_t*	Rigid portable	30 April 2019	07 December 2021
[266]	KR	Reaction force adjusting device of exoskeleton system and variable stiffness actuator using the same	*PExo_t*	Rigid portable	30 March 2018	16 January 2020

Table A1. *Cont.*

Cite	Country Code	Title	Type of Treatment	Type of Structure	Application Date	Publication Date
[267]	US	Safety monitoring and control system and methods for a legged mobility exoskeleton device	*PExo_t*	Rigid portable	18 November 2016	03 August 2021
[268]	KR	Elastic type sole assembly in wearable robot absorbing impact and detecting ground reaction force	*PExo_p*	Rigid portable	16 September 2014	10 May 2016
[269]	JP	Foot mounting structure of joint motion assist device	*PExo_p*	Rigid portable	09 August 2013	08 June 2016
[270]	US	Adaptable robotic gait trainer	*PExo_t*	Rigid non-portable	20 April 2018	05 October 2021
[271]	KR	Reaction force adjusting device and method using variable stiffness actuator of exoskeleton system	*PExo_t*	Rigid portable	30 March 2018	16 January 2020
[272]	US	Exosuit systems and methods	*PExo_t*	Suit	28 November 2018	07 December 2021
[273]	CN	A kind of ectoskeleton stopping means	*PExo_t*	Rigid portable	19 May 2016	30 January 2018
[274]	ES	System to assist walking	*PExo_t*	Rigid portable	17 October 2016	11 September 2018
[275]	CN	Electro–hydraulic hybrid driving exoskeleton device	*PExo_t*	Rigid portable	25 May 2018	07 April 2020
[276]	CN	Pneumatic weight-reducing walking power-assisted robot	*PExo_t*	Rigid portable	30 August 2018	16 June 2020
[277]	DE	Mobility system	*PExo_t*	Rigid non-portable	17 August 2016	05 October 2017
[278]	KR	Gait assist robot for rehabilitation training with lift device	*PExo_t*	Rigid portable with wheels	24 October 2019	03 December 2021
[279]	KR	Robot device for upper and lower extremity rehabilitation	*PExo_t*	Rigid non-portable	21 April 2020	30 September 2021
[280]	CN	Device and method for realizing cooperative motion of weight-reducing vehicle and lower-limb robot through communication	*PExo_t*	Rigid portable with wheels	18 November 2019	21 September 2021
[281]	CN	Counter weight type lower-limb rehabilitation robot	*PExo_t*	Rigid non-portable	21 October 2019	16 November 2021

Table A1. *Cont.*

Cite	Country Code	Title	Type of Treatment	Type of Structure	Application Date	Publication Date
[282]	CN	Lower-limb exoskeleton robot with overload slipping function	*PExo_p*	Rigid portable	26 April 2019	08 October 2021
[283]	CN	Device and method for assisting lower-limb robot to transfer gravity center by aid of weight reduction vehicle	*PExo_t*	Rigid portable with wheels	02 December 2019	19 November 2021
[284]	CN	Anti-falling walking aid vehicle for lower-limb rehabilitation training and rehabilitation training method	*PExo_t*	Rigid non-portable	12 July 2019	21 July 2020
[285]	KR	Patient weight burden reduction device of walking rehabilitation training robot	*PExo_t*	Rigid portable with wheels	24 October 2019	26 October 2021
[286]	CN	Exoskeleton joint self-locking mechanism, knee joint and bionic rehabilitation robot	*PExo_p*	Rigid portable	29 September 2020	05 February 2021
[287]	CN	Lower-limb exoskeleton ankle joint based on telecentric mechanism	*PExo_p*	Rigid portable	23 October 2018	02 March 2021
[288]	KR	Lower extremity exoskeleton robotic device	*PExo_t*	Rigid portable	20 May 2021	24 August 2021
[289]	CN	Lower-limb exoskeleton capable of being used for both wheel and leg	*PExo_t*	Rigid portable with wheels	01 July 2019	11 May 2021
[290]	CN	Auxiliary dual-purpose outer limb robot for human body movement	*PExo_t*	Rigid portable	08 July 2019	08 September 2020
[291]	KR	Rehabilitation walking method considering patient's rom characteristics and system thereof	*PExo_t*	Rigid portable	27 December 2017	09 September 2019
[292]	JP	Lower limbs of the exoskeleton with low power consumption	*PExo_t*	Rigid portable	19 September 2018	17 March 2021
[293]	CN	Novel self-balancing ectoskeleton robot	*PExo_t*	Rigid portable	23 April 2021	01 February 2022
[294]	KR	Rehabilitation robot	*PExo_t*	Rigid non-portable	24 July 2018	26 February 2020

Table A1. *Cont.*

Cite	Country Code	Title	Type of Treatment	Type of Structure	Application Date	Publication Date
[295]	KR	Leg opening joint of walking exoskeleton and walking exoskeleton comprising the same	*PExo_t*	Rigid portable	19 November 2018	15 November 2019
[296]	CN	Motion decoupling parallel driving type exoskeleton robot ankle joint	*PExo_p*	Rigid portable	23 April 2021	25 January 2022
[297]	CN	Locking-free hip adjusting device	*PExo_p*	Rigid portable	23 December 2020	09 April 2021
[298]	CN	Constant-force human body suspension system for rehabilitation training	*PExo_t*	Rigid non-portable	06 August 2019	29 September 2020
[299]	CN	Control method of hydraulic system of knee joint rehabilitation robot	*PExo_p*	Rigid portable	24 August 2017	21 July 2020
[300]	CN	Walking-aid boots	*PExo_p*	Rigid portable	20 September 2017	07 April 2020
[301]	KR	Walking assistance system	*PExo_t*	Rigid non-portable	18 February 2020	17 January 2022
[302]	KR	Wearable suit control method	*PExo_t*	Suit	15 November 2019	10 August 2021
[303]	CN	Power-assisted exoskeleton control method, power-assisted exoskeleton control system and computer equipment	*PExo_t*	Rigid portable	15 November 2019	29 October 2021
[304]	CN	Variable-rigidity knee joint exoskeleton robot based on shape–memory alloy	*PExo_p*	Rigid portable	03 August 2020	14 January 2022
[305]	KR	Motion assist apparatus	*PExo_p*	Rigid portable	25 July 2018	16 September 2020
[306]	RU	Exoskeleton	*PExo_t*	Rigid portable	15 April 2021	26 November 2021
[307]	KR	Two-leg walking assistant system for boarding type	*PExo_t*	Rigid portable	02 December 2013	03 March 2015
[308]	US	Soft inflatable exosuit for knee rehabilitation	*PExo_t*	Suit	31 July 2018	01 March 2022
[309]	KR	Walk assistance and fall prevention wearable suit	*PExo_t*	Suit	15 November 2019	07 March 2022

Table A1. *Cont.*

Cite	Country Code	Title	Type of Treatment	Type of Structure	Application Date	Publication Date
[310]	KR	Size-adjustable pelvis unit and wearable walking robot comprising the same	*PExo_p*	Rigid portable	16 March 2016	09 August 2017
[311]	CN	A kind of human foot's bionic exoskeleton system	*PExo_t*	Rigid portable	22 September 2016	28 August 2018
[312]	KR	Apparatus and method for observing feedback force of wearable exoskeleton system	*PExo_t*	Rigid portable	13 February 2018	01 October 2019
[313]	JP	Measurement system, measurement method, and program	*PExo_t*	Rigid portable	18 May 2017	02 June 2021
[314]	KR	Devices designed to be positioned near joints and systems incorporating such devices	*PExo_p*	Rigid portable	28 October 2019	09 December 2021

Table A2. PatentScope platform.

Cite	Country Code	Title	Type of Treatment	Type of Structure	Application Date	Publication Date
[315]	US	Lower extremity exoskeleton for gait retraining	*PExo_t*	Rigid portable	26 October 2015	02 June 2016
[27]	US	Lower extremity exoskeleton for gait retraining	*PExo_t*	Rigid portable	28 September 2012	29 August 2013
[316]	WO	Lower extremity exoskeleton for gait retraining	*PExo_t*	Rigid portable	28 September 2012	04 April 2013
[317]	US	Pneumatic lower extremity gait rehabilitation training system	*PExo_t*	Rigid non-portable	12 September 2016	15 March 2018
[318]	WO	Powered orthotic system for cooperative overground rehabilitation	*PExo_t*	Rigid portable	13 March 2014	02 October 2014
[319]	IN	Robotic exoskeleton assisted (locomotion) rehabilitation system "rears"	*PExo_t*	Rigid portable with wheels	14 December 2017	01 February 2019
[320]	WO	Active sling for the motion neurological rehabilitation of lower limbs, system comprising such sling and process for operating such system	*PExo_t*	Rigid portable	20 February 2013	19 September 2013

Table A2. *Cont.*

Cite	Country Code	Title	Type of Treatment	Type of Structure	Application Date	Publication Date
[321]	EP	Active orthosis for the motion neurological rehabilitation of lower limbs, system comprising such orthosis and process for operating such system	*PExo_t*	Rigid portable	20 February 2013	21 January 2015
[322]	WO	Interface for adjusting the motion of a powered orthotic device through externally applied forces	*PExo_t*	Rigid portable	15 January 2014	24 July 2014
[323]	EP	Interface for adjusting the motion of a powered orthotic device through externally applied forces	*PExo_t*	Rigid portable	15 January 2014	25 November 2015
[324]	EP	Powered orthotic system for cooperative overground rehabilitation	*PExo_t*	Rigid portable	13 March 2014	20 January 2016
[325]	EP	Assistive flexible suits, flexible suit systems, and methods for making and control thereof to assist human mobility	*PExo_t*	Suit	03 December 2014	19 October 2016
[326]	WO	Interactive exoskeleton robotic knee system	*PExo_p*	Rigid portable	01 March 2016	17 November 2016
[327]	US	Admittance shaping controller for exoskeleton assistance of the lower extremities	*PExo_p*	Rigid portable	05 December 2017	12 April 2018
[328]	US	Assistive flexible suits, flexible suit systems, and methods for making and control thereof to assist human mobility	*PExo_t*	Suit	03 December 2014	20 July 2017
[329]	WO	Bio-inspired adaptive impedance-based controller for human–robot interaction and method	*PExo_t*	Rigid non-portable	26 August 2019	27 February 2020
[330]	US	Apparatus and system for limb rehabilitation	*PExo_t*	Rigid portable with wheels	29 November 2018	20 June 2019
[28]	US	Lower extremity robotic rehabilitation system	*PExo_t*	Rigid non-portable	05 October 2012	10 April 2014
[331]	US	Admittance shaping controller for exoskeleton assistance of the lower extremities	*PExo_p*	Rigid portable	25 June 2015	18 February 2016
[332]	US	Modular and minimally constraining lower-limb exoskeleton for enhanced mobility and balance augmentation	*PExo_t*	Rigid portable	04 October 2017	22 August 2019
[333]	WO	Modular and minimally constraining lower-limb exoskeleton for enhanced mobility and balance augmentation	*PExo_t*	Rigid portable	04 October 2017	12 April 2018
[334]	EP	Soft exosuit for assistance with human motion	*PExo_t*	Suit	30 May 2014	13 April 2016

Table A2. *Cont.*

Cite	Country Code	Title	Type of Treatment	Type of Structure	Application Date	Publication Date
[335]	EP	Soft exosuit for assistance with human motion	*PExo_t*	Suit	30 May 2014	17 February 2021
[336]	US	Methods of enhancing the or training rehabilitation of an exoskeleton wearer	*PExo_t*	Rigid portable	11 November 2015	22 March 2018
[337]	WO	Apparatus and system for limb rehabilitation	*PExo_t*	Rigid portable with wheels	29 November 2018	20 June 2019
[338]	US	Bio-inspired standing balance controller for a full-mobilization exoskeleton	*PExo_p*	Rigid portable	16 July 2020	21 January 2021
[339]	WO	Methods for improved user mobility and treatment	*PExo_t*	Rigid portable	27 May 2021	02 December 2021
[340]	WO	Exoskeleton ankle robot	*PExo_p*	Rigid portable	01 March 2016	17 November 2016
[341]	US	Powered medical device and methods for improved user mobility and treatment	*PExo_t*	Rigid portable	27 May 2021	02 December 2021
[342]	WO	Methods of enhancing the rehabilitation or training of an exoskeleton wearer	*PExo_t*	Rigid portable	11 November 2015	19 May 2016
[343]	EP	Soft wearable muscle assisting device	*PExo_t*	Suit	21 December 2017	02 September 2020
[344]	EP	Exoskeleton	*PExo_t*	Rigid portable	11 November 2015	20 September 2017
[345]	WO	Soft wearable muscle assisting device	*PExo_p*	Suit	21 December 2017	05 July 2018
[346]	US	Human movement research, therapeutic, and diagnostic devices, methods, and systems	*PExo_t*	Rigid non-portable	21 April 2015	02 February 2017
[347]	US	Robotic management system for limb rehabilitation	*PExo_t*	Rigid portable with wheels	30 October 2019	04 June 2020
[348]	US	Interface for adjusting the motion of a powered orthotic device through externally applied forces	*PExo_t*	Rigid portable	15 January 2014	10 December 2015
[349]	US	Control system for movement reconstruction and/or restoration for a patient	*PExo_t*	Rigid non-portable	13 November 2019	14 May 2020

Table A2. *Cont.*

Cite	Country Code	Title	Type of Treatment	Type of Structure	Application Date	Publication Date
[350]	US	Data logging and third-party administration of a mobile robot	*PExo_t*	Rigid portable	27 May 2021	02 December 2021
[351]	US	Apparatus comprising a support system for a user and its operation in a gravity assist mode	*PExo_t*	Rigid non-portable	17 August 2017	16 September 2021
[352]	US	Integrated platform to monitor and analyze individual progress in physical and cognitive tasks	*PExo_t*	Rigid portable	18 July 2016	14 January 2020
[353]	EP	Soft exosuit for assistance with human motion	*PExo_t*	Suit	17 September 2013	22 July 2015
[354]	EP	Movement assistance device	*PExo_t*	Rigid portable	17 June 2013	22 April 2015
[355]	WO	Movement assistance device	*PExo_t*	Rigid portable	17 June 2013	19 December 2013
[356]	US	Exoskeleton and master	*PExo_t*	Rigid portable	27 June 2017	01 August 2019
[357]	US	Powered orthotic system for cooperative overground rehabilitation	*PExo_t*	Rigid portable	13 March 2014	04 February 2016
[358]	EP	Exoskeleton for assisting human movement	*PExo_t*	Rigid portable	25 November 2015	04October 2017
[359]	WO	Data logging and third-party administration of a mobile robot	*PExo_t*	Rigid portable	27.05.2021	02 December 2021
[360]	US	Reconfigurable exoskeleton	*PExo_t*	Rigid portable	11 December 2013	10 December 2015
[32]	US	Powered gait assistance systems	*PExo_t*	Rigid portable	31 July 2017	30 September 2021
[361]	CA	Apparatus and method for restoring voluntary control of locomotion in neuromotor impairments	*PExo_t*	Rigid non-portable	-	05 December 2013
[362]	AU	Apparatus and method for restoring voluntary control of locomotion in neuromotor impairments	*PExo_t*	Rigid non-portable	29 May 2013	11 December 2014
[363]	WO	Apparatus and method for restoring voluntary control of locomotion in neuromotor impairments	*PExo_t*	Rigid non-portable	29 May 2013	05 December 2013

Table A2. *Cont.*

Cite	Country Code	Title	Type of Treatment	Type of Structure	Application Date	Publication Date
[364]	EP	Apparatus for restoring voluntary control of locomotion in neuromotor impairments	$PExo\_t$	Rigid non-portable	29 May 2013	08 April 2015
[365]	EP	Reconfigurable exoskeleton	$PExo\_t$	Rigid portable	11 December 2013	21 October 2015
[366]	WO	Methods of communication exoskeleton and control	$PExo\_t$	Rigid portable	14 April 2016	20 October 2016
[367]	EP	Soft exosuit for assistance with human motion	$PExo\_t$	Suit	17 September 2013	17 March 2021
[368]	US	Soft exosuit for assistance with human motion	$PExo\_t$	Suit	17 September 2013	25 June 2015
[369]	EP	Apparatus for restoring voluntary control of locomotion in neuromotor impairments	$PExo\_t$	Rigid non-portable	29 May 2013	08 November 2017
[370]	US	Soft exosuit for assistance with human motion	$PExo\_t$	Suit	17 March 2015	12 November 2015
[371]	US	Power-assist with adjustable lower-limb exoskeleton robot stiffness joints	$PExo\_t$	Rigid portable	10 January 2019	29 October 2020
[372]	WO	Mobility assistance devices with automated assessment and adjustment control	$PExo\_t$	Rigid portable	07 February 2018	27 September 2018
[373]	US	Soft exosuit for assistance with human motion	$PExo\_t$	Suit	13 April 2016	04 August 2016
[374]	US	Exoskeleton ankle robot	$PExo\_p$	Rigid portable	21 June 2015	17 November 2016
[375]	US	Mobility assistance devices with automated assessment and adjustment control	$PExo\_t$	Rigid portable	07 February 2018	27 February 2020
[376]	EP	Method and apparatus for providing economical, portable deficit-adjusted adaptive assistance during movement phases of an impaired ankle	$PExo\_p$	Rigid portable	20 June 2016	15 September 2021
[377]	WO	Method and apparatus for providing economical, portable deficit-adjusted adaptive assistance during movement phases of an impaired ankle	$PExo\_t$	Rigid portable	20 June 2016	29 December 2016
[378]	WO	Customizable orthotic/prosthetic braces and lightweight modular exoskeleton	$PExo\_p$	Rigid portable	14 June 2017	21 December 2017
[31]	US	Soft exosuit for assistance with human motion	$PExo\_t$	Suit	13 August 2020	11 February 2021

Table A2. *Cont.*

Cite	Country Code	Title	Type of Treatment	Type of Structure	Application Date	Publication Date
[379]	WO	Powered gait assistance systems	*PExo_t*	Rigid portable	31 July 2017	01 February 2018
[380]	US	Customizable orthotic/prosthetic braces and lightweight modular exoskeleton	*PExo_p*	Rigid portable	14 June 2017	31 October 2019
[381]	IN	Method and apparatus for providing economical portable deficit adjusted adaptive assistance during movement phases of an impaired ankle	*PExo_p*	Rigid portable	13 December 2017	16 March 2018
[382]	US	Hybrid terrain- adaptive lower-extremity systems	*PExo_p*	Rigid portable	31 December 2018	11 July 2019
[383]	US	Robotic system for simulating a wearable device and method of use	*PExo_t*	Rigid non-portable	20 December 2012	20 June 2013
[384]	US	Kinetic sensing, signal generation, feature extraction, and pattern recognition for control of autonomous wearable leg devices	*PExo_p*	Rigid portable	08 November 2017	10 October 2019
[385]	US	Open-loop control for exoskeleton motor	*PExo_t*	Rigid portable	19 March 2021	23 September 2021
[386]	WO	Kinoped lower extremity performance improvement, injury prevention, and rehabilitation system	*PExo_t*	Rigid non-portable	04 September 2020	11 March 2021
[387]	EP	System for assisting walking	*PExo_p*	Rigid portable	18 November 2016	26 September 2018
[388]	CA	Methods of communication exoskeleton and control	*PExo_t*	Rigid portable	14 April 2016	20 October 2016
[389]	US	Movement assistance device	*PExo_t*	Rigid portable	17 June 2013	21 May 2015
[390]	US	Hybrid terrain-adaptive lower-extremity systems	*PExo_p*	Rigid portable	24 September 2013	20 March 2014
[391]	US	Interactive exoskeleton robotic knee system	*PExo_p*	Rigid portable	21 June 2015	17 November 2016

Table A2. *Cont.*

Cite	Country Code	Title	Type of Treatment	Type of Structure	Application Date	Publication Date
[392]	US	Hybrid terrain-adaptive lower-extremity systems	*PExo_p*	Rigid portable	11 April 2016	13 October 2016
[393]	US	Implementing a stand-up sequence using a lower-extremity prosthesis or orthosis	*PExo_t*	Rigid portable	23 September 2013	20 March 2014
[394]	US	Hybrid terrain-adaptive lower-extremity systems	*PExo_t*	Rigid portable	30 July 2018	25 April 2019
[395]	WO	Apparatus comprising a support system for a user and its operation in a gravity assist mode	*PExo_t*	Rigid non-portable	17 August 2017	22 February 2018
[396]	US	Orthopedic device including protruding members	*PExo_p*	Rigid portable	10 April 2015	02 February 2017
[29]	US	Exoskeleton device and control system	*PExo_p*	Rigid portable	25 May 2017	10 May 2018
[397]	US	Cloud-based control system and method enabling interactive clinical care using a powered mobility assistance device	*PExo_t*	Rigid portable	03 September 2019	16 December 2021
[398]	US	Hybrid terrain-adaptive lower-extremity systems	*PExo_p*	Rigid portable	28 April 2016	18 August 2016
[399]	US	Methods of communication exoskeleton and control	*PExo_t*	Rigid portable	14 April 2016	05 April 2018
[400]	US	Exoskeleton device and control system	*PExo_p*	Rigid portable	24 February 2021	02 December 2021
[401]	US	Hybrid terrain-adaptive lower-extremity systems	*PExo_p*	Rigid portable	30 November 2016	30 March 2017
[402]	US	Soft exosuit for assistance with human motion	*PExo_t*	Suit	30 May 2014	21 April 2016
[403]	US	Systems, methods, and devices for assisting walking for developmentally delayed toddlers	*PExo_p*	Rigid portable	05 February 2015	01 December 2016
[404]	CA	Method and apparatus for providing economical, portable deficit-adjusted adaptive assistance during movement phases of an impaired ankle	*PExo_p*	Rigid portable	-	29 December 2016

Table A2. *Cont.*

Cite	Country Code	Title	Type of Treatment	Type of Structure	Application Date	Publication Date
[405]	US	Torque control methods for an exoskeleton device	*PExo_p*	Rigid portable	25 May 2017	30 November 2017
[406]	US	Patient aid devices, particularly for mobile upper extremity support in railed devices such as parallel bars and treadmills	*PExo_p*	Rigid portable	23 October 2017	15 February 2018
[407]	WO	Autonomous mobile support system for the robotic mobility-impaired	*PExo_t*	Rigid portable with wheels	18 August 2020	06 May 2021
[408]	US	Wearable robot and control method thereof	*PExo_t*	Rigid portable	02 December 2014	16 July 2015
[409]	US	Patient aid devices, particularly for mobile upper extremity support in railed devices such as parallel bars and treadmills	*PExo_p*	Rigid portable	21 May 2015	26 November 2015
[410]	WO	Wearable devices for protecting against musculoskeletal injuries and enhancing performance	*PExo_p*	Rigid portable	15 February 2019	22 August 2019
[411]	WO	Exosuit systems and methods	*PExo_t*	Suit	28 November 2018	06 June 2019
[412]	US	Exosuit systems and methods	*PExo_t*	Suit	28 November 2018	30 May 2019
[413]	US	System and method for the regeneration of at least one severed nerve conduit	*PExo_t*	Rigid non-portable	20 July 2018	15 November 2018
[414]	US	Exosuit load bearing distribution systems	*PExo_t*	Suit	28 November 2018	30 May 2019
[415]	US	Orthosis leg and orthosis	*PExo_t*	Rigid portable	24 June 2015	08 June 2017
[416]	CN	Lower extremity exoskeleton control method and apparatus	*PExo_t*	Rigid non-portable	16 October 2019	21 February 2020
[417]	EP	Esoskeleton equipped with electro-or magneto- rheological fluid type semi-active joints	*PExo_t*	Rigid portable	29 December 2017	06 November 2019
[30]	WO	Esoskeleton equipped with electro-or magneto- rheological fluid type semi-active joints	*PExo_t*	Rigid portable	29 December 2017	05 July 2018
[418]	US	Low profile exoskeleton	*PExo_t*	Rigid portable	05 November 2015	12 May 2016
[419]	WC	System for movement control	*PExo_p*	Rigid portable	15 May 2020	26 November 2020

Table A3. Lens platform.

Cite	Country Code	Title	Type of Treatment	Type of Structure	Application Date	Publication Date
[420]	EP	Powered orthotic system for cooperative overground rehabilitation	$PExo\_t$	Rigid portable	13 March 2014	11 August 2021
[421]	US	Movement assistance device	$PExo\_t$	Rigid non-portable	20 June 2017	06 October 2020
[422]	US	Hybrid terrain-adaptive lower-extremity systems	$PExo\_p$	Rigid portable	30 June 2018	03 March 2020
[423]	EP	System for assisting walking	$PExo\_t$	Rigid portable	18 November 2016	04 November 2020
[424]	US	Optimal design of a lower-limb exoskeleton or orthosis	$PExo\_t$	Rigid portable	16 December 2014	18 February 2020
[78]	US	Powered lower-limb devices and methods of control thereof	$PExo\_t$	Rigid portable	03 November 2017	07 December 2021
[425]	US	Leg orthosis and orthosis	$PExo\_p$	Rigid portable	24 June 2015	23 November 2021
[11]	US	Orthopedic device including protruding members	$PExo\_p$	Rigid portable	10 April 2015	15 December 2020
[426]	US	Soft exosuit for assistance with human motion	$PExo\_t$	Suit	30 May 2014	24 November 2020
[427]	US	Portable human exoskeleton system	$PExo\_t$	Rigid portable	09 February 2016	16 February 2021
[428]	US	Motorized limb assistance device	$PExo\_p$	Rigid portable	02 May 2017	30 June 2020
[429]	US	Legged robotic device utilizing modifiable linkage mechanism	$PExo\_t$	Rigid portable	05 May 2015	08 September 2020
[430]	EP	Esoskeleton equipped with electro-or magneto-rheological fluid type semi-active joints	$PExo\_t$	Rigid portable	29 December 2017	07 April 2021
[40]	US	Low profile exoskeleton	$PExo\_t$	Rigid portable	05 November 2015	18 February 2020
[213]	US	System and method for the regeneration of at least one severed nerve conduit	$PExo\_t$	Rigid non-portable	20 July2018	02 February 2021
[431]	US	Regulation of autonomic control of bladder voiding after a complete spinal cord injury	$PExo\_t$	Rigid non-portable	21 August 2015	25 August 2020
[432]	EP	Movement assistance device	$PExo\_t$	Rigid portable	17 June 2013	08 January 2020

References

1. Yagn, N. Apparatus for Facilitating Walking, Running, and Jumping. U.S. Patent 420,179, 28 January 1890. Available online: https://patentimages.storage.googleapis.com/0f/e0/2c/161bea0a876b81/US420179.pdf (accessed on 15 January 2022).
2. De la Tejera, J.A.; Bustamante-Bello, R.; Ramirez-Mendoza, R.A.; Izquierdo-Reyes, J. Systematic Review of Exoskeletons towards a General Categorization Model Proposal. *Appl. Sci.* **2021**, *11*, 76. [CrossRef]
3. Barjuei, E.S.; Ardakani, M.M.G.; Caldwell, D.G.; Sanguineti, M.; Ortiz, J. On the Optimal Selection of Motors and Transmissions for a Back-Support Exoskeleton. In Proceedings of the IEEE International Conference on Cyborg and Bionic Systems (CBS), Munich, Germany, 19–20 September 2019; Available online: https://ieeexplore.ieee.org/document/9114465 (accessed on 15 January 2022).
4. Chen, G.; Chan, C.K.; Guo, Z.; Yu, H. A review of lower extremity assistive robotic exoskeletons in rehabilitation therapy. *Crit. Rev. Biomed. Eng.* **2013**, *41*, 343–363. [CrossRef] [PubMed]
5. Colombo, G.; Joerg, M.; Schreier, R.; Dietz, V. Treadmill training of paraplegic patients using a robotic orthosis. *J. Rehabil. Res. Dev.* **2000**, *37*, 693–700. [PubMed]
6. Carpino, G.; Pezzola, A.; Urbano, M.; Guglielmelli, E. Assessing effectiveness and costs in robot-mediated lower limbs rehabilitation: A meta-analysis and state of the art. *J. Healthc. Eng.* **2018**, *2018*, 7492024. [CrossRef]
7. Bruni, M.F.; Melegari, C.; De Cola, M.C.; Bramanti, A.; Bramanti, P.; Calabrò, R.S. What does best evidence tell us about robotic gait rehabilitation in stroke patients: A systematic review and meta-analysis. *J. Clin. Neurosci.* **2018**, *48*, 11–17. [CrossRef]
8. Rodríguez-León, J.F.; Chaparro-Rico, B.D.M.; Russo, M.; Cafolla, D. an Autotuning Cable-Driven Device for Home Rehabilitation. *J. Healthc. Eng.* **2021**, *2021*, 6680762. [CrossRef]
9. Dong, K.H.; Su, H.K.; Sil, P.J.; Ill, J.H. Lower Extremity Exoskeleton Robotic Device. KR102292983B1, 24 August 2021. Available online: https://patents.google.com/patent/KR102292983B1/en (accessed on 15 January 2022).
10. Yan, T.; Cempini, M.; Oddo, C.M.; Vitiello, N. Review of assistive strategies in powered lower-limb orthoses and exoskeletons. *Robot. Auton. Syst.* **2015**, *64*, 120–136. [CrossRef]
11. Meng, W.; Liu, Q.; Zhou, Z.; Ai, Q.; Sheng, B.; Xie, S.S. Recent development of mechanisms and control strategies for robot-assisted lower limb rehabilitation. *Mechatronics* **2015**, *31*, 132–145. [CrossRef]
12. Rupal, B.S.; Singla, A.; Virk, G.S. Lower limb exoskeletons: A brief review. In Proceedings of the Conference on Mechanical Engineering and Technology (COMET2016), IIT (BHU), Varanasi, India, 15–17 January 2016; pp. 130–140.
13. Chen, B.; Ma, H.; Qin, L.Y.; Gao, F.; Chan, K.M.; Law, S.W. Recent developments and challenges of lower extremity exoskeletons. *J. Orthop. Transl.* **2016**, *5*, 26–37. [CrossRef]
14. Shi, D.; Zhang, W.; Zhang, W.; Ding, X. A review on lower limb rehabilitation exoskeleton robots. *Chin. J. Mech. Eng.* **2019**, *32*, 74. [CrossRef]
15. Sanchez-Villamañan, M.C.; Gonzalez-Vargas, J.; Torricelli, D.; Moreno, J.C.; Pons, J.L. Compliant lower limb exoskeletons: A comprehensive review on mechanical design principles. *J. NeuroEng. Rehabil.* **2019**, *16*, 55. [CrossRef]
16. Koch, M.A.; Font-Llagunes, J.M. Lower-Limb Exosuits for Rehabilitation or Assistance of Human Movement: A Systematic Review. *Appl. Sci.* **2021**, *11*, 8743. [CrossRef]
17. Kian, A.; Widanapathirana, G.; Josephet, A.M.; Lai, D.T.H.; Begg, R. Application of wearable sensors in actuation and control of powered ankle exoskeletons: A Comprehensive Review. *Sensors* **2022**, *22*, 2244. [CrossRef] [PubMed]
18. Yang, X.; She, H.; Lu, H.; Fukuda, T.; Shen, Y. State of the Art: Bipedal Robots for Lower Limb Rehabilitation. *Appl. Sci.* **2017**, *7*, 1182. [CrossRef]
19. Tijjani, I.; Kumar, S.; Boukheddimi, M. A Survey on Design and Control of Lower Extremity Exoskeletons for Bipedal Walking. *Appl. Sci.* **2022**, *12*, 2395. [CrossRef]
20. Meda-Gutiérrez, J.R.; Zúñiga-Avilés, L.A.; Vilchis-González, A.H.; Ávila-Vilchis, J.C. Knee Exoskeletons Design Approaches to Boost Strength Capability: A Review. *Appl. Sci.* **2021**, *11*, 9990. [CrossRef]
21. Zhou, J.; Yang, S.; Xue, Q. Lower limb rehabilitation exoskeleton robot: A review. *Adv. Mech. Eng.* **2021**, *13*, 16878140211011862. [CrossRef]
22. Rupal, B.S.; Rafique, S.; Singla, A.; Singla, E.; Isaksson, M.; Virk, G.S. Lower-limb exoskeletons: Research trends and regulatory guidelines in medical and non-medical applications. *Int. J. Adv. Robot. Syst.* **2017**, *14*, 1729881417743554. [CrossRef]
23. Contreras-Vidal, J.L.; Bhagat, N.A.; Brantley, J.; Cruz-Garza, J.G.; He, Y.; Manley, Q.; Nakagome, S.; Nathan, K.; Tan, S.H.; Zhu, F.; et al. Powered exoskeletons for bipedal locomotion after spinal cord injury. *J. Neural Eng.* **2016**, *13*, 031001. [CrossRef]
24. Baud, R.; Manzoori, A.R.; Ijspeert, A.; Bouri, M. Review of control strategies for lower-limb exoskeletons to assist gait. *J. NeuroEng. Rehabil.* **2021**, *18*, 119. [CrossRef]
25. Shorter, K.A.; Xia, J.; Hsiao-Wecksler, E.T.; Durfee, W.K.; Kogler, G.F. Technologies for powered ankle-foot orthotic systems: Possibilities and challenges. *IEEE/ASME Trans. Mech.* **2013**, *18*, 337–347. [CrossRef]
26. Aviv Reuven, S.; Rosenfeld, A. Publication patterns' changes due to the COVID-19 pandemic: A longitudinal and short-term scientometric analysis. *Scientometrics* **2021**, *126*, 6761–6784. [CrossRef] [PubMed]
27. Unluhisarcikli, O. Lower Extremity Exoskeleton for Gait Retraining. U.S. Patent 20130226048, 29 August 2013.
28. Jianjuen, H. Lower Extremity Robotic Rehabilitation System. U.S. Patent 20140100491, 10 April 2014.
29. Walsh, C.J. Orthopedic Device Including Protruding Members. U.S. Patent 20170027735, 2 February 2017.

30. Cheng, W. Lower Extremity Exoskeleton Control Method and ApparatusShenzhen Milebot Robot. Cn-110812127, 21 February 2020.

31. Anindo, R. Method and Apparatus for Providing Economical, Portable Deficit-Adjusted Adaptive Assistance During Movement Phases of an Impaired Ankle. Ep-3878603, 15 September 2021.

32. Zoss, A. Powered Orthotic System for Cooperative Overground Rehabilitation. U.S. Patent 20160030201, 4 February 2016.

33. Goffer, A. Gait Device with a Crutch. U.S. Patent 10137050-B2, 27 November 2018.

34. Johnson, C.L. Patient Aid Devices, Particularly for Mobile Upper Extremity Support in Railed Devices Such as Parallel Bars and Treadmills. U.S. Patent 10617906-B2, 14 April 2020.

35. Lee, J.M.; Kim, H.; Kim, S.J.; Choi, J.; Hwang, Y.; Chung, S.; Kim, Y.; Choi, J. Recognition Method of Human Walking Speed Intention from Surface Electromyogram Signals of Plantar Flexor and Walking Speed Control Method of a Lower-Limb Exoskeleton Robot. U.S. Patent 10434027-B2, 8 October 2019.

36. Hyun, Y.S.; Baek, L.G.; Gyeong, L.C. Walk Assist Apparatus. Kr-101368817-B1, 3 March 2014.

37. Kim Hyun, I.; Kim Hyun, A. Wearable Crutch with Lap Joints. Kr-101530525-B1, 23 June 2015.

38. Tong, K.Y.; Ockenfeld, C.U.; Yeung, L.F.; Ho, S.K.; Wai, H.W.; Pang, M.K. Interactive Exoskeleton Robotic Knee System. U.S. Patent 10390973-B2, 27 August 2019.

39. Won, Y.J.; Hun, P.S. Active Type Step Assistance Apparatus. Kr-101471856-B1, 12 December 2014.

40. LaChappelle, E.J.; Blue, A. Low Profile Exoskeleton. U.S. Patent 10561564-B2, 18 February 2020.

41. Lee, L.W.; Chiang, H.H. Multi-Function Lower Limb Ambulation Rehabilitation and Walking Assist Device. U.S. Patent 9789023-B1, 17 October 2017.

42. Walsh, C.; Asbeck, A.T.; Ye, D.; Bujanda, I.G.; De Rossi, S.M.M. For Aiding in the Soft Machine Armor of Human Motion. Cn-105263448-B, 4 May 2018.

43. Asbeck, A.T.; Bujanda, I.G.; Ding, Y.; Dyer, R.J.; Larusson, A.F.; Quinlivan, B.T.; Schmidt, K.; Wagner, D.; Walsh, C.J.; Wehner, M. Soft Exterior Protector for Aiding in Human Motion. Cn-104869969-B, 9 June 2017.

44. Kyung, S.W.; Young, L.H. Gait Assistant Robot. Kr-101656695-B1, 13 September 2016.

45. Lee, J.W.; Kim, H.Y. Robot for Assisting User to Walk. Kr-101979938-B1, 20 May 2019.

46. Legaz, J.G.; Herruzo, B.B. System to Assist Walk. Es-2618285-B1, 6 April 2018.

47. Lee, J.W.; Kim, H.Y. Robot for Assisting User to Walk with Lower Body Exoskeleton. Kr-101869968-B1, 21 June 2018.

48. Cao, H.; MO, S.; Zhu, J.; Wang, Y.; Jiang, J.; Chen, Y.; Cui, J.; Zhang, Y. Rehabilitation Type Lower Limb Exoskeleton. Cn-109044742-B, 13 March 2020.

49. Tong, K.Y.; Yeung, L.F.; Ockenfeld, C.U.; Ho, S.K.; Wai, H.W.; Pang, M.K. Exoskeleton Ankle Robot. U.S. Patent 10426637-B2, 1 October 2019.

50. Borisoff, J.; Rafer, V. Mobility System Including an Exoskeleton Assembly Releasably Supported on a Wheeled Base. U.S. Patent 9849048-B2, 26 December 2017.

51. Won, K.S.; Hwi, Y.L. Rehabilitation Robot of Legs, Boarding and Driving Method Thereof. Kr-101955925-B1, 12 March 2019.

52. Linon, R. Support Structure. Es-2759298-T3, 8 May 2020.

53. Ang, X.; Gan, Z. Wearable Lower Limb Exoskeleton Driven by Lasso Artificial Muscles. Cn-107468487-B, 3 November 2020.

54. Armada, E.G.; Cestari Soto, M.J.; Sanz Merodio, D.; Carrillo De Hijes, X. Exo Skeleton for Human Movement Assistance. Es-2575255-B1, 6 April 2017.

55. Yang, C.; Wang, H.; Yang, W.; Ma, Z.; Wei, Q.; Zhao, Y. A Kind of Lower Limb Rehabilitation Exoskeleton System and Its Walking Control Method. Cn-108379038-B, 9 July 2019.

56. Mun, J.H.; Youn, S.H.; Joo, S.B.; Sim, T.Y.; Choi, A.R. Wearable Apparatus for Support Holding Posture. Kr-101774152-B1, 4 September 2017.

57. Lee, C.S.; Li, I.H.; Chiang, H.H.; Lee, L.W.; Hong, F.C.; Chen, R.P.; Chen, W.G. Pneumatic Lower Extremity Gait Rehabilitation Training System. U.S. Patent 10292892-B2, 21 May 2019.

58. Walsh, C.J.; Goldfield, E.C.; Song, S.E.; Park, E. Systems, Methods, and Devices for Assisting Walking for Developmentally-Delayed Toddlers. U.S. Patent 10278883-B2, 7 May 2019.

59. Si, G.; Xu, F.; Li, G.; Huang, W.; Chu, M. Variable-Rigidity Flexible Driver for Exoskeleton Type Lower Limb Rehabilitation Robot. Cn-108309688-B, 6 December 2019.

60. Cook, D.L. Actuation System for a Joint. U.S. Patent 9498354-B2, 22 November 2016.

61. Yang, C.; Yang, W.; Xu, L.; Wang, H.; Wei, Q.; Ma, Z. Lower Limb Rehabilitation Training Exoskeleton System and Its Walking Control Method and Hip Joint Structure. Cn-108392378-B, 20 September 2019.

62. Jang, W.Y.; Lee, S.H.; Lee, J.H. Weight Bearing Brace. Kr-101997997-B1, 8 July 2019.

63. Taek, J.I.; Park, Y.D. Training System for Leg Rehabilatation Having Saparated Treadmill. Kr-101275030-B1, 17 June 2013.

64. Wei, X.; Cao, W.; Wei, W.; Cheng, G.; Lu, H.; Yu, H.; Hu, B.; Meng, Q. Wearable Lower Limb Exoskeleton Rehabilitation Robot. Cn-107811805-B, 21 February 2020.

65. Unluhisarcikli, O.; Mavroidis, C.; Bonato, P.; Pietrusisnki, M.; Weinberg, B. Lower Extremity Exoskeleton for Gait Retraining. U.S. Patent 9198821-B2, 1 December 2015.

66. Guo, C.; Xiao, X.; Zhang, C.; Zhou, J. A Kind of Bionical Lower Limb Exoskeleton Robot Driven Based on Rope. Cn-106956243-B, 2 August 2019.

67. Zou, Y.; Wang, N.; Liu, K.; Li, J.; Geng, X.; Huang, Y. Movable Parallel Flexible Cable Driven Lower Limb Rehabilitation Robot and Implementation Method Thereof. Cn-108606907-B, 18 February 2020.
68. Wu, X.; Peng, A.; Wang, C.; Chen, C.; Liu, D.; Feng, W. Wearable Lower Limb Exoskeleton Robot. Cn-106109186-B, 14 August 2018.
69. Accoto, D.; Sergi, F.; Carpino, G.; Tagliamonte, N.L.; Galzerano, S.; Di Palo, M.; Guglielmelli, E. Robotic Device for Assistance and Rehabilitation of Lower Limbs. Ep-2906172-B1, 21 December 2016.
70. Aguirre-Ollinger, G.; Nagarajan, U.; Goswami, A. Admittance Shaping Controller for Exoskeleton Assistance of the Lower Extremities. U.S. Patent 9907722-B2, 6 March 2018.
71. Lu, T.; Tao, X.; Chang, H.; Yi, J. Exoskeleton-Type Moves Walking Rehabilitation Training Device and Method. Cn-105456004-B, 1 February 2019.
72. Gao, Y.; Liu, Y.; Zhu, Y. Based on Rope-Pulley Mechanism Drive Lacking Lower Limb Assistance Exoskeleton Robot. Cn-107137207-B, 17 May 2019.
73. Sun, K.C.; Tsai, Y.J.; Wu, C.H.; Hu, J.S. Method for Estimating Posture of Robotic Walking Aid. Ep-3173191-B1, 17 March 2021.
74. Herr, H.M.; Casler, R.; Han, Z. Hybrid Terrain-Adaptive Lower-Extremity Systems. U.S. Patent 8419804-B2, 16 April 2013.
75. Shi, X.; Li, Z.; Lin, S.; Zhu, J.; Liao, Z. A Kind of Dedicated Power-Assisted Healing Robot of Single Lower Limb Individuals with Disabilities. Cn-107411939-B, 27 September 2019.
76. Noda, T.; Morimoto, J. Actuator Device, Power Assist Robot and Humanoid Robot. Jp-5906506-B1, 20 April 2016.
77. Chung, H.I.; Kweon, H.D.; Ha, K.S.; Park, H. Monitoring System of Walking Balance for Lower Limb Rehabilitation. Kr-102014162-B1, 26 August 2019.
78. Grandmaison, C.; Sensinger, J. Powered Lower Limb Devices and Methods of Control Thereof. U.S. Patent 11191653-B2, 7 December 2021.
79. Liu, H.; Liu, G. A Kind of Artificial Intelligence Motion's Auxiliary Equipment. Cn-105213155-B, 29 March 2017.
80. Noda, T.; Nakata, Y.; Ishiguro, H.; Morimoto, J. Actuator Device, Humanoid Robot and Power Assist Device. Jp-6035590-B2, 30 November 2016.
81. Herr, H.M.; Kuan, J.Y. Robotic System for Simulating a Wearable Device and Method of Use. U.S. Patent 9498401-B2, 22 November 2016.
82. Agrawal, S.; Mankala, K.K.; Banala, S. Passive Swing Assist Leg Exoskeleton. U.S. Patent 8900167-B2, 2 December 2014.
83. Smith, F.M. Forward or Rearward Oriented Exoskeleton. Ep-2942044-B1, 19 January 2022.
84. Kapure, K.; Mudgal, G.; Shah, U.; Goyal, K.K.; Dahiya, S. Apparatus and System for Limb Rehabilitation. U.S. Patent 10492977-B2, 3 December 2019.
85. Goldfarb, M.; Murray, S. Movement Assist Device. Es-2773853-T3, 15 July 2020.
86. Han, J.W.; Kim, Y.H.; Min, D.M.; Lee, J.H. Sitting Type Walking Rehabilitation Robot. Cn-107708641-B, 12 January 2021.
87. Agrawal, S.K.; Vashista, V.; Kang, J.; Jin, X. Human Movement Research, Therapeutic, and Diagnostic Devices, Methods, and Systems. Ep-3133998-B1, 3 July 2019.
88. Chen, B.; Zi, B.; Wang, D.; Wang, Z.; Qian, S. Hip Joint Rehabilitation Exoskeleton Based on Multifunctional Driver and Motion Control Method Thereof. Cn-109172289-B, 23 February 2021.
89. Kim, J.Y.; Moon, K.; Shim, Y.B.; Lee, J.S. Wearable Robot and Control Method Thereof. U.S. Patent 10201437-B2, 12 February 2019.
90. Roh, C.H. Wearable Robot and Method for Controlling the Same. U.S. Patent 10556335-B2, 11 February 2020.
91. Huang, J.; Huang, Z.; Tu, X.; Zhang, H.; Xiong, C. A Kind of Link-Type Lower Limb Exoskeleton Rehabilitation Robot. Cn-107260483-B, 26 June 2018.
92. Tsai, Y.J.; Teng, M.C. Exoskeleton Robot. Jp-6729870-B2, 29 July 2020.
93. Amundson, K.; Harding, N.; Stryker, J. Interface for the Movement by Externally Applied Force Motivation of Adjustment Orthopedic Appliance. Cn-104936570-B, 9 October 2018.
94. Roy, A.; Forrester, L.W.; Macko, R.F. Method and Apparatus for Providing Deficit-Adjusted Adaptive Assistance During Movement Phases of an Impaired Joint. U.S. Patent 9943459-B2, 17 April 2018.
95. Little, R. Walking Aid. Cn-105899177-B, 11 September 2020.
96. Fu, C.; Chang, Y.; Wang, W. A Kind of Quasi—Passive Knee Ankle-Joint Coupling Lower Limb Exoskeleton and Its Control Method. Cn-107126348-B, 18 June 2019.
97. Roy, A.; Macko, R.F. Method and Apparatus for Providing Economical, Portable Deficit-Adjusted Adaptive Assistance During Movement Phases of an Impaired Ankle. Ca-2989421-C, 24 March 2020.
98. Julin, A.; Hughes, M. Ball Screw and Tensile Member Exoskeleton Joint Actuation Device. Ep-3243606-B1, 16 February 2022.
99. Robot for Assisting User to Walk. Kr-101988644-B1, 12 June 2019.
100. Zoss, A.; Evans, J.; Sandler, R.; Harding, N.; Julin, A.; Lubin, J.; Heanue, T.; Fairbanks, D.; Stryker, J. Reconfigurable Ectoskeleton Cn-104869970-D, 29 September 2017.
101. Einav, O. Gait Rehabilitation Methods and Apparatuses. U.S. Patent 8888723-B2, 18 November 2014.
102. Smith, J.; Bhugra, K. Methods of Operating an Exoskeleton for Gait Assistance and Rehabilitation. U.S. Patent 9545353-B2, 17 January 2017.
103. Agrawal, S.; Banala, S. Powered Orthosis. U.S. Patent 8147436-B2, 3 April 2012.
104. Kim, S.W.; Chung, S.G.; Kim, H.C.; Beom, J.W.; Nam, H.S.; Lee, C.W. A Treatment Device for Hemiplegia. Kr-101511427-B1, 10 April 2015.

105. Farris, R.; Clausen, M.; Wilson, E. Wearable Robotic Device. U.S. Patent 11141342-B2, 12 October 2021.
106. Kazerooni, H.; Amundson, K.; Angold, R.; Harding, N. An Exoskeleton and Method for Controlling a Swing Leg of the Exoskeleton. Ep-2346447-B1, 4 September 2019.
107. Sankai, Y. Wearing Tool for Measuring Biological Signal, and Wearing-Type Motion Assisting Device. Kr-101226169-B1, 24 January 2013.
108. Jeon, D.Y.; Moon, H.S.; Kwak, S.U.; Lee, J.H. Wheelchair Walking Assist Robot. Jp-5400890-B2, 29 January 2014.
109. Yuan, B.; Jiang, F.; Jiang, M.; Li, J.; Sun, Z. Reduction Exoskeleton Joint and Exoskeleton Power Assisting Device Thereof. Cn-106109181-B, 14 April 2020.
110. Patton, J.L.; Peshkin, M.A.; Sulzer, J.S. Cable Driven Joint Actuator and Method. U.S. Patent 9597217-B2, 21 March 2017.
111. Walsh, C.J.; Asbeck, A.T.; Yarri, M.W.; Cochran, J.C.; De Rossi, S.M.M. Orthopedic Device Including Protruding Members. U.S. Patent 10864100-B2, 15 December 2020.
112. Chen, W.; Wu, S.; Zhou, T.; Xiong, C. A Kind of Passive Exoskeleton Device of Hip Joint Based on Energy Timesharing Regulation. Cn-108095980-B, 22 November 2019.
113. Zoss, A.; Musick, D.; Harding, N. Methods of Enhancing the Rehabilitation or Training of an Exoskeleton Wearer. U.S. Patent 10576008-B2, 3 March 2020.
114. Lee, J.W.; Kim, H.Y. Robot for Assisting User to Walk. Kr-101979937-B1, 20 May 2019.
115. Nakashima, I.; Manabe, S. Leg Support Device. Ep-2583657-B1, 18 February 2015.
116. Lear, M.C.; Witherspoon, K.G.; Grant, M.; Kernbaum, N.I.; Mahoney, R.; Tayson-Frederick, M.L.; Fielding, L.C.; Riggs, V.; Shahoian, E.; Hogue, M.E. Systems and Methods for Assistive Exosuit System. U.S. Patent 10926123-B2, 23 February 2021.
117. Amundson, K.; Angold, R.; Harding, N.; Kazerooni, H. Device and Method for Reducing a Person's Oxygen Consumption During a Regular Walk by Using a Load-Bearing Exoskeleton. Es-2549004-T3, 22 October 2015.
118. Homayoon, K.; Kurt, A.; Nathan, H. Device and Method for Decreasing Energy Consumption of a Person by Use of a Lower Extremity Exoskeleton. Cn-102088933-B, 18 March 2015.
119. Smith, F.; Olivier, M. Walking Robot System That Regenerates Energy. Jp-6927912-B2, 1 September 2021.
120. Zhang, J.; Yu, T. It Is Single to Drive Bionical Gait Rehabilitation Training Robot System. Cn-106420271-B, 30 November 2018.
121. Jacobsen, S.C.; Olivier, M.X.; Maclean, B.J. Control Logic for Biomimetic Joint Actuators. U.S. Patent 8731716-B2, 20 May 2014.
122. Smith, F. Legged Robotic Device Utilizing Modifiable Linkage Mechanism. Kr-101778025-B1, 13 September 2017.
123. Dong, W.; Chen, C.; Du, Z.; Mao, W. Lower Limb Exoskeleton Robot System Based on Man-Machine Terminal Interaction. Cn-109223456-B, 13 October 2020.
124. Bouton, C.E.; Annetta, N.; Friedenberg, D.A.; Sharma, G. Systems for Neural Bridging of the Nervous System. U.S. Patent 10857358-B2, 08 December 2020.
125. Wang, X.; Jiang, C. Use the Lower Limb Exoskeleton Robot Control Method of Air Bag Sensor. Cn-105796286-B, 6 April 2018.
126. Morbi, A.; Ahmadi, M.; Beranek, R. Control System and Device for Patient Assist. Ca-2867484-C, 9 October 2018.
127. Angold, R.; Harding, N.H.; Kazerooni, H. Semi-Motorized Exoskeleton of the Lower Extremities. Es-2491218-T3, 5 September 2014.
128. Stephen, J.; Marc, O. Contact Displacement Actuator System. Cn-101489732-B, 28 January 2015.
129. Menga, G. System for Controlling a Robotic Device during Walking, in Particular for Rehabilitation Purposes, and Corresponding Robotic Device. Ep-2497610-B1, 22 October 2014.
130. Lee, J.W.; Kim, H.Y. Robot for Assisting User to Walk. Kr-102334854-B1, 3 December 2021.
131. Ye, Z.; Liu, P.; Lu, J.Y.; Yue, J. Lower Limb Rehabilitation Walking-Aid Robot Supporting Omnidirectional Movement and Control Method. Cn-107149539-B, 24 March 2020.
132. Ou, M.; Jiegao, W. Apparatus and Method for Reduced-Gravity Simulation. U.S. Patent 8152699-B1, 10 April 2012.
133. Zoss, A.; Swift, T.; Berg, A.; Strausser, K.; John, E.S. Powered Orthopedic System for Cooperative Above-Ground Rehabilitation. Es-2886134-T3, 16 December 2021.
134. Huang, J.; Yan, Q.; Cao, H. Walking Stick Type Autonomous Falling Protection Rehabilitation Walking-Aid Robot. Cn-107693314-B, 14 July 2020.
135. Song, Z.; Yan, P.; Hu, Y.; Tian, C.; Jia, C.; Deng, X. The Ectoskeleton Wheelchair Integrated Mobile Auxiliary Robot of Telescopic. Cn-106420203-B, 20 April 2018.
136. Wang, B.; Xu, D.; Wang, Z.; Liu, G.; Yang, W.; Wang, Y.; Hu, F. A Kind of Link Joint Integrated Hydraulic Driving Ectoskeleton. Cn-105686930-B, 20 November 2018.
137. Song, S.; Zhao, M.; Zou, Z.; Bai, P.; Chen, W.; Sun, J. Collapsible Mobile Lower Limb Exoskeleton. Cn-105686927-B, 11 July 2017.
138. Zhao, K.; Zhang, L.; Yi, W. Lower Limb Rehabilitation Robot Based on Bidirectional Neural Interface. Cn-109199786-B, 30 July 2021.
139. Zhang, J.; Collins, S.H. Torque Control Methods for an Exoskeleton Device. U.S. Patent 10555865-B2, 11 February 2020.
140. Park, H.S.; Lee, D.W.; Jung, J.Y. Sensor System for a User's Intention Following and Walk Supporting Robot. Kr-101171225-B1, 6 August 2012.
141. Ahn, S.H.; Kwon, Y.D.; Shim, Y.B.; Suk, J.Y. Walking Assist Robot and Control Method Thereof. U.S. Patent 9655805-B2, 23 May 2017.

142. Yang, W.; Xu, L.; Cao, B.; Peng, Z.; Canjun, Y. Lower Limb Exoskeleton System with Actively Adjustable Leg Rod Length and Control Method Thereof. Cn-111805511-B, 14 September 2021.
143. Wu, Q.; Zhu, Y.; Chen, B.; Chen, X.; Xu, D. Variable-Rigidity Lower Limb Exoskeleton Power-Assisted Robot. Cn-110652425-B, 17 December 2021.
144. Huang, L.; Chen, C.; Liu, X.; Lu, X.; Sun, H. Wearable Lower Limb Healing Robot Based on Ectoskeleton. Cn-105997441-B, 23 January 2018.
145. Wang, X.; Xi, R.; Zhu, Z. A Kind of Passive Energy Storage Foot Mechanism for Lower Limb Assistance Exoskeleton. Cn-105616113-B, 20 October 2017.
146. Yang, J.; Li, Z. Lower Limb Rehabilitation Robot Movement Intention Reasoning Method. Cn-109953761-B, 22 October 2021.
147. Wang, J.; Zhang, K.; Guo, X.; Li, M. Portable Ankle Joint Rehabilitation Robot Based on Active Intention Control. Cn-107997929-B, 1 September 2020.
148. Chen, W.; Wei, T.; Zhou, T.; Wu, S.; Xiong, C. A Kind of Passive Exoskeleton Device of Hip Knee Doublejointed Based on Clutch Timesharing Regulation. Cn-109528451-B, 8 October 2019.
149. Zhang, D.; Gui, K. Exoskeleton Hybrid Control System and Method for Lower Limb Walking Aid Machine. Cn-106112985-B, 22 May 2020.
150. Patoglu, V. Reconfigurable Ankle Exoskeleton Device. U.S. Patent 8366591-B2, 5 February 2013.
151. Belforte, G.; Eula, G.; Appendino, S.; Geminiani, G.C.; Zettin, M. Active Orthosis for the Neurological Rehabilitation of the Movement of the Lower Limbs, a System Comprising Said Orthosis and a Process to Put Said System into Operation. Es-2666382-T3, 04 May 2018.
152. Wu, C.; Ye, J.; Chen, G.; Zhang, X.; Hu, G.; Guo, D. Lower Limb Exoskeleton Control Method and Device. Cn-110812127-B, 4 January 2022.
153. Zhang, L. A Kind of Wearable Flexible Lower Limb Exoskeleton Based on Negative Pressure Rotary Pneumatic Artificial-Muscle. Cn-108542718-B, 26 July 2019.
154. Zhang, L.; Jiang, Y.; Xie, C.; Jiang, X.; Chen, H.; Zhang, Z.; Li, G.; Ma, J. A Kind of Portable Waist Hunting Gear. Cn-105232293-B, 11 July 2017.
155. Ping, Y. A Kind of Unweighting Walking Rehabilitation Training Robot. Cn-105726272-B, 16 October 2018.
156. Zhang, L.; Huang, Q.; Wang, W.; Wang, Z. A Kind of Wearable Flexible Knee Joint Robotic Exoskeleton Equipment Based on Gait. Cn-106420279-B, 9 April 2019.
157. Zhang, X.; Zheng, W.; Xiao, P.; Dong, Z.; Li, Y.; Zuo, G.; Cheng, Y. Semi-Automatic Bone Installations of Pulling Together. Cn-105919779-B, 9 March 2018.
158. Vitiello, N.; Giovacchini, F.; Cempini, M.; Fantozzi, M.; Moise, M.; Muscolo, M.; Cortese, M. Actuation System for Hip Joint Orthosis. Ru-2708223-C2, 4 December 2019.
159. Dong, W.; Du, Z.; Chen, C.; Mao, W. Lower Limb Exoskeleton Heterogeneous Knee Joint Based on Parallel Elastomers. Cn-109528453-B, 15 June 2021.
160. Katherine, G.W.; Alexander, S.K. Flexible Body Harness. Cn-210277443-U, 10 April 2020.
161. Ye, J.; Chen, G.; Xu, F.; Jiang, L.; Feng, J.; Zhang, X.; Hu, G.; Liu, S.; Li, Y. Lower Limb Training Rehabilitation Apparatus. U.S. Patent 11166866-B2, 9 November 2021.
162. Chen, B.; Wang, B. Motion Control Method Suitable for Exoskeleton Robot. Cn-111604890-B, 25 May 2021.
163. Hou, L.; Zhou, Y.; Wang, L.; Qiu, J.; Cheng, H.; Zhao, E. Anti-Falling System Based on Lower Limb Exoskeleton Robot. Cn-108836759-B, 29 December 2020.
164. Du, F.; Mu, X.; Chen, J.; Jiang, Z.; Ma, Z.; Zhao, Z.; Zhao, X.; Guo, H.; Song, G.; Wang, J. A Kind of Lower Limb Exoskeleton Robot. Cn 106137687-B, 5 September 2017.
165. Zhang, L.; Huang, Q. A Kind of Unpowered Wearable Auxiliary Walking Servomechanism. Cn-106491318-B, 25 December 2018.
166. A Kind of Sufficient Isomorphism Deformation Type Wheelchair Exoskeleton Robot of Wheel. Cn-108309593-B, 3 December 2019.
167. Zhu, Y.; Hua, Y.; Zhang, G.; Li, M. Overload Slipping Mechanism of Lower Limb Exoskeleton Robot. Cn-109998859-B, 8 October 2021.
168. Huang, Z.; Huang, Y. The External Bone Robot of Hemiparalysis Recovery Type. Cn-106063760-B, 12 October 2018.
169. Chen, Y.; Qiang, C.; Cao, Y.; Yang, M.; Sun, F. A Kind of Exoskeleton Robot. Cn-106580637-B, 28 December 2018.
170. Sankai, Y. Life Activity Detection Device and Life Activity Detection System Cyberdyne. Jp-6742196-B2, 19 August 2020.
171. Han, C.S.; Lee, S.; Choi, Y.S.; Soonwoong, H.; Kim, B.; Hwang, S.H. Walking Assistance Apparatus and Operation Method of the Same. Kr-101991588-B1, 20 June 2019.
172. Tang, S.; Cao, J.; Wang, G.; Li, J.; Guo, Z. Human Lower Limb Assisting Device. Cn-109464264-B, 18 September 2020.
173. Umashankar, N.; Ambarish, G. Control Design Framework for Resistant Exoskeleton. Jp-6800634-B2, 16 December 2020.
174. Li, M.; Wang, X.; Ren, G.; Wang, D.; Sun, C.; Yu, K.; He, J. Auxiliary Exercise System and Lower Limb Exoskeleton Control Method. Cn-110292506-B, 18 May 2021.
175. Cheng, H.; Wang, F.; Li, Z.; Qiu, J.; Yin, Z. A Kind of the Lower Limb Exoskeleton Training Method and System of the Triggering of Mental Imagery Pattern Brain-Computer Interface. Cn-105708587-B, 3 November 2020.
176. Hou, L.; Zhou, Y.; Zhao, E.; Qiu, J.; Cheng, H.; Huang, P.; Chen, C. Human Motion Intention Recognition Control Device and Control Method. Cn-109498375-B, 25 December 2020.

177. Ji, Y.H.; Shin, D.B.; Hwang, S.H.; Han, Y.H.; Lee, S.C.; Jang, H.Y.; Han, C.S. Ankle Module for Gait Rehabilitation Robot. Kr-101912920-B1, 28 December 2018.
178. Kap, H.S.; Jong, B.H.; Jongil, L.; Kyon, M.Y.; Seokjae, L.; Sohn, D.S. Soft Exo Suit for Fall Prevention and Gait Assistance. Kr-102027069-B1, 30 September 2019.
179. Ming, D.; Jiang, S.; Wang, Z.; Xu, M.; Zhao, X.; Qi, H.; Zhou, P. The Ectoskeleton Walk Help System Driven with Functional Muscle Electric Stimulation. Cn-106334265-B, 8 January 2019.
180. De Sapio, V.; Goldfarb, S.E.; Ziegler, M. Integrated Platform to Monitor and Analyze Individual Progress in Physical and Cognitive Tasks. U.S. Patent 10532000-B1, 14 January 2020.
181. Shen, C.; Zhu, X.; Yin, M.; Cheng, H.; Huang, R.; Xue, H. Human Motion Trend Detection Device and Detection Method Based on Force Sensor. Cn-105596018-B, 28 July 2020.
182. Peng, A.; Wu, X.; Chen, C.; Wang, C.; Liu, D.; Feng, W. Exoskeleton Robot Leg Exercise System. Cn-106236517-B, 7 September 2018.
183. He, F.; Huang, H.; Liu, Y.; Zhou, X.; Liu, L. Online Step Generation Control System for Exoskeleton Robot Contralateral Training. Cn-112220650-B, 16 April 2021.
184. Zhang, G.; Liu, H.; Wang, X.; Gao, K.; Chen, S. A Kind of Detachable Recovery Set for Lower Limbs and Control Method. Cn-106726369-B, 30 April 2019.
185. Park, K.T.; Nam, B.; Son, J.K.; Yu, S.; Lee, W. Adaptive Assistive And/or Rehabilitative Device and System. Ep-3524218-B1, 8 September 2021.
186. Son, J.K.; Nam, B.; Park, K.T.; Yu, S.; Lee, W. Wearable Assistive Device That Efficiently Delivers Assistive Force. U.S. Patent 10792209-B2, 6 October 2020.
187. Tüttemann, M.; Vogel, C. Leg Straightening Device and Straightening Device. Jp-6647225-B2, 14 February 2020.
188. Zhang, D.; Gui, K. The Lower Limb Rehabilitation Ectoskeleton Control System and Method That Subject Dominates. Cn-106214427-B, 24 May 2019.
189. Song, K.T.; Ko, C.L. Walking Rehabilitation Robot System. Tw-I702977-B, 1 September 2020.
190. Gong, G.C.; Lee, J.M.; Kim, Y.S.; Yoon, S.H.; Kwon, C.M.; Im, M.J. Ankle Assist Apparatus. Kr-102018436-B1, 5 September 2019.
191. In, N.; Yang, Z.; Sun, Y.; Wang, Y.; Fang, Y. A Kind of Single Rope Towards Gait and Balance Rehabilitation Training Suspends Active Loss of Weight System in Midair. Cn-106236519-B, 10 July 2018.
192. Jinglun, S.; Yanfang, W.; Huai, H.; Bin, Y. Convalescence Device Walking Trigger Control Method Based on Trunk Centre-of Gravity Shift. Cn-105326627-B, 13 April 2018.
193. Xiaodong, Z.; Gui, Y.; Guangyue, L.; Liangliang, L.; Qinyi, S.; Runlin, D.; Hanzhe, L. Brain-Myoelectricity Fusion Small-World Neural Network Prediction Method for Human Lower Limb Movement. Cn-111714339-B, 7 September 2021.
194. Wang, C.; Duan, L.; Liu, Q.; Shen, Y.; Shang, W.; Lin, Z.; Sun, T.; Xia, J.; Sun, Z.; Chen, X.; et al. Assisted Rehabilitation Training Robot. Jp-6555790-B2, 7 August 2019.
195. Lee, D.-C. Wearable Soft Exoskeleton Apparatus. Kr-101814676-B1, 4 January 2018.
196. Zhi, L.; Wenkang, Z.; Jianmin, L.; Qiang, W.; Hongliang, L.; Li, W. Walking Aid for Hemiplegia Patients. Cn-108670732-B, 30 June 2020.
197. Sawicki, J.T.; Laubscher, C.A.; Farris, R.J.; Etheridge, S.J.-S. Drive Device for Motorized Orthosis. Es-2858316-T3, 30 September 2021.
198. Choi, J.D.; Kwon, Y.D.; Kim, G.R.; Kim, J.M.; Shim, Y.B.; Ha, T.S. Wearable Robot and Control Method for the Same. Kr-102146363-B1, 20 August 2020.
199. Arzanpour, S.; Sadeqi, S.; Bourgeois, S.; Park, J.W. System and Device for Guiding and Detecting Motions of 3-Dof Rotational Target Joint. U.S. Patent 11135120-B2, 5 October 2021.
200. Patlogu, V. Ankle Treatment and Exoskeleton Measurement Device Not Making Contact with Ground, Capable of Being Worn and Capable of Being Reconstructed. Cn-203662948-U, 25 June 2014.
201. Tao, H. A Kind of Unpowered Walking Power-Assisted Flexible Exoskeleton Device. Cn-108354785-B, 20 September 2019.
202. Xinyu, G.; Yinbo, L.; Renhao, L.; Qinghui, M.; Linhong, J. Lower Limb Exoskeleton Driver. Cn-110559162-B, 29 September 2020.
203. Nakashima, I. Walking Training Apparatus and State Determination Method. Ep-3238686-B1, 30 September 2020.
204. Kim, J.-H.; Baek, Y.-S.; Son, B.-J. Shoe Module for Detecting Walking Phase, Method, Gait Analysis System and Active Walking Assist Device Using the Same. Kr-101648270-B1, 16 August 2016.
205. Kim, D.-H.; Seo, G.-H.; Yoon, S.-H.; Seo, J.-H.; Heo, Y.-S. Muscle Rehabilation Training Method Using Walking-Assistive Robot. Kr-101556747-B1, 1 October 2015.
206. Wang, X.; Qu, D.; Di, P.; Li, G.; Song, J.; Liu, S. Method for Controlling Man-Machine Interactive Motion of Lower Limb Exoskeleton Based on Joint Stress. Cn-111358667-B, 11 February 2022.
207. Qin, T.; Jin, C.; Wu, K.; Wei, C.; Qiu, J.; Wen, J.; Meng, X. Pelvic Auxiliary Walking Rehabilitation Training Robot. Cn-110812130-B, 9 November 2021.
208. Li, J.; Li, Y.; Liu, J.; Jiang, S.; Zhao, H.; Wang, X.; Liao, X. Auxiliary Standing Device and Auxiliary Standing Mechanism. Cn-110279562-B, 28 December 2021.
209. Lee, H.D. Tendon Device for Suit Type Robot for Assisting Human with Physical Strength. U.S. Patent 10799381-B2, 13 October 2020.
210. Wang, C.; Zhang, C.; Chen, B. Pneumatic Waist Assistance Exoskeleton Robot. Cn-110385692-B, 8 September 2020.

211. Ansi, P.; Wu, X.; Wang, C.; Chen, C.; Feng, W. Exoskeleton Robot Line Winding Driving Hip Joint. Cn-106236518-B, 14 August 2018.
212. Smith, J.A.; Bhugra, K. Method and System for Control and Operation of Motorized Orthotic Exoskeleton Joints. U.S. Patent 10278885-B1, 7 May 2019.
213. Podschun, T. System and Method for the Regeneration of At Least One Severed Nerve Conduit. U.S. Patent 10905877-B2, 2 February 2021.
214. Jeon, D.Y.; Hwang, B.S.; Jeon, J.H. Robot for Assistance Exoskeletal Power. Kr-101363850-B1, 18 February 2014.
215. Liu, K.; Luo, X.; Ji, S.; Sun, Z.; Xu, H. Sit and Stand and Go Multi-Functional Motion Auxiliary Robot. Cn-108992259-B, 1 May 2020.
216. Nam, B.; Park, K.T.; Son, J.K.; Yu, S.; Lee, W. Wearable Assistive Device Performing Protection Operation for Drive System. Ep-3539526-B1, 10 November 2021.
217. Xia, N. A Kind of Exoskeleton Robot Follow-Up Control Device. Cn-105078708-B, 31 May 2017.
218. Lim, B.; Park, Y.; Seo, K. Walking Assistance Method and Apparatuses. U.S. Patent 11109778-B2, 7 September 2021.
219. Sandler, R.; Strausser, K.; Fiedler, M.; Amundso, K.; Brown, D.; Smith, R.; Sweeney, M.D.; Angold, R.; Mccaffrey, N.; Edmonds, D.; et al. Methods of Exoskeleton Communication and Control. U.S. Patent 10694948-B2, 30 June 2020.
220. Laffranchi, M.; D'angella, S.; Uboldi, P.; Saglia, J.A.; Sanfilippo, C. Foot for a Robotic Exoskeleton for Assisted Walking of Persons Suffering from Locomotor Disorders. Ep-3322395-B1, 17 July 2019.
221. Threlfall, J. External Structure Holder Device. Cn-105473116-B, 9 January 2018.
222. Liao, W.-H.; Chen, B.; Qin, L.; Zhao, X.; Ma, H. Magneto-Rheological Series Elastic Actuator. U.S. Patent 10626944-B2, 21 April 2020.
223. Yoon, J.W.; Pyo, S.H. A Knee-Orthosis to Assist the Gait by Support the Knee-Joint. Kr-101330343-B1, 15 November 2013.
224. Qiu, J.; Deng, Q.; Yue, C.; Lin, X.; Zheng, X.; Chen, Y. A Kind of Plantar Pressure Measuring Device and Method for Ectoskeleton Control. Cn-105662419-B, 12 October 2018.
225. Lerner, Z.F. Exoskeleton Device. U.S. Patent 11034016-B2, 15 June 2021.
226. Lerner, Z.F. Exoskeleton Device. U.S. Patent 11090801-B2, 17 August 2021.
227. Wang, J.; Zhang, X.; Xiao, D.; Dong, K.; Gu, J. Power-Source-Free Knee Joint Mechanism. Cn-111419652-B, 29 September 2020.
228. Ming, D.; Jiang, S.; Wang, Z.; Xu, M.; Zhao, X.; Qi, H.; Zhou, P. A Kind of Ectoskeleton Walk Help System Driven with Functional Muscle Electric Stimulation. Cn-106377838-B, 5 April 2019.
229. Hong, M.B.; Yoon, Y.H.; Kim, D.J. Robot for Lower Limb with Multi-Link Type Knee Joint and Method for Controlling the Same. Kr-101828256-B1, 13 February 2018.
230. Wu, X.; Peng, A.; Wang, C.; Wang, Z.; Wu, G.; Liu, D.; Wu, C.; Liao, Y.; Liang, G. A Kind of Hip Joint Structure of Wearable Exoskeleton Robot. Cn-105877973-B, 21 September 2018.
231. Liu, D.; Du, Y.; Cong, M.; Zhang, J.; Yang, J. Variable-Rigidity Lower Limb Exoskeleton Robot Based on Shape Memory Alloy. Cn-111823217-B, 4 January 2022.
232. Wang, Z.; Yu, S.; Zhao, N.; Zhang, X.; Meng, D.; Zhang, K. Lower Limb Exoskeleton Inverse Motion Analysis Method Under Random Road Surface Condition. Cn-106548033-B, 19 May 2020.
233. Fu, C.; Liu, J. A Kind of Light-Duty Ankle-Joint Ectoskeleton. Cn-106625605-B, 5 February 2019.
234. Liu, Y.; Zhang, G.; Song, R.; Li, Y. Wearable Metatarsophalangeal Joint Walking Power Assisting Device. Cn-111671624-B, 27 August 2021.
235. Zhang, L.; Wang, Z.; Cai, K.; Wang, W.; Huang, Q. Flexible Exoskeleton Robot Assisting Movement of Hip Joint and Knee Joint. Cn-108938340-B, 26 June 2020.
236. Song, W.K.; Kwon, S.C.; Ko, B.W. Walking Pattern Training and Intension Analysis System Through Complex Stimulus, and Method Thereof. Kr-101911221-B1, 28 December 2018.
237. Nam, B.; Park, K.T.; Son, J.K.; Yu, S.; Lee, W. Belt for Effective Wearing and Wearable Assistive Device Having the Same. U.S. Patent 10603786-B2, 31 March 2020.
238. Ivanov, A.V.; Tolstov, K.M.; Kostenko, A.A.; Kuznetsov, A.S.; Komarov, P.A. Femal Link of an Active Foot Orthosis. Ru-171262-U1, 25 May 2017.
239. Park, H.S.; Kim, C.J. Gait Rehabilitation Apparatus. Kr-102130886-B1, 6 July 2020.
240. Shi, J.; Wang, Y.; Huang, H.; Zhao, W. The Walking Trigger Control Method of Convalescence Device Based on Foot Pressure Sensor. Cn-105342808-B, 15 May 2018.
241. Han, C.S.; Lee, S.; Choi, Y.S.; Soonwoong, H.; Kim, B.; Hwang, S.H. Walking Assistance Apparatus and Operation Method of the Same. Kr-101981402-B1, 28 August 2019.
242. Yamamoto, K.; Toshiyuki, K. Joint Motion Assist Device. Jp-6105091-B2, 29 March 2017.
243. Daba, H. Walking Training Apparatus and Method of Controlling the Same. Ep-3360529-B1, 23 September 2020.
244. Chen, L.; Song, X.; Ma, S.; Wang, J. Control Method of Lower Limb Exoskeleton Robot. Cn-109276415-B, 22 December 2020.
245. Dong, W.; Chen, C.; Du, Z.; Mao, W. Sole Human-Computer Interaction Measuring Device Based on Multi-Source Information Fusion. Cn-109222984-B, 6 July 2021.
246. Handrail Type Intelligent Tumble Protection Walking Aid Rehabilitation Robot. Cn-109157381-B, 4 August 2020.

247. Wang, C.; Liao, Y.; Wu, X.; Peng, A.; Wu, C.; Zhang, S.; Hu, X.; Li, C. Support Member and the Self-Adapting Seat Device with the Support Member. Cn-105342811-B, 31 October 2017.
248. Jin, H.; Yang, S.; Jin, L.; Mei, L.; Hu, R. Lower Limb Power Assisting Device. Cn-110192966-B, 8 February 2022.
249. Han, C.S.; Lee, S.; Choi, Y.S.; Soonwoong, H.; Kim, B.; Hwang, S.H. Walking Assistance Apparatus and Operation Method of the Same. Kr-101981403-B1, 28 August 2019.
250. Shi, J.; Wang, Y.; Zhao, W.; Yang, N.; She, H. A Kind of Convalescence Device Speed of Travel Control Method Rocked Based on Trunk. Cn-105287164-B, 5 January 2018.
251. Lin, X.; Deng, Q.; Yin, Z.; Qiu, J.; Chen, Y.; Zhou, C. A Kind of Adaptive Ectoskeleton Knee Joint Support Plate Unlocked. Cn-106074094-B, 15 June 2018.
252. He, J.; Luo, Z.; Liao, Y. Standing Mode Control Method of Exoskeleton Mechanical Leg Rehabilitation System. Cn-106726357-B, 22 September 2020.
253. Chen, L.; Li, S.; Zhang, Y.; Zhang, C. Electromyographic Signal Collection Position Choosing Method Based on Complex Network. Cn-105769186-B, 29 May 2018.
254. Liang, W. The Bionical Dynamic Knee Joint System in the Wearable List Source of One Kind and Its Control Method. Cn-108175639-B, 30 August 2019.
255. Little, R.; Shishbaradaran, J.D. Exoskeleton and Mounting Arrangement. Ep-3501473-B1, 25 November 2020.
256. Borisoff, J.; Dougall, R.; Janzen, E.; Mattie, J. Active Arm Passive Leg Exercise Machine with Guided Leg Movement. U.S. Patent 11083924-B2, 10 August 2021.
257. Arzanpour, S.; Wook, P.J.; Page, L.; Bazhanov, N.; Dehghani, H. Self-Supported Device for Guiding Motions of a Target Joint. Ca-3098479-C, 8 June 2021.
258. Li, T.; Li, M. A Kind of Wearable Leg Power Brace of Self-Regulation. Cn-105997438-B, 12 February 2019.
259. Ji, L.; Ma, Q. A Wheeled Drive Self-Balancing Power Ectoskeleton of Sole for Spinal Cord Injury Patient. Cn-107174488-B, 20 March 2020.
260. Nam, B.; Park, K.T.; Son, J.K.; Yu, S.; Lee, W. Controlling Position of Wearable Assistive Device Depending on Operation Mode. Ep-3539527-B1, 17 November 2021.
261. Nigamadyanov, N.R.; Lukyanov, V.I.; Ivanova, G.I.; Tsykunov, M.B. Modular Orthopedic Apparatus. Ru-2659132-C1, 28 June 2018.
262. Nasiri, R.; Ahmadabadi, M.N.; Ahmadi, A. Methods and Systems for an Exoskeleton to Reduce a Runners Metabolic Rate. U.S. Patent 10549138-B2, 4 February 2020.
263. Shuai, M. Telescopic Structure and Exoskeleton Robot with Same. Cn-109431752-B, 2 February 2021.
264. Dalley, S.; Farris, R.; Murray, S. Advanced Gait Control System and Methods Enabling Continuous Walking Motion of a Powered Exoskeleton Device Parker-Hannifin Corporation. Ep-3750166-B1, 5 January 2022.
265. Van Den Bogert, A.J.; Miller, D.; Doris, R.; Moody, R.; Grimes, D., Jr. Exoskeleton System. U.S. Patent 11191694-B2, 7 December 2021.
266. Lee, S.R.; Kang, O.H.; Yun, J.H.; Yi, H. Reaction Force Adjusting Device of Exoskeleton System and Variable Stiffness Actuator Using the Same. Kr-102067127-B1, 16 January 2020.
267. Dalley, S.A.; Farris, R.; Etheridge, S.; Morrison, S.; Murray, S. Safety Monitoring and Control System and Methods for a Legged Mobility Exoskeleton Device. U.S. Patent 11077556-B2, 3 August 2021.
268. Hong, M.B.; Kim, D.J.; Lee, S.H.; Seo, C.H.; Wang, J.H. Elastic Type Sole Assembly in Wearable Robot Absorbing Impact and Detecting Ground Reaction Force. Kr-101619297-B1, 10 May 2016.
269. Yamamoto, K.; Ishii, M. Foot Mounting Structure of Joint Motion Assist Device. Jp-5930354-B2, 8 June 2016.
270. Sulzer, J.; Shin, S.Y. Adaptable Robotic Gait Trainer. U.S. Patent 11135119-B2, 5 October 2021.
271. Lee, S.R.; Kang, O.H.; Yun, J.H.; Yi, H. Reaction Force Adjusting Device and Method Using Variable Stiffness Actuator of Exoskeleton System. Kr-102067124-B1, 16 January 2020.
272. Mahoney, R.; Gadway, C.; Hogue, M.E.; Kernbaum, N.I.; Lear, M.C.; Stolee-Smith, H.; Tayson-Frederick, M.L.; Toh, R.; Vincent, R.; Weber, T.; et al. Exosuit Systems and Methods. U.S. Patent 11192237-B2, 7 December 2021.
273. Dai, Y.; Deng, C.; Tang, K.; Tang, P. A Kind of Ectoskeleton Stopping Means. Cn-105856195-B, 30 January 2018.
274. Lugris, A.U.; Cuadrado, A.J.; Font, L.J.M.; Clos, C.D.; Alonso, S.F.J.; Romero, S.F. System to Assist Walking. Es-2663899-B2, 11 September 2018.
275. Quan, L.; Ge, L.; Liu, B.; Zhang, X.; Wang, B.; Li, Z. Electro-Hydraulic Hybrid Driving Exoskeleton Device. Cn-108670731-B, 07 April 2020.
276. Wu, H.; Sun, J.; Pan, W.; Liu, X.; Tian, J. Pneumatic Weight-Reducing Walking Power-Assisted Robot. Cn-109124999-B, 16 June 2020.
277. Stefan, F.; Gussen, U.; Petri, F. Mobility System. De-102016215399-B3, 5 October 2017.
278. Lee, J.W.; Kim, H.Y. Gait Assist Robot for Rehabilitation Training with Lift Device. Kr-102334851-B1, 3 December 2021.
279. Lee, J.C.; Bae, Y.H. Robot Device for Upper and Lower Extremity Rehabilitation. Kr-102305924-B1, 30 September 2021.
280. Sheng, Y.; Huang, H.; Feng, L.; Chen, Z.; Qian, Y. Device and Method for Realizing Cooperative Motion of Weight-Reducing Vehicle and Lower Limb Robot Through Communication. Cn-110916970-B, 21 September 2021.
281. Li, B.; Hu, C.; Huang, Q.; Zhou, Y.; Zhang, Y. Counter Weight Type Lower Limb Rehabilitation Robot. Cn-110613587-B, 16 November 2021.

282. Li, Z.; Li, D.; Xu, H. Lower Limb Exoskeleton Robot with Overload Slipping Function. Cn-109998860-B, 8 October 2021.
283. Sheng, Y.; Huang, H.; Feng, L.; Chen, Z.; Qian, Y. Device and Method for Assisting Lower Limb Robot to Transfer Gravity Center by Aid of Weight Reduction Vehicle. Cn-110946742-B, 19 November 2021.
284. Pei, Z.; Pei, P.; Gu, H.; Liu, H.; Tang, Z. Anti-Falling Walking Aid Vehicle for Lower Limb Rehabilitation Training and Rehabilitation Training Method. Cn-110353952-B, 21 July 2020.
285. Lee, J.W.; Kim, H.Y. Patient Weight Burden Reduction Device of Walking Rehabilitation Training Robot. Kr-102317401-B1, 26 October 2021.
286. Wang, J.; Wang, T.; Xiao, D.; Gu, J.; Li, C. Exoskeleton Joint Self-Locking Mechanism, Knee Joint and Bionic Rehabilitation Robot. Cn-111920651-B, 5 February 2021.
287. Dong, W.; Du, Z.; Chen, C.; Mao, W. Lower Limb Exoskeleton Ankle Joint Based on Telecentric Mechanis. Cn-109528440-B, 2 March 2021.
288. Kweon, H.D.; Ha, K.S.; Park, J.S.; Jeung, H.I. Lower Extremity Exoskeleton Robotic Device. Kr-102292983-B1, 24 August 2021.
289. Wang, X.; Cheng, W.; Du, F. Lower Limb Exoskeleton Capable of Being Used for Both Wheel and Leg. Cn-110393656-B, 11 May 2021.
290. Huang, J.; Huo, J.; Zhang, M.; Xiong, C.; Xiao, X. Auxiliary Dual-Purpose Outer Limb Robot for Human Body Movement. Cn-110292510-B, 8 September 2020.
291. Youngwoo, K. Rehabilitation Walking Method Considering Patient's Rom Characteristics and System Thereof. Kr-102020129-B1, 9 September 2019.
292. Geffard, F.; Ponsort, D. Lower Limbs of the Exoskeleton with Low Power Consumption. Jp-6845378-B2, 17 March 2021.
293. Wu, X.; He, Y.; Liu, J.; Li, J.; Ma, Y.; Li, F.; Sun, J.; Dong, Y.; Lian, P. Novel Self-Balancing Ectoskeleton Robot. Cn-113181009-B, 1 February 2022.
294. Yoon, J.W.; Sabaapour, M.; Lee, H.S. Rehabilitation Robot. Kr-102081911-B1, 26 February 2020.
295. Kwon, S.C.; Kim, K.H. Leg Opening Joint of Walking Exoskeleton and Walking Exoskeleton Comprising the Same. Kr-102037316-B1, 15 November 2019.
296. Liu, J.; Wu, X.; He, Y.; Li, J.; Li, F.; Ma, Y.; Cao, W.; Wang, D.; Sun, J.; Lian, P. Motion Decoupling Parallel Driving Type Exoskeleton Robot Ankle Joint. Cn-113183176-B, 25 January 2022.
297. Wang, J.; Wang, T.; Xiao, D.; Li, C.; Gu, J. Locking-Free Hip Adjusting Device. Cn-112245239-B, 9 April 2021.
298. Zhang, W.; Su, Y.; Xiao, B.; Shao, Y.; Ding, X. Constant-Force Human Body Suspension System for Rehabilitation Training. Cn-110507516-B, 29 September 2020.
299. Zhang, J.; Hu, K.; Zhu, Y.; Shen, L.; Yang, J.; Xiong, L.; Li, J. Control Method of Hydraulic System of Knee Joint Rehabilitation Robot. Cn-107669447-B, 21 July 2020.
300. Gao, W. Walking-Aid Boots. Cn-107661193-B, 7 April 2020.
301. Park, S.S.; Jung, S.H.; Park, K.T. Walking Assistance System. Kr-102352537-B1, 17 January 2022.
302. Seo, K.H.; Han, J.B.; Lee, J.I.; Yang, K.M.; Sin, H.S.; Lee, S.J.; Sohn, D.S. Wearable Suit Control Method. Kr-102288425-B1, 10 August 2021.
303. Zhu, H.; Zhang, B.; Qiu, P. Power-Assisted Exoskeleton Control Method, Power-Assisted Exoskeleton Control System and Computer Equipment. Cn-112809651-B, 29 October 2021.
304. Liu, D.; Du, Y.; Cong, M.; Zhang, J.; Yang, J. Variable-Rigidity Knee Joint Exoskeleton Robot Based on Shape Memory Alloy. Cn-111823218-B, 14 January 2022.
305. Hong, M.-B.; Yoon, Y.-H. Motion Assist Apparatus. Kr-102156837-B1, 16 September 2020.
306. Andrey Valerievich Borisov, A.V.; Borisova, V.L.; Konchina, L.V.; Kulikova, M.G.; Maslova, K.S. Exoskeleton. Ru-2760527-C1, 26 November 2021.
307. Ryong, L.S.; Young, L.C.; Hyun, K.O.; Hwan, K.I. Two-Leg Walking Assistant System for Boarding Type. Kr-101497673-B1, 3 March 2015.
308. Polygerinos, P.; Sridar, S.; Maruyama, T.; St. Clair, C.; Kwasnica, C. Soft Inflatable Exosuit for Knee Rehabilitation. U.S. Patent 11259980-B2, 1 March 2022.
309. Ho, S.K.; Boo, H.J.; Il, L.J.; Mo, Y.K.; Seob, S.H.; Jae, L.S.; Seop, S.D. Walk Assistance and Fall Prevention Wearable Suit. Kr-102370567-B1, 7 March 2022.
310. Hwan, K.K.; Hun, C.K. Size-Adjustable Pelvis Unit and Wearable Walking Robot Comprising the Same. Kr-101765547-B1, 9 August 2017.
311. Jiongming, S.; Bicong, Z.; Jiali, B.; Jiayin, H.; Mengqi, A.; Ganggang, Z. A Kind of Human Foot's Bionic Exoskeleton System. Cn-106344344-B, 28 August 2018.
312. Ryong, L.S.; Hyun, K.O.; Hwan, Y.J.; Hak, Y. Apparatus and Method for Observing Feedback Force of Wearable Exoskeleton System. Kr-102002635-B1, 1 October 2019.
313. Takashi, I.; Kenichiro, F.; Takeo, N. Measurement System, Measurement Method, and Program. Jp-6879301-B2, 2 June 2021.
314. Larose, P.; Denninger, M.; Looking, B.J.-P.; Plante, J. S.; Veronno, C.; Grenir, J. Devices Designed to Be Positioned Near Joints and Systems Incorporating Such Devices. Kr-102336585-B1, 9 December 2021.
315. Unluhisarcikli, O. Lower Extremity Exoskeleton for Gait Retraining. U.S. Patent 20160151176, 2 June 2016
316. Unluhisarcikli, O. Lower Extremity Exoskeleton for Gait Retraining. Wo/2013/049658, 4 April 2013.

317. Lee, C.-S. Pneumatic Lower Extremity Gait Rehabilitation Training System. U.S. Patent 20180071580, 15 March 2018.
318. Zoss, A. Powered Orthotic System for Cooperative Overground Rehabilitation Ekso Bionics. Wo/2014/159857, 2 October 2014.
319. Mudgal, G. Robotic Exoskeleton Assisted (Locomotion) Rehabilitation System "Rears"Bionic Yantra Private LimitedMudgal. In-201741045051, 1 February 2019.
320. Belforte, G. Active Sling for the Motion Neurological Rehabilitation of Lower Limbs, System Comprising Such Sling and Process for Operating Such System. Wo/2013/136351, 19 September 2013.
321. Belforte, G. Active Orthosis for the Motion Neurological Rehabilitation of Lower Limbs, System Comprising Such Orthosis and Process for Operating Such System. Ep-2825146, 21 January 2015.
322. Amundson, K. Interface for Adjusting the Motion of a Powered Orthotic Device Through Externally Applied ForcesEkso Bionics. Wo/2014/113456, 24 July 2014.
323. Amundson, K. Interface for Adjusting the Motion of a Powered Orthotic Device Through Externally Applied Forces. Ep-2945590, 25 November 2015.
324. Zoss, A. Powered Orthotic System for Cooperative Overground Rehabilitation. Ep-2968052, 20 January 2016.
325. De Rossi, S.M.M. Assistive Flexible Suits, Flexible Suit Systems, and Methods for Making and Control Thereof to Assist Human Mobility. Ep-3079581, 19 October 2016.
326. Tong, K.-Y. Interactive Exoskeleton Robotic Knee System. Wo/2016/180074, 17 November 2016.
327. Aguirre-Ollinger, G. Admittance Shaping Controller for Exoskeleton Assistance of the Lower Extremities. U.S. Patent 20180098907, 12 April 2018.
328. De Rossi, S.M.M. Assistive Flexible Suits, Flexible Suit Systems, and Methods for Making and Control Thereof to Assist Human Mobility. U.S. Patent 20170202724, 20 July 2017.
329. Ortlieb, A.L. Bio-Inspired Adaptive Impedance Based Controller for Human-Robot Interaction and Method. Wo/2020/039409, 27 February 2020.
330. Kaustubh, K. Apparatus and System for Limb Rehabilitation. U.S. Patent 20190183715, 20 June 2019.
331. Aguirre-Ollinger, G. Admittance Shaping Controller for Exoskeleton Assistance of the Lower Extremities. U.S. Patent 20160045385, 18 February 2016.
332. Ortlieb, A.L. Modular and Minimally Constraining Lower Limb Exoskeleton for Enhanced Mobility and Balance Augmentation. U.S. Patent 20190254908, 22 August 2019.
333. Ortlieb, A.L. Modular and Minimally Constraining Lower Limb Exoskeleton for Enhanced Mobility and Balance Augmentation. Wo/2018/065913, 12 April 2018.
334. Walsh, C. Soft Exosuit for Assistance with Human Motion. Ep-3003231, 13 April 2016.
335. Walsh, C. Soft Exosuit for Assistance with Human Motion. Ep-3777677, 17 February 2021.
336. Zoss, A. Methods of Enhancing the or Training Rehabilitation of an Exoskeleton Wearer. U.S. Patent 20180078442, 22 March 2018.
337. Kapure, K. Apparatus and System for Limb Rehabitation. Bionic Yantra Private Limited Kapure. Wo/2019/116093, 20 June 2019.
338. Vouga, T. Bio-Inspired Standing Balance Controller for a Full-Mobilization Exoskeleton. U.S. Patent 20210015694, 21 January 2021.
339. Swartz, A. Methods for Improved User Mobility and Treatment. Roam Robotics. Wo/2021/243056, 2 December 2021.
340. Tong, K.-Y. Exoskeleton Ankle Robot. Wo/2016/180073, 17 November 2016.
341. Swartz, A. Powered Medical Device and Methods for Improved User Mobility and Treatment Roam Robotics. U.S. Patent 20210370495, 2 December 2021.
342. Zoss, A. Methods of Enhancing the Rehabilitation or Training of an Exoskeleton Wearer. Wo/2016/077442, 19 May 2016.
343. Reiner, R. Soft Wearable Muscle Assisting Device. Ep-3701927, 2 September 2020.
344. Zoss, A. Exoskeleton. Ep-3217942, 20 September 2017.
345. Reiner, R. Soft Wearable Muscle Assisting Device. Wo/2018/122106, 5 July 2018.
346. Agrawal, S.K. Human Movement Research, Therapeutic, and Diagnostic Devices, Methods, and Systems. U.S. Patent 20170027803, 2 February 2017.
347. Kapure, K. Robotic Management System for Limb Rehabilitation. U.S. Patent 20200170876, 4 June 2020.
348. Amundson, K. Interface for Adjusting the Motion of a Powered Orthotic Device Through Externally Applied Forces. U.S. Patent 20150351991, 10 December 2015.
349. Caban, M. Control System for Movement Reconstruction And/or Restoration for a Patient. U.S. Patent 20200147382, 14 May 2020.
350. Stuart, R. Data Logging and Third-Party Administration of a Mobile Robot Roam. U.S. Patent 20210369542, 2 December 2021.
351. Von Zitzewitz, J. Apparatus Comprising a Support System for a User and Its Operation in a Gravity Assist Mode. U.S. Patent 20210283001, 16 September 2021.
352. De Sapio, V. Integrated Platform to Monitor and Analyze Individual Progress in Physical and Cognitive Tasks. U.S. Patent 10532000, 14 January 2020.
353. Asbeck, A.T. Soft Exosuit for Assistance with Human Motion. Ep-2895133, 22 July 2015.
354. Goldfarb, M. Movement Assistance Device. Ep-2861198, 22 April 2015.
355. Goldfarb, M. Movement Assistance Device. Wo/2013/188868, 19 December 2013.
356. Reese, M. Exoskeleton and Master. U.S. Patent 20190232485, 1 August 2019.
357. García, A.E. Exoskeleton for Assisting Human Movement. Ep-3225363, 4 October 2017.
358. Stuart, R. Data Logging and Third-Party Administration of a Mobile Robot Roam Robotics. Wo/2021/242991, 2 December 2021.

359. Zoss, A. Reconfigurable Exoskeleton. U.S. Patent 20150351995, 10 December 2015.
360. Bulea, T. Powered Gait Assistance Systems. U.S. Patent 20210298984, 30 September 2021.
361. Courtine, G. Apparatus and Method for Restoring Voluntary Control of Locomotion in Neuromotor Impairments. Ca-2874101, 5 December 2013.
362. Courtine, G. Apparatus and Method for Restoring Voluntary Control of Locomotion in Neuromotor Impairments. Au-2013269175, 11 December 2014.
363. Courtine, G. Apparatus and Method for Restoring Voluntary Control of Locomotion in Neuromotor Impairments. Wo/2013/179230, 5 December 2013.
364. Courtine, G. Apparatus for Restoring Voluntary Control of Locomotion in Neuromotor Impairments. Ep-2854939, 8 April 2015.
365. Zoss, A. Reconfigurable Exoskeleton. Ep-2931204, 21 October 2015.
366. Sandler, R. Methods of Communication Exoskeleton and Control. Wo/2016/168463, 20 October 2016.
367. Thomas, A.A. Soft Exosuit for Assistance with Human Motion. Ep-3791834, 17 March 2021.
368. Walsh, C. Soft Exosuit for Assistance with Human Motion. U.S. Patent 20150173993, 25 June 2015.
369. Courtine, G. Apparatus for Restoring Voluntary Control of Locomotion in Neuromotor Impairments. Ep-3241586, 8 November 2017.
370. Asbeck, A.T. Soft Exosuit for Assistance with Human Motion. U.S. Patent 20150321339, 12 November 2015.
371. Wang, B. Power-Assist with Adjustable Lower Limb Exoskeleton Robot Stiffness Joints. U.S. Patent 20200337934, 29 October 2020.
372. Dalley, S. Mobility Assistance Devices with Automated Assessment and Adjustment Control. Wo/2018/175004, 27 September 2018.
373. Walsh, C. Soft Exosuit for Assistance with Human Motion. U.S. Patent 20160220438, 4 August 2016.
374. Tong, K.-Y. Exoskeleton Ankle Robot. U.S. Patent 20160331557, 17 November 2016.
375. Dalley, S. Mobility Assistance Devices with Automated Assessment and Adjustment Control. U.S. Patent 20200060921, 27 February 2020.
376. Roy, A. Method and Apparatus for Providing Economical, Portable Deficit-Adjusted Adaptive Assistance During Movement Phases of an Impaired Ankle. Wo/2016/209770, 29 December 2016.
377. Contreras-Vidal, J.-L. Customizable Orthotic/Prosthetic Braces and Lightweight Modular Exoskeleton. Wo/2017/218661, 21 December 2017.
378. Walsh, C. Soft Exosuit for Assistance with Human Motion. U.S. Patent 20210039248, 11 February 2021.
379. Bulea, T.C. Powered Gait Assistance Systems. Wo/2018/023109, 1 February 2018.
380. Contreras-Vidal, J.-L. Customizable Orthotic/Prosthetic Braces and Lightweight Modular Exoskeleton. U.S. Patent 20190328604, 31 October 2019.
381. Roy, A. Method and Apparatus for Providing Economical Portable Deficit Adjusted Adaptive Assistance during Movement Phases of an Impaired Ankle. In-201717044844, 16 March 2018.
382. Casler, R.J., Jr. Hybrid Terrain- Adaptive Lower-Extremity Systems. U.S. Patent 20190209348, 11 July 2019.
383. Herr, H.M. Robotic System for Simulating a Wearable Device and Method of Use. U.S. Patent 20130158444, 20 June 2013.
384. Herr, H.M. Kinetic Sensing, Signal Generation, Feature Extraction, and Pattern Recognition for Control of Autonomous Wearable Leg Devices. U.S. Patent 20190307583, 10 October 2019.
385. Lerner, Z.F. Open-Loop Control for Exoskeleton Motor. U.S. Patent 20210291355, 23 September 2021.
386. Foucault, A. Kinoped Lower Extremity Performance Improvement, Injury Prevention, and Rehabilitation System. Wo/2021/046482, 11 March 2021.
387. Legaz, J.G. System for Assisting Walking. Ep-3378446, 26 September 2018.
388. Sandler, R. Methods of Communication Exoskeleton and Control. Ca-2982778, 20 October 2016.
389. Goldfarb, M. Movement Assistance Device. U.S. Patent 20150142130, 21 May 2015.
390. Miller Herr, H.M. Hybrid Terrain-Adaptive Lower-Extremity Systems. U.S. Patent 20140081424, 20 March 2014.
391. Tong, K.-Y. Interactive Exoskeleton Robotic Knee System. U.S. Patent 20160331560, 17 November 2016.
392. Herr, M.H. Hybrid Terrain-Adaptive Lower-Extremity Systems. U.S. Patent 20160296348, 13 October 2016.
393. Herr, M.H. Implementing a Stand-Up Sequence Using a Lower-Extremity Prosthesis or Orthosis. U.S. Patent 20140081420, 20 March 2014.
394. Herr, M.H. Hybrid Terrain-Adaptive Lower-Extremity Systems. U.S. Patent 20190117415, 25 April 2019.
395. Von Zitzewitz, J. Apparatus Comprising a Support System for a User and Its Operation in a Gravityassist Mode. Wo/2018/033591, 22 February 2018.
396. Witte, K.A. Exoskeleton Device and Control System. U.S. Patent 20180125738, 10 May 2018.
397. Dalley, S.A. Cloud-Based Control System and Method Enabling Interactive Clinical Care Using a Powered Mobility Assistance Device. U.S. Patent 20210386611, 16 December 2021.
398. Herr, H.M. Hybrid Terrain-Adaptive Lower Extremity Systems. U.S. Patent 20160235557, 18 August 2016.
399. Sandler, R. Methods of Communication Exoskeleton and Control. U.S. Patent 20100092536, 5 April 2018.
400. Witte, K.A. Exoskeleton Device and Control System. U.S. Patent 20210369537, 2 December 2021.
401. Casler, R.J., Jr. Hybrid Terrain-Adaptive Lower-Extremity Systems. U.S. Patent 20170086991, 30 March 2017.
402. Walsh, C.J. Soft Exosuit for Assistance with Human Motion. U.S. Patent 20160107309, 21 April 2016.
403. Walsh, C.J. Systems, Methods, and Devices for Assisting Walking for Developmentally-Delayed Toddlers. U.S. Patent 20160346156, 1 December 2016.

404. Method and Apparatus for Providing Economical, Portable Deficit-Adjusted Adaptive Assistance During Movement Phases of an Impaired Ankle. Ca-2989421, 29 December 2016.
405. Zhang, J. Torque Control Methods for an Exoskeleton Device. U.S. Patent 20170340506, 30 November 2017.
406. Johnson, C.L. Patient Aid Devices, Particularly for Mobile Upper Extremity Support in Railed Devices Such as Parallel Bars and Treadmills. U.S. Patent 20180043202, 15 February 2018.
407. Lonner, J.H. Autonomous Mobile Support System for the Robotic Mobility-Impaired Ambulatus Robotics. Wo/2021/086471, 6 May 2021.
408. Ji, Y.K. Wearable Robot and Control Method Thereof. U.S. Patent 20150196403, 16 July 2015.
409. Johnson, C.L. Patient Aid Devices, Particularly for Mobile Upper Extremity Support in Railed Devices Such as Parallel Bars and Treadmills. U.S. Patent 20150335940, 26 November 2015.
410. Bujanda, I.G. Wearable Devices for Protecting Against Musculoskeletal Injuries and Enhancing Performance. Wo/2019/161232, 22 August 2019.
411. Mahoney, R. Exosuit Systems and Methods Seismic. Wo/2019/108655, 6 June 2019.
412. Mahoney, R. Exosuit Systems and Methods Seismic. U.S. Patent 20190160651, 30 May 2019.
413. Trutz, P. System and Method for the Regeneration of At Least One Severed Nerve Conduit. U.S. Patent 20180326210, 15 November 2018.
414. Mahoney, R. Exosuit Load Bearing Distribution Systems. U.S. Patent 20190160652, 30 May 2019.
415. Markus, T. Orthosis Leg and Orthosis. U.S. Patent 20170156963, 8 June 2017.
416. Scattareggia, M.S. Esoskeleton Equipped with Electro-or Magneto—Rheological Fluid Type Semi-Active Joints. Ep-3562443, 6 November 2019.
417. Scattareggia, M.S. Esoskeleton Equipped with Electro-or Magneto—Rheological Fluid Type Semiactive Joints. Wo/2018/122886, 5 July 2018.
418. Easton, J.L.C. Low Profile Exoskeleton. U.S. Patent 20160128890, 12 May 2016.
419. Varghese, R.J. System for Movement Control. Wo/2020/234184, 26 November 2020.
420. Zoss, A. Powered Orthotic System for Cooperative Overground Rehabilitation. Ep-2968052 B1, 11 August 2021.
421. Goldfarb, M. Movement Assistance Device. U.S. Patent 10792210 B2, 6 October 2020.
422. Herr, H.M. Hybrid Terrain-Adaptive Lower-Extremity Systems. U.S. Patent 10575971 B2, 3 March 2020.
423. García, L.J. System for Assisting Walking. Ep-3378446 B1, 4 November 2020.
424. Herr, H.M. Optimal Design of a Lower Limb Exoskeleton or Orthosis. U.S. Patent 10561563 B2, 18 February 2020.
425. Tuttemann, M. Leg Orthosis and Orthosis. U.S. Patent 11179288 B2, 23 November 2021.
426. Walsh, C.J. Soft Exosuit for Assistance with Human Motion. U.S. Patent 10843332 B2, 24 November 2020.
427. Huang, C.-M. Portable Human Exoskeleton System. U.S. Patent 10918558 B2, 16 February 2021.
428. Herr, H. Motorized Limb Assistance Device. U.S. Patent 10695256 B2, 30 June 2020.
429. Smith, F.M. Legged Robotic Device Utilizing Modifiable Linkage Mechanism. U.S. Patent 10766133 B2, 8 September 2020.
430. Scattareggia, M.S. Esoskeleton Equipped with Electro-or Magneto—Rheological Fluid Type Semi-Active Joints. Ep-3562443 B1, 7 April 2021.
431. Edgerton, V.R. Regulation of Autonomic Control of Bladder Voiding After a Complete Spinal Cord Injury. U.S. Patent 10751533 B2, 25 August 2020.
432. Goldfarb, M. Movement Assistance Device. Ep-2861198 B1, 8 January 2020.

Review

HAL Training in Spinal Cord Injured Patients: A Narrative Review of 10 Years Experience

Alexis Brinkemper [1,*], Dennis Grasmücke [2], Emre Yilmaz [1], Thomas Armin Schildhauer [1] and Mirko Aach [2]

[1] Department of General and Trauma Surgery, BG University Hospital Bergmannsheil, 44789 Bochum, Germany
[2] Department of Spinal Cord Injuries, BG University Hospital Bergmannsheil, 44789 Bochum, Germany
* Correspondence: alexis.brinkemper@bergmannsheil.de; Tel.: +49-(0)-234-302-6918

Abstract: To provide a summary and overview of the use of Hybrid Assistive Limb in spinal cord injured patients over the past 10 years. A review of the literature was performed via Web of Science and PubMed using the search terms "Hybrid assistive limb" or "HAL" or "wearable robot" or "exoskeleton" and "SCI" or "spinal cord injury" by two of the authors. Relevant articles were then studied in full text. Our review of the literature found 21 articles that met the inclusion criteria of this narrative review including 344 participants. Articles were sorted into two general categories: (1) clinical trials, and (2) single-case or two-case reports. The vast majority of patients improved functionally, showing increased walking distances, walking speeds, and endurance. In addition, a variety of other advances were described, such as temporary decrease in spasticity, improvement in bladder and bowel management, pain reduction, and change in muscle activity. Even though there is no uniform application of HAL training in people living with SCI the current study situation suggests that many patients could benefit from this innovative training within their means.

Keywords: spinal cord injury; exoskeleton; Hybrid Assistive Limb; rehabilitation; review

Citation: Brinkemper, A.; Grasmücke, D.; Yilmaz, E.; Schildhauer, T.A.; Aach, M. HAL Training in Spinal Cord Injured Patients: A Narrative Review of 10 Years Experience. *Appl. Sci.* **2023**, *13*, 1369. https://doi.org/10.3390/app13031369

Academic Editors: Ionuţ Daniel Geonea and Cristian Copilusi Petre

Received: 30 November 2022
Revised: 10 January 2023
Accepted: 17 January 2023
Published: 20 January 2023

1. Introduction

In 2016, globally, there were 0.93 million new cases of spinal cord injury (SCI) with age-standardized incidence rates of 13 per 100,000 [1]. The number of prevalent cases of SCI was 27.04 million [1] and thus represents a substantial portion of the global injury burden. Individuals who have experienced SCI find themselves in a situation where even previously simple things become insuperable barriers. Fortunately, a lot has happened in the treatment of SCI in recent years. Robotic therapy approaches and exoskeletons have increasingly found their way into SCI centers around the world. Most exoskeletons have a similar appearance from the outside. There is an external support frame along the lower extremities, a type of hip belt or attachment to the hip, and a shoe device, either for use with the patient's own shoes or over shoes belonging to the system. However, the systems differ in their application and clinical objective. Some systems are assistive, and some are rehabilitative. There is also a difference in the control mechanism. Here there is joystick control or posture control and in one case, electromyography (EMG) based support. Our narrative review focuses on this EMG controlled system, the Hybrid Assistive Limb (HAL; Cyberdyne, Inc., Tsukuba, Japan). The HAL is a wearable robot suit that senses a patient's voluntary actions, such as real-time myoelectric potential, foot pressure, and joint angle, and assists hip and knee joint movement [2]. Figure 1 shows the HAL. Initially, HAL was developed in order to physically support a wearer's daily activities and heavy work [3] but it has also been used from the earliest times for welfare and patient training. The HAL was used in the rehabilitation of acute [4] and chronic [5] stroke patients, patients with neuromuscular diseases [6], and SCI [7–27]. To the best of our knowledge, there is currently no paper that concisely presents the previous studies in patients with SCI and helps to provide an overview of the different training approaches, patient collectives and the results achieved. Therefore, the purpose of this paper is to summarize and present a review of previous studies on HAL training in patients with SCI.

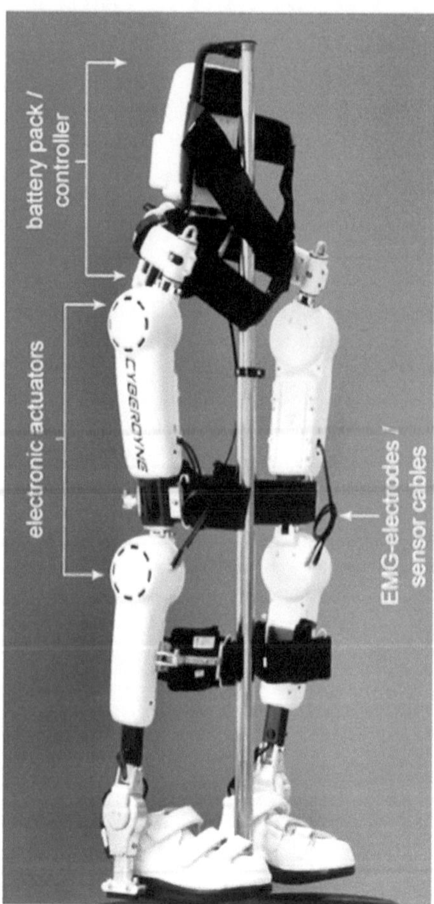

Figure 1. HAL®—Hybrid assistive limb exoskeleton (Cyberdyne Inc.).

2. Materials and Methods

A narrative review of the literature was conducted with electronic searches for eligible studies using the databases Web of Science, PubMed, and Scopus on 15 September 2022 and 6 January 2023 using the following search term combinations in the titles: "Hybrid assistive limb" or "HAL" AND "SCI", "Hybrid assistive limb" or "HAL" AND "spinal cord injury", "Hybrid assistive limb" or "wearable robot" AND "SCI", "Hybrid assistive limb" or "wearable robot" AND "spinal cord injury", "Hybrid assistive limb" or "exoskeleton" AND "SCI", "Hybrid assistive limb" or "exoskeleton" AND "spinal cord injury", "HAL" or "wearable robot" AND "SCI", "HAL" or "wearable robot" AND "spinal cord injury", "HAL" or "exoskeleton" AND "SCI", "HAL" or "exoskeleton" AND "spinal cord injury". The search was limited to articles that were published in (or subsequently translated to) English or German. No restriction on date published was enforced. All study designs were included. Two of the authors screened the titles and then abstracts independently and studies were considered relevant if they addressed any clinical application of the HAL system in people living with SCI. In case of disagreement, a discussion was held to reach consensus on which studies should be reviewed in full text. If necessary, a third author was consulted to make the final decision. Studies reporting the use of HAL in other conditions, such as post-stroke or multiple sclerosis, were excluded as well as studies reporting use of other exoskeleton systems. Reference lists of the involved studies were manually searched for further articles. A narrative review rather than a systematic review was undertaken to report a broad overview of the use of the HAL system in the field of rehabilitative SCI treatment. Although the system has been available for a long time, there are currently no randomized controlled trials, which the authors believe are essential for a systematic review. Articles were sorted into 2 general categories for discussion: (1) clinical trials, and (2) single-case or two-case reports.

3. Results

In their paper from 2005 Suzuki et al. [7] proposed for the first time an algorithm to estimate human intentions related with walking in order to comfortably and safely support a paraplegia patient's walk with the HAL robot suit. This work presents mainly technical data of algorithm development and therefore was not considered further in this review. Nevertheless, it should be mentioned here, as it has set the stage for further application studies.

A total of 1940 studies were found using the search criteria described above. After excluding studies that had a completely different topic but were still listed, either used exoskeletons other than the HAL system, or used the HAL device for other conditions, and after removing all duplicates, 27 published articles were identified. Studies only reporting technology data (3) and reviews on general HAL application (3) were excluded. Twenty-one relevant articles, as deemed so by the authors, were then studied in full text (Figure 2). Overall, the studies discuss results from 344 participants, with some of the same subjects included in different studies.

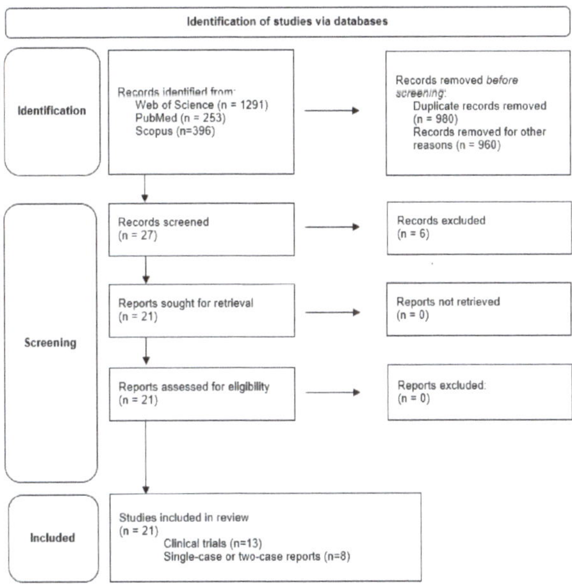

Figure 2. Presentation of the results of the literature search.

Of these 21 articles, 13 are reports about clinical trials using HAL [8–20] and eight are single-case [2,22–25] or two-case reports [21,26,27]. Ten articles originate from Germany [9–13,16,17,20–22], ten from Japan [2,8,14,15,18,19,23,24,26,27], and one from the USA [25]. Years of publication of these articles range from 2013 to 2021. The underlying experimental series started in part as early as 2012, which means that we present a survey of the last ten years in this review. To the best of our knowledge this is the first ever review summarizing experiences with HAL in people living with SCI.

Study protocols differed in terms of training type (treadmill or walker), training time (20 to 40 min of net HAL walking training), time period over which training was performed (2 to 52 weeks), number of training sessions per week (2 to 5), resulting total training sessions (8 to approximately 180), and assessments performed. The patients included in these studies represent the full range of spinal cord injuries. There are traumatic and non-traumatic injuries, chronic and acute patients, complete and incomplete paralysis, lesion levels from C2 to L4, American spinal injury association impairment scale (AIS) A to D, and patients with and without spastic motor behavior. The age of the participants varies from 16 to 74 years. Some studies conducted by the same research groups (Ruhr-University Bochum Group [9–13,16,17,20–22], Keio University School of Medicine Group [15,18,19], University of Tsukuba and University of Tsukuba Hospital Group [8,14,23,24,26,27]) have partial overlap in their study populations.

3.1. Clinical Trials

In 2013, Kubota et al. [8] published a paper summarizing findings of previous studies on robotic devices used in clinical settings with the conclusion that robot-assisted rehabilitation is controversial. They introduced the newly developed HAL and had the intention to investigate the feasibility of HAL rehabilitation training for patients with limited mobility. The study collective consisted of 12 stroke patients, four patients with musculoskeletal diseases, eight SCI patients, and 14 other diseases. The SCI patients were all in a chronic stage (1–6 years since SCI) and included four paraplegia as well as four tetraplegia. Four of them were incomplete, one complete, one had spina bifida, one central cervical cord injury, and one spinal dural arteriovenous fistula. Patients performed 16 HAL sessions in eight weeks using a walking device or treadmill for 90 min (net walking time 20 min). One of these patients dropped out in course of the study due to neuropathic pain and for one patient the baseline measurement was not assessed, as he was unable to ambulate with any assistance. Therefore, only six SCI patients were included in the statistical analysis. In this study, the functional ambulation results were presented for the whole group of participants included in the statistical analysis ($n = 27$). Since the informative value about this diverse patient collective, in the view of the authors of the review, is very limited, we refrain from presenting it here in detail. However, significant improvements in gait speed, number of steps, and cadence were observed, as assessed by the 10-Meter-walking-test (10MWT). Improved Timed-Up-and-Go test (TUG) and Berg balance scale (BBS) results were also observed, but they were not statistically significant [8]. The only SCI specific result that was given is a large effect size in gait speed (Cohan d = 0.78), which leads the authors to conclude that training effectiveness in SCI patients can be expected. The conclusion of this study states that HAL training is feasible, safe and effective in patients with limited mobility [8].

Results that are more detailed are provided by a study from Aach et al. [9] involving eight patients in chronic stage (1–19 years since SCI) after traumatic SCI. Included patients had incomplete paraplegia (AIS B/C/D, $n = 4$) or complete paraplegia (AIS A, $n = 4$) with zones of partial preservation (ZPP) and performed treadmill training using HAL five times per week for 90 days (mean number of sessions 51.75 ± 5.6). Treadmill associated results improved over all patients for walking speed and walking time, but no information was provided on the statistical significance except for ambulated distance, which increased from 195.88 ± 166.71 m to 954.13 ± 380.35 m ($p < 0.05$). Functional assessments were the 10MWT, TUG, 6-Minute-Walk-Test (6MWT), walking index for SCI II (WISCI II) score, and lower extremity motor score (LEMS). The authors state that although the WISCI II score did not improve statistically, three patients showed functional improvements in gait abilities. For the 10MWT a significant increase in gait speed similar to Kubota et al. [8] and decrease in number of steps were found as well as a decreased time for the TUG and increased LEMS. Similarly, they reported significant improvement for the 6MWT, although it should be noted that only three participants were able to take part in the baseline measurement (with an average walking distance of 187 ± 162.2 m) and the five others who did not participate (scored as 0 m) decimated the total baseline data to 70.1 ± 130 m. However, the subgroup who performed the 6MWT at baseline improved their walking distance to 287.3 ± 229.4 [9]. In this study, information on statistical significance was not given in detail, it was only stated that the p value is <0.05. Thus, it remains unknown whether the improvements were highly significant or whether the results were just at the threshold of significance. In addition to these results, the authors reported on seven participants with increased muscle volume as measured by the circumferences 10/20 cm cranial of the inner knee joint gap and 15 cm distal of it. One patient lost muscle volume. Furthermore, one participant with spinal spasticity was able to reduce Ashworth scale from four to two for a few hours after training. One patient switched from AIS B to C after treatment.

Seven of these eight patients from the Aach et al. [9] study were also part of a study by Sczesny-Kaiser et al. [10] to investigate whether body weight supported treadmill training with HAL affects cortical excitability in the primary somatosensory cortex (S1) in SCI patients, as measured by paired-pulse somatosensory evoked potentials (ppSEP) stimulated above the level of injury. These seven patients were supplemented by four additional patients (8.5 months to 17 years since SCI) performing the same intervention protocol as described for a total of eleven participants. No electrophysiological results of these eleven patients were published until then. Nerve stimulation was conducted successively on both sides with a block electrode placed on the wrist. ppSEP was assessed before and after the HAL training period and compared with a healthy control group (n = 11). Additionally, electroneurography (ENG), somatosensory evoked potentials (SEP) of the tibial nerves, and motor evoked potentials (MEP) of anterior tibial muscles were performed before and after the training period. The authors present a statistical difference between SCI patients and healthy control group at baseline. After intervention, this difference was no longer present. Comparison of patient's data pre- and post-training showed a significant statistical training effect. ENG, SEP of tibial nerves and MEP showed no significant differences after training. Functional outcomes consequently to the previous study from Aach et al. [9] were statistical significance for 10MWT-speed, 6MWT, TUG, LEMS. These results show that SCI patients had increased cortical excitability in the hand area of S1 prior to training and that this was normalized following 12 weeks HAL training accompanied by significant improvements in walking abilities [10]. However, a correlation analysis revealed no relationship between the extent of changes in excitability and time since injury, functional walking parameters, or LEMS. The authors, anyway, conclude that walking improvements may be related to a renewed S1 and M1 representation of impaired lower extremities likely related to the recruitment and more effective use of remaining somatosensory afferent pathways and corticospinal tracts [10]. The missing correlation between functional abilities and changes of excitability in S1 might indicate that either other cortical areas or even a complex supraspinal network is required for walking rehabilitation [10].

In 2017, Jansen et al. [11] published a proof of concept to the pilot study from Aach et al. [9] and expanded the collective to 21 chronic SCI (1–19 years since SCI) patients. All of them improved significantly in respect to treadmill associated data (walking distance, speed) and in functional tests assessed by standardized over-ground walking tests (10MWT, TUG, 6MWT) without HAL, which therefore confirms findings of the pilot study [9,11]. In addition to these results, this study revisits the issue of spasticity and states that in eight subjects with spastic motor behavior intervention led to a temporary relief. This was already discussed for one patient in the pilot study [9]. Furthermore, the authors indicate that 18 of 21 patients reported improvements of their bowel and bladder dysfunctions after the training period with two of them able to discontinue self-catheterization [11]. However, both spasticity and bowel and bladder improvements were not supported by statistical data.

A question often discussed in the SCI literature is in which patient population improvements are most likely to occur and in which less so. According to Grasmücke et al. [12] it is mainly postulated that the neurological and ambulatory recovery depends on a patient's lesion level and age, as well as their initial AIS grade whereby low AIS grade and age $\geq$ 50 years seem to be negative predictors for functional recovery following SCI. In order to investigate this in relation to HAL training a study was set up with the aim to compare the functional outcomes of patients with chronic SCI (1 to 20 years since SCI) in four injury-level groups and additionally analyze results as a function of age (<50 years vs. $\geq$50 years) [12]. The four groups were classified as (1) incomplete tetraplegia (n = 13, C2–8, AIS C and D), (2) incomplete paraplegia with spastic motor behavior (n = 15, T2–12, AIS C and D), (3) complete paraplegia with an absence of spastic motor behavior (n = 18, T11–L4, AIS A and ZPP from L3 to S1), and (4) incomplete paraplegia with an absence of spastic motor behavior (n = 9, T12–L3, AIS C and D) [12]. A total of 55 participants were involved. As part of the Ruhr-University Bochum group, the same training program (12 weeks, 5 units/week) was completed and the results of the initial eight patients from Aach et al. [9] and in part from Jansen et at. [11] were also included. As seen in the other studies [9–11], the cohort improved in treadmill related and functional parameters. In contrast, however, to the assumptions made in the literature, the lesion level and patients' age were not significantly associated with training-related functional improvements. No significant differences between subgroups were found. Subgroup 2 showed slightly less improvement in treadmill walking speed, treadmill distance and 6MWT, but these differences were statistically not significant [12]. Concerning patients' age only 10MWT-speed showed a significantly larger improvement in the younger group compared to the older group. Another interesting finding of this study was that most improvements occurred during the 4th and 10th week of the training period [12]. From the results demonstrated here, the authors conclude that older age or lesion level seem to be less important in terms of the extent of functional improvement following HAL training. Nevertheless, incomplete SCI lesions of the thoracic spine, including spastic motor behavior, appear to be a non-significant negative predictor for training-related improvements [12].

To examine long-term data of variable-frequency HAL training after the initial training period, it was again Jansen et al. [13] who followed up the pilot study of Aach et al. [9] with one year of HAL training. All eight chronic SCI participants continued HAL training after their initial training period of 12 weeks and were divided into two groups. Group 1 (n = 4) continued with 3–5 training sessions per week for a further 40 weeks and group 2 (n = 4) continued with one session per week. In the first 12 weeks, the participants had 51.75 $\pm$ 5.6 training sessions on average. During the second training period, the mean number of training sessions in group 1 was 126.8 $\pm$ 7.9 and 32.3 $\pm$ 3.3 in group 2. Results were given both for the first training period (repetition of Aach et al. [9] results) and after one year of treatment. In order not to be redundant, we here only discuss new results after one year. Concerning treadmill related parameters (speed, walking time, and distance) no significant improvements were demonstrated between 12 and 52 weeks results for evaluation as a total group (n = 8), as individual groups (n = 4/4), and for comparison between subgroups. The same was seen in the functional testing. Considering all eight patients together, irrespective of continuous training frequency during the following 40 weeks, there were no significant changes for 10MWT, cadence, number of steps, step length, TUG, 6MWT, or WISCI II. Taking into consideration the two subgroups separately, there was no difference between 12 and 52 weeks and no difference between the groups. These certainly surprising results indicate that the initial functional gain after 3 months of daily HAL training was stable and consistent over a 1-year period [13].

A completely new approach to HAL therapy was described by Shimizu et al. [14] in their 2017 study. HAL support is predominantly based on EMG signals from the lower extremity or weight shift over the sensor shoes. Since such muscle signals are usually undetectable in patients with complete paraplegia or tetraplegia and controlled loading of the feet is rare, the researchers decided to use upper extremity muscle activity as a trigger of lower extremity movement. This is based on the assumption that the lower extremities move synchronously and almost simultaneously with the contralateral upper extremities during natural locomotion [14]. The authors hypothesized that triggering lower limb motion by upper limb muscle activity was feasible to generate voluntary gait and furthermore, that this may activate paralyzed lower limb muscles. Muscle activities from the anterior and posterior deltoid were used for contralateral hip flexion and extension and activities from the biceps and triceps brachii for contralateral knee flexion and extension [14]. Patients underwent ten sessions of walker assisted HAL training protocol for 60–90 min (net HAL training time 30 min) once or twice a month ($n = 2$) or twice a week ($n = 2$). In two patients who were able to flex their hips (before or along the intervention) a knee extension training using hip flexor activation was performed additionally in order to regain active knee extension. EMG was used to evaluate muscle activity of the tensor fasciae latae (TFL) and quadriceps femoris (Quad) [14]. Prior to intervention none of the participants showed apparent activation in Quad and only one in TFL. However, gait phase dependent activity of the lower limb muscles was seen during voluntarily triggered ambulation driven by upper limb muscle activities and in some cases active contraction in Quad was observed after knee extension session using the hip flexor as trigger [14]. Therefore, the authors conclude that this method is a feasible option for patients with severe paraplegia or tetraplegia. Although these results are certainly exciting, it is difficult to generalize. In this study, only four patients were examined and they had different training frequencies. Furthermore, in two of the four patients, the upper body controlled training was started after three and five sessions, respectively, and another mode of the HAL accompanied with heavy assistance from three therapists was used beforehand to familiarize the patients with the use of such a device. One of the four patients received additional physiotherapy during the training phase. Nevertheless, with this study the authors contribute to the understanding of possible applications of the HAL even in severely injured patients. In addition to the results already mentioned, the patients improved their walking distance and three out of four patients showed improvements in Ashworth score, which is in line with the results of other studies [9,11].

A question similar in approach to Grasmücke et al. [12] was pursued by Okawara et al. [15], who asked whether HAL training is appropriate for all severities of SCI. They stated that in most of the preliminary studies patients already had a certain walking function before the treatment participated. They set out to investigate the effect of training in relation to the severity of SCI. For this purpose, 20 chronic patients were recruited and divided into a low walking ability group ($n = 8$) and a high walking ability group ($n = 12$) based on their WISCI II scores at baseline. Twenty training sessions with a frequency of 2–5 per week were performed on the treadmill. Gait performance on treadmill (speed, distance, and time) was significantly lower in the low walking ability group while perceived exertion was higher. Functional testings' (10MWT time, 10MWT-speed, 10MWT number of steps, TUG) improved in the higher walking ability group significantly while in the low walking ability group none of the participants were able to complete the assessments at either point. BBS score improved significantly following training in the high walking ability group but not in the low walking ability group. Barthel Index (BI) and Functional Independence Measure (FIM) as assessment of activities of daily living (ADL) were unchanged for both groups, indicating that there was no effect of training regardless of baseline walking ability [15]. The authors summed up the results such that patients with a WISCI II score below six barely benefited from training. They contrast these results with Jansen et al. [11] in whose collective, which improved significantly, there was also one patient with a WISCI II score of one. In Jansen's study, however, individual results were not reported, so it is not clear whether this patient also improved. Interestingly, despite the sparse results in the slow walking ability group, 19 of the 20 patients reported subjective improvement after training and all patients were either fully or slightly satisfied.

The issue of bladder and bowel dysfunctions raised by Jansen et al. [11] is revisited by Brinkemper et al. [16] in a retrospective study. Here, 35 patients with acute ($n = 13$) or chronic ($n = 22$) SCI completed two standardized and established gastroenterological questionnaires (Cleveland Clinic Constipation Scoring System (CCCS), Wexner Score) and a self-developed questionnaire asking for bladder function before and after completing the training. The study protocol corresponds to that of the other Ruhr-University Bochum studies and functional results of the patients were already published. Comparisons were made for all patients and the group was divided into subgroups of acute and chronic patients. Within these subgroups, further subgroups were formed according to high or low baseline scores. Wexner Score, a questionnaire to assess the severity of fecal incontinence, over all patients and for the group of chronic patients decreased significantly. For the group of acute patients Wexner Score decreased insignificantly. Patients from both groups with higher baseline scores could decrease significantly. CCCS, a questionnaire to assess the severity of constipation, reduced non-significantly for all patients, the group of acute, and the group of chronic patients. For the subgroup of chronic patients with higher baseline scores, CCCS decreased missing out significance. The self-developed questionnaire showed an improvement in bladder function in 28.24% of all patients, 31.43% of chronic patients, and 23.08% of acute patients. The authors conclude that there is a trend toward improvement in bladder and bowel function after HAL training, with patients with higher baseline scores in Wexner Score and CCCS seeming to benefit more and see this as confirming the results of studies [28–31] using other exoskeletons. However, the mixed population of acute and chronic patients and different injury levels as well as the retrospective study approach remains a limitation of this study.

Another study based on the Ruhr-University Bochum protocol was published by Brinkemper et al. [17]. This time, however, the focus was not on functional improvements of the patients but on whether the functional improvements were accompanied by physiological improvements in gait. For this purpose, a 3D gait analysis was performed on 15 patients (acute $n = 5$, chronic $n = 10$) using an inertial measurement unit sensor system. While all functional parameters improved as expected, physiological differences were also found in some phases of gait cycle showing improved knee extension during initial contact and increased maximum hip extension prior to swing phase. Furthermore, all joint angles showed a larger range of motion and those findings were accompanied by significant improvement in all spatiotemporal and gait phase parameters. Thus, according to the authors, this study was the first to demonstrate improved gait physiology after HAL training in SCI patients.

In addition to the frequently addressed ability to walk in patients with SCI, the ability to maintain a seated position for an extended period of time also plays an important role. To evaluate the effect of HAL training on trunk muscle strength, Okawara et al. [18] conducted a study in nine chronic (1–10 years since SCI) patients. The HAL intervention consisted of 20 training sessions of approximately 40 min net HAL training time each. Trunk muscle strength was defined as the ability to maintain a seated posture in four directions (anterior, posterior, left lateral, and right lateral) and measured pre- and post-training [18]. An examiner pushed the participant's torso using a dynamometer until the participant could no longer maintain the seated posture. The patients were part of the collective from Okawara et al. [15] where the functional recovery results have already been published. Trunk muscle strengths from baseline to after 20 training sessions improved in all directions, however, statistical significance was only shown in the change of lateral trunk muscle strength. Surprisingly for the authors, the changes in anterior, posterior, and lateral trunk muscle strength showed significant positive correlations with age at baseline, indicating that older adult subjects with chronic SCI experienced greater improvements in trunk muscle strength [18]. This contrasts somewhat with the results of Grasmücke et al. [12] who induced a lower level of improvement in gait function (10MWT-speed) in older adult patients than in younger patients during the chronic phase of SCI.

A study by Sawada et al. [19], which is based on the same study population as Okawara et al. [15], deals with the influence of HAL training on quality of life (QOL). Since paraplegia affects one's entire life, QOL suffers from it and requires follow-up in the course of rehabilitation. Patients performed the already known 20 sessions HAL protocol and were assessed by short Form−36 questionnaire Japanese version (SF-36v2), WISCI II, FIM motor score, and the Neuropathic Pain Symptom Inventory (NPSI) self-questionnaire. As with Okawara et al. [15] patients were divided into a low walking ability group and a high walking ability group. Baseline values in SF-36v2 of the participants were lower compared to healthy individuals and were not improved in the course of the training. Likewise, WISCI II, FIM motor score, and NPSI did not show significant differences, irrespective of baseline walking abilities. However, a correlation analysis showed positive correlation between FIM motor score and some subscales of SF-36v2 and a negative correlation between NPSI and the change in subscale Vitality and Mental Health, leading the authors to conclude that in their protocol those with higher functional independence and lower pain at pre intervention were more likely to improve [19]. The authors see an explanation for the generally low improvement in QOL in the unchanged pain score, which in turn can be explained by an already low pain prevalence at baseline.

The largest study population in a clinical trial with HAL to date is provided by the study of Zieriacks et al. [20]. This involved 121 patients with acute ($n = 47$) and chronic ($n = 74$) SCI. Partial results of this collective have been published previously [9–13,16,17]. The aim was to assess whether HAL training is advantageous for acute and chronic participants and if length of time post injury affects the outcome. Significant improvements in treadmill-related parameters (time, distance) and all functional parameters (10MWT, 6MWT, WISCI II, LEMS) were reported for the whole group as well as for the subgroups of acute and chronic patients. However, significant differences between acute and chronic participants' outcomes were found in 6MWT, LEMS, and WISCI II, showing that chronic participants improved significantly less than acute participants did. Nevertheless, since both subgroups showed significant improvements individually, the authors concluded that HAL training is advantageous for both acute and chronic patients and that there is no time related cut-off threshold following SCI for effectiveness [20]. Table 1 summarizes the clinical trials studies details and results.

Table 1. Spinal cord injury clinical trials studies characteristics.

Author	Number of Patients	Acute/Chronic	Lesion Levels	Training type	Training Time (Minutes)	Training Period (Weeks)	Training Sessions per Week	Total Training Sessions	Baseline Measurement	Results	Patients Involved in Other Studies
Kubota et al. [8]	8 (6)	chronic	4 incomplete, 1 complete, 1 spina bifida, 1 central cervical cord injury, 1 spinal dural arteriovenous fistula	Sitting, Walker, Treadmill	20	8	-	16	Yes	large effect size in gait speed (Cohan d = 0.78)	No
Aach et al. [9]	8	chronic	T8-L2, AIS A–D	Treadmill	30	12	5	60	Yes	improved 10MWT and LEMS, decreased number of steps in 10MWT and time in TUG, improved 6MWT, increased muscle volume	No
Sczesny-Kaiser et al. [10]	11	1 acute, 10 chronic	T8-L1, AIS A–C	Treadmill	30	12	5	60	Yes	PpSEPs increased in SCI patients compared to controls at baseline. Following training, ppSEPs were increased from baseline and no longer significantly differed from controls; Improved 10MWT-speed, 6MWT, TUG, LEMS	7 of [9]
Jansen et al. [11]	2	chronic	C4-L3, AIS A–D	Treadmill	30	12	5	60	Yes	improved treadmill associated data (walking distance, speed) and in 10MWT, TUG, 6MWT	Eight from [9]
Grasmücke et al. [12]	55	chronic	C2-L4, AIS A–D	Treadmill	30	12	5	60	Yes	influences of SCI lesion level and age on functional outcome were nonsignificant	Eight from [9] and partly from [11]
Jansen et al. [13]	8	chronic	T8-L2, AIS A–D	Treadmill	30	12 52	5 (n = 8) 1 (n = 4); 3-5 (n = 4)	52 32 127	Yes	functional improvements after 12 weeks, but no difference between 12 and 52 weeks for neither group	Eight from [9]
Shimizu et al. [14]	4	chronic	C6-T11, AIS A–B	Walker	30	-	2/week (n = 2) 1–2/month (n = 2)	10	Yes	decreased modified Ashworth scale (in n = 3); gait phase dependent activity of the lower limb muscles during voluntarily triggered ambulation driven by upper limb muscle activities	No
Okawara et al. [15]	20	chronic *	C-L, AIS A–D	Treadmill	40	-	2–5	20	Yes	improvement in 10MWT, TUG, and BBS; Walking ability improved in participants with high walking ability at baseline but not in participants with low walking ability	No
Brinkemper et al. [16]	35	acute and chronic	AIS A–D	Treadmill	30	12	5	60	Yes (retrospective)	decreased Wexner score over all patients and for group of chronic patients; improvement in bladder function in self-developed questionnaire	Partly from [12]
Brinkemper et al. [17]	15	acute and chronic	C3/4-L3, AIS A, C, D	Treadmill	30	12	5	60	Yes	improved 10MWT, TUG, 6MWT, WISCI II; significant improvement in all spatiotemporal and gait phase parameters	Partly from [12]

Table 1. *Cont.*

Author	Number of Patients	Acute/Chronic	Lesion Levels	Training type	Training Time (Minutes)	Training Period (Weeks)	Training Sessions per Week	Total Training Sessions	Baseline Measurement	Results	Patients Involved in Other Studies
Okawara et al. [18]	9	chronic	C5-T12, AIS A–D	Treadmill	40	-	2–5	20	Yes	trunk muscle strengths improved in all directions, but statistical significance was only shown in the change of lateral trunk muscle strength; anterior, posterior and lateral trunk muscle strength showed significant positive correlations with age	Partly from [15]
Sawada et al. [19]	19	chronic *	C-L, AIS A–D	Treadmill	40	-	2–5	20	Yes	in SF-36v2, mean values of all subscales lower than those in healthy individuals; none of the measures showed significant improvement, even in individuals with some residual walking ability; correlation analysis between the baselines WISCI-II, FIM, or NPSI values and mean SF-36v2 subscale changes throughout the training; the baseline FIM motor score was positively correlated with the mean changes in Role Emotional and Mental Health; NPSI was negatively correlated with the mean change in Vitality and Mental Health.	Partly from [15]
Zieriacks et al. [20]	121	acute and chronic	C2-L4, AIS A, C, D	Treadmill	30	12	5	60	Yes	all participants significantly improved in the functional assessments performed without the exoskeleton; significant differences between acute and chronic participants' outcomes were found in 6MWT, LEMS, and WISCI II showing that chronic participants improved significantly lesser than acute participants; since both subgroups showed significant improvements individually, HAL training is advantageous for both acute and chronic patients; there is no time related cut-off threshold following SCI for effectiveness	Partly from [12]

* Defined here as >6 months after injury.

3.2. Single-Case or Two-Case Reports

In addition to clinical trials in more or less large collectives, there are some publications that are single case or two case reports on selected issues.

For example, Cruciger et al. [21] published the pain-related data of two subjects from the collective of the Aach et al. [9] study individually. Those chronic patients (both AIS A with ZPP, 10 and 19 years since SCI) had therapy-resistant chronical neuropathic pain. During the course of HAL training, pain symptoms improved in both patients to the extent that medication could be reduced. A brief increase in pain in response to the lower dosage was followed by a renewed reduction in pain during training. Patients changed from a permanent medication to an on-demand medication. The average pain intensity, measured by numerical pain rating scale (NRS-11; 0 = no pain, 10 = worst pain possible), decreased from 4.3 at the first training week to 0.6 after 12 weeks. Furthermore, an improvement on health related QOL in the SF-36 questionnaire was found reflecting the impact of the intervention. According to the authors in a one-year follow up, both patients reported neither recurrence of pain nor need for medication [21]. These results are in contrast to those of Sawada et al. [19] who found no changes in pain and QOL in their study. However, the collective in Sawada et al. [19] also showed no motor functional improvements while in the study by Cruciger et al. [21], two patients improved in terms of 10MWT and LEMS. This could potentially have had an impact on the QOL data

It was also Cruciger et al. [22] who first reported the use of HAL training in an acutely injured SCI patient in a single case report. The patient had an incomplete motor T10 lesion (AIS C) and began training 77 days after the accident. After 12 weeks HAL training there was recovery of motor functions and walking abilities as shown by increase in WISCI II from 8 to 18 and conversion to AIS D.

Ikumi et al. [23] wanted to find out if HAL application was possible in a severely affected chronic complete tetraplegic patient and subjected him to 10 HAL trainings in 5 weeks. To perform the treadmill training, two physicians and one therapist were needed. Walking distance and time increased from 25.2 m and 7.6 min to 148.3 m and 15 min and the modified Ashworth scale decreased from 15.3 to 5.75 points after HAL training and lasted for up to 30 min [23]. Therefore, the authors conclude feasibility and efficiency of HAL rehabilitation for tetraplegic patients.

Another special case is reported by Shimizu et al. [24] in their presentation of a man whose condition deteriorated due to a spinal dural arteriovenous fistula 19 years after the initial SCI. He underwent surgery and his neurological as well as motor status improved again. Six months post-operatively his improvement had plateaued. This was the point where HAL training started for ten sessions during three months. The patient improved in terms of gait speed and cadence in 10MWT, International Standards for Neurological and Functional Classification of Spinal Cord Injury (ISNSCI) motor score (14 to 16), WISCI II (7 to 12), manual muscle testing and to some extent in muscle activities measured via EMG. Even though the improvements shown here are not enormous, they do show a further development after the performance increase had initially reached a plateau.

The very first application of HAL in the USA is described by Yilmaz et al. [25]. This involved six patients with neurologic motor deficits of various etiologies who underwent 60 HAL sessions in 12 weeks. Among others, one patient was post spinal cord infarction following pulmonary embolism. The aim of this study was to investigate the impact of HAL training on QOL. The spinal cord infarction patient showed improvements in distance on treadmill and in a physical function mobility score. However, she showed a worse QOL after training in a six months follow up. The authors note that this is probably due to the very high expectations for the novel therapy, which is an extremely interesting point that has not been mentioned in previous studies. In the other patients of the collective, there was no noteworthy improvement in QOL despite good mobility improvements, which suggests that QOL is determined by more diverse factors [25]. This result contrasts with that of Cruciger et al. [21], whose two patients improved in health related QOL, but is confirmed by the study of Sawada et al. [19]

HAL training was completed as a contiguous training phase in most studies. Kanazawa et al. [2], however, chose an alternating design of conventional gait training (A) and HAL training (B) in their single-case study of a patient with chronic thoracic SCI. A similar protocol has already been practiced in stroke patients. The patient completed a total 25-week program with a switch from conventional gait training to HAL training every 5 weeks. In total, 15 weeks of conventional gait training and 10 weeks of HAL (A1-B1-A2-B2-A3) were performed, with three sessions per week. Accompanying physiotherapy took place permanently. Global parameters such as 2-min-walk-test, WISCI II, and BBS improved from the beginning to 25 weeks while spasticity did not. Walking speed, stride length, and cadence improved after phase A but not B [2]. During phase B a decrease in speed was seen, which could, according to the authors, reflect fatigue and changes in lower limb spasticity after using HAL. However, the alternating training program is considered potentially successful, although it remains unclear which phases have which share in the overall result.

As in Yilmaz et al. [25] a two-case study by Watanabe et al. [26] included patients after spinal cord infarction. Both patients were in a very early acute phase (14 and 7 days after infarction) at the start of training. In addition to conventional physical therapy, they received 7–8 HAL sessions 3–4 times per week. Improvements in LEMS, WISCI II, comfortable gait speed, stride, cadence, BI, FIM, and further assessments were observed in both study participants [26]. Therefore, the authors state that gait treatment using HAL in people with acute spinal cord infarction is possible and may be beneficial.

Soma et al. [27] presents in their study a severely affected case with a patient who was diagnosed with cervical spondylotic myelopathy, underwent surgery, and was rediagnosed with traumatic cervical SCI (incomplete tetraplegic, AIS C) after a fall. He received HAL training 2–3 times per week for 13 sessions and additional physical therapy. Improvements were observed in gait speed, step length, and cadence based on a 10MWT, ISNCSCI motor score, and WISCI II score [27]. The authors state, as Okawara et al. [15] did before, that in other studies, such as that of Aach et al. [9], only patients who were already able to perform high-performance testing at baseline, such as the 6MWT, participated and it can be seen from their results that even severely affected patients benefit from HAL training. However, it must be noted that also in the mentioned Aach et al. [9] study only three out of eight patients were able to perform the 6MWT at baseline. After the training, all eight were able to do so. Regardless of this, a notable improvement of a severely affected patient is reported here, who, beyond the functional tests, also showed kinematic improvements, e.g., in hip extension, which is consistent with the results of Brinkemper et al. [17]. Table 2 summarizes single-case or two-case reports studies details and results.

Table 2. Spinal cord injury single-case or two-case reports studies characteristics.

Author	Number of Patients	Acute/Chronic	Lesion Levels	Training Type	Training Time (Minutes)	Training Period (Weeks)	Training Sessions per Week	Total Training Sessions	Baseline Measurement	Results	Patients Involved in Other Studies
Crucige et al. [21]	2	chronic	T11, AIS A; L2, AIS A	Treadmill	30	12	5	60	Yes	improvements in motor functions and walking abilities, significant reduction in pain severity and improvements in all HRQoL domains	In [9]
Cruciger t al. [22]	1	acute	T10, AIS C	Treadmill	30	12	5	60	Yes	increase in WISCI II score from 8 to 18; conversion to AIS D	No
Ikumi et al. [23]	1	chronic	C4, AIS A	Treadmill	60	5	2	10	Yes	walking distance and time Increased; mAs score decreased after HAL training	No
Shimizu et al. [24]	1	chronic	L3, AIS C	Walker	20	5	2	10	Yesy	improved 10MWT; after the intervention, the right Gmax and both Quad muscles were activated in stance phase rhythmically according to the gait cycle; ISNCSCI motor score improved; WISCI II score improved	No
Yilmaz et al [25]	6(1)	acute	spinal cord infarction following pulmonary embolism	Treadmill	30	12	5	60	Yes	improvements in distance on treadmill and in a physical function mobility score; worse QOL after training in six months follow up	No
Kanazawa et a [2]	1	chronic	T7, AIS C	Walker	30	25	3	-	Yes	walking ability of patients with a chronic SCI improved, over a short period by combining gait training, using HAL-assisted and conventional gait training and physical therapy	No
Watanabe et al. 26]	2	acute	T8-T10, AIS C; T12-L1, AIS D	Walker	20–30	2	3–4	8	Yes	improvements in LEMS, WISCI II, Comfortable gait speed, stride, cadence, BI, FIM, mRS, and joint angles	No
Soma et al. [27	1	chronic*	C7, AIS C	Walker	15–20	8	2–3	13	Yes	improvements in gait speed, step length, and cadence based on a 10MWT, ISNCSCI motor score, and WISCI II score	No

* Defined here as >6 months after injury.

4. Discussion

The studies shown in this review on the use of HAL in patients with SCI demonstrate a wide range of use from acute patients to chronic patients and from patients with well-preserved residual functionality to severely affected patients. With the exception of skin reddening at the site of the electrodes, leg cuffs or shoes, no other adverse events or injuries were reported. The vast majority of patients improved functionally, showing increased walking distances, walking speeds, and endurance. In addition, a variety of other advances were described, such as temporary decrease in spasticity, improvement in bladder and bowel management, pain reduction, and change in muscle activity. Nevertheless, no clear recommendation can be made as to which patient population is most likely to benefit from this innovative training. It seems that in principle its application is feasible and effective in many types of SCI, but whether it is superior to other training and therapies remains unclear. To date, there are no published controlled randomized trials with firstly, other similar exoskeleton systems and secondly, conventional therapy of the same extent and frequency. This is imperative to differentiate the potential benefits of HAL therapy from others and should be done in the future.

Accessibility to HAL training is mentioned in only a few studies. Okawara et al. [18] states that in Germany, HAL training is only funded for patients within the scope of the employers' liability insurance association, and in Japan, government support is provided for patients with neuromuscular diseases. Thus, access to training for a wide range of patients still seems to be a problem which should be solved. In addition to the insurance and thus financial hurdle, the availability for people of different constitutions must also be mentioned. Limitations for the application of this exoskeletal system, as with other manufacturers, are restrictions on patient's height and weight, contractures, insufficient cardiopulmonary capacity, and spasticity that does not decrease with the use of the system.

The following limitations of the present work should be considered. The underlying selective literature search remains incomplete, and a possible bias cannot be excluded with certainty. Some of the authors of the review are also authors of included studies. The studies found are partly limited in their methodological quality. HAL was developed in Japan and is most widely used there. There might be studies published in Japanese journals that were not included in this review. Seven of the included studies [8,9,14,23,24,26,27] list Professor Yoshiyuki Sankai, the inventor of HAL and CEO of Cyberdyne Inc. as a co-author. However, most state that Professor Sankai and Cyberdyne were not involved in conducting the study, data collection and analysis, writing, or submission. Different versions of HAL may have been used in the early studies than in the later studies. It was not always documented which other therapies were made and if medication was changed in the course of the training. Many of the studies presented have mixed collectives including chronic, acute, incomplete, and complete SCI.

5. Conclusions

Currently, there is no uniform application regarding the appropriate collective, as well as the type and frequency of training. Nevertheless, the current study situation suggests that many patients with paraplegia could benefit from qualified use within their means. Although exoskeletal training in SCI patients is still controversial, it offers unprecedented opportunities for patients and therapists. Future studies should focus on a comparison of HAL with other therapies as mentioned above and develop an optimal standardized therapy in terms of training weeks, training units per week, and training duration per unit.

Author Contributions: Conceptualization, A.B., D.G. and E.Y.; methodology, A.B., D.G. and E.Y.; formal analysis, A.B. and D.G.; data curation, A.B. and M.A.; writing—original draft preparation, A.B.; writing—review and editing, D.G. and E.Y.; visualization, A.B.; supervision, T.A.S. and M.A.; project administration, M.A. All authors have read and agreed to the published version of the manuscript.

Funding: This research received no external funding.

Institutional Review Board Statement: Not applicable.

Informed Consent Statement: Not applicable.

Data Availability Statement: Not applicable.

Acknowledgments: We acknowledge support by the Open Access Publication Funds of the Ruhr-Universität Bochum.

Conflicts of Interest: The authors declare no conflict of interest.

References

1. James, S.L.; Theadom, A.; Ellenbogen, R.G.; Bannick, M.S.; Montjoy-Venning, W.; Lucchesi, L.R.; Abbasi, N.; Abdulkader, R.; Abraha, H.N.; Adsuar, J.C. Global, regional, and national burden of traumatic brain injury and spinal cord injury, 1990–2016: A systematic analysis for the global burden of disease study. *Lancet Neurol.* **2019**, *18*, 56–87. [CrossRef] [PubMed]

2. Kanazawa, A.; Yoshikawa, K.; Koseki, K.; Takeuchi, R.; Mutsuzaki, H. A Consecutive 25-Week Program of Gait Training, Using the Alternating Hybrid Assistive Limb (HAL®) Robot and Conventional Training, and its Effects on the Walking Ability of a Patient with Chronic Thoracic Spinal Cord Injury: A Single Case Reversal Design. *Medicina* **2019**, *55*, 746. [CrossRef] [PubMed]

3. Okamura, J. EMG-based prototype powered assistive system for walking aid. In Proceedings of the Asian Symposium on Industrial Automation and Robotics, Bangkok, Thailand, 6–7 May 1999; pp. 229–234.

4. Ueba, T.; Hamada, O.; Ogata, T.; Inoue, T.; Shiota, E.; Sankai, Y. Feasibility and Safety of Acute Phase Rehabilitation After Stroke Using the Hybrid Assistive Limb Robot Suit. *Neurol. Med. Chir.* **2013**, *53*, 287–290. [CrossRef] [PubMed]

5. Kawamoto, H.; Kamibayashi, K.; Nakata, Y.; Yamawaki, K.; Ariyasu, R.; Sankai, Y.; Sakane, M.; Eguchi, K.; Ochiai, N. Pilot study of locomotion improvement using hybrid assistive limb in chronic stroke patients. *BMC Neurol.* **2013**, *13*, 141. [CrossRef] [PubMed]

6. Sczesny-Kaiser, M.; Kowalewski, R.; Schildhauer, T.; Aach, M.; Jansen, O.; Grasmücke, D.; Güttsches, A.-K.; Vorgerd, M.; Tegenthoff, M. Treadmill Training with HAL Exoskeleton—A Novel Approach for Symptomatic Therapy in Patients with Limb-Girdle Muscular Dystrophy—Preliminary Study. *Front. Neurosci.* **2017**, *11*, 449. [CrossRef] [PubMed]

7. Suzuki, K.; Kawamura, Y.; Hayashi, T.; Sakurai, T.; Hasegawa, Y.; Sankai, Y. Intention-Based Walking Support for Paraplegia Patient. In Proceedings of the 2005 IEEE International Conference on Systems, Man and Cybernetics, Waikoloa, HI, USA, 10–12 October 2005; Volume 3, pp. 2707–2713.

8. Kubota, S.; Nakata, Y.; Eguchi, K.; Kawamoto, H.; Kamibayashi, K.; Sakane, M.; Sankai, Y.; Ochiai, N. Feasibility of Rehabilitation Training With a Newly Developed Wearable Robot for Patients With Limited Mobility. *Arch. Phys. Med. Rehabil.* **2013**, *94*, 1080–1087. [CrossRef] [PubMed]

9. Aach, M.; Cruciger, O.; Sczesny-Kaiser, M.; Höffken, O.; Meindl, R.C.; Tegenthoff, M.; Schwenkreis, P.; Sankai, Y.; Schildhauer, T. Voluntary driven exoskeleton as a new tool for rehabilitation in chronic spinal cord injury: A pilot study. *Spine J.* **2014**, *14*, 2847–2853. [CrossRef]

10. Sczesny-Kaiser, M.; Höffken, O.; Aach, M.; Cruciger, O.; Grasmücke, D.; Meindl, R.; Schildhauer, T.A.; Schwenkreis, P.; Tegenthoff, M. HAL® exoskeleton training improves walking parameters and normalizes cortical excitability in primary somatosensory cortex in spinal cord injury patients. *J. Neuroeng. Rehabil.* **2015**, *12*, 68. [CrossRef]

11. Jansen, O.; Grasmuecke, D.; Meindl, R.C.; Tegenthoff, M.; Schwenkreis, P.; Sczesny-Kaiser, M.; Wessling, M.; Schildhauer, T.; Fisahn, C.; Aach, M. Hybrid Assistive Limb Exoskeleton HAL in the Rehabilitation of Chronic Spinal Cord Injury: Proof of Concept; the Results in 21 Patients. *World Neurosurg.* **2018**, *110*, e73–e78. [CrossRef]

12. Grasmücke, D.; Zieriacks, A.; Jansen, O.; Fisahn, C.; Sczesny-Kaiser, M.; Wessling, M.; Meindl, R.C.; Schildhauer, T.A.; Aach, M. Against the odds: What to expect in rehabilitation of chronic spinal cord injury with a neurologically controlled Hybrid Assistive Limb exoskeleton. A subgroup analysis of 55 patients according to age and lesion level. *Neurosurg. Focus* **2017**, *42*, E15. [CrossRef]

13. Jansen, O.; Schildhauer, T.; Meindl, R.C.; Tegenthoff, M.; Schwenkreis, P.; Sczesny-Kaiser, M.; Grasmücke, D.; Fisahn, C.; Aach, M. Functional Outcome of Neurologic-Controlled HAL-Exoskeletal Neurorehabilitation in Chronic Spinal Cord Injury: A Pilot With One Year Treatment and Variable Treatment Frequency. *Glob. Spine J.* **2017**, *7*, 735–743. [CrossRef] [PubMed]

14. Shimizu, Y.; Kadone, H.; Kubota, S.; Suzuki, K.; Abe, T.; Ueno, T.; Soma, Y.; Sankai, Y.; Hada, Y.; Yamazaki, M. Voluntary Ambulation by Upper Limb-Triggered HAL® in Patients with Complete Quadri/Paraplegia Due to Chronic Spinal Cord Injury. *Front. Neurosci.* **2017**, *11*, 649. [CrossRef] [PubMed]

15. Okawara, H.; Sawada, T.; Matsubayashi, K.; Sugai, K.; Tsuji, O.; Nagoshi, N.; Matsumoto, M.; Nakamura, M. Gait ability required to achieve therapeutic effect in gait and balance function with the voluntary driven exoskeleton in patients with chronic spinal cord injury: A clinical study. *Spinal Cord* **2020**, *58*, 520–527. [CrossRef] [PubMed]

16. Brinkemper, A.; Grasmücke, D.; Yilmaz, E.; Reinecke, F.; Schildhauer, T.A.; Aach, M. Influence of Locomotion Therapy With the Wearable Cyborg HAL on Bladder and Bowel Function in Acute and Chronic SCI Patients. *Global Spine J.* **2021**, Epub *ahead of print*. [CrossRef]

17. Brinkemper, A.; Aach, M.; Grasmücke, D.; Jettkant, B.; Rosteius, T.; Dudda, M.; Yilmaz, E.; Schildhauer, T.A. Improved Physiological Gait in Acute and Chronic SCI Patients After Training With Wearable Cyborg Hybrid Assistive Limb. *Front. Neurorobot.* **2021**, *15*, 116. [CrossRef]

18. Okawara, H.; Tashiro, S.; Sawada, T.; Sugai, K.; Matsubayashi, K.; Kawakami, M.; Nori, S.; Tsuji, O.; Nagoshi, N.; Matsumoto, M.; et al. Neurorehabilitation using a voluntary driven exoskeletal robot improves trunk function in patients with chronic spinal cord injury: A single-arm study. *Neural Regen. Res.* **2022**, *17*, 427–432. [CrossRef]

19. Sawada, T.; Okawara, H.; Matsubayashi, K.; Sugai, K.; Kawakami, M.; Tashiro, S.; Nori, S.; Tsuji, O.; Nagoshi, N.; Matsumoto, M.; et al. Influence of body weight-supported treadmill training with voluntary-driven exoskeleton on the quality of life of persons with chronic spinal cord injury: A pilot study. *Int. J. Rehabilitation Res.* **2021**, *44*, 343–349. [CrossRef]

20. Zieriacks, A.; Aach, M.; Brinkemper, A.; Koller, D.; Schildhauer, T.A.; Grasmücke, D. Rehabilitation of Acute Vs. Chronic Patients With Spinal Cord Injury With a Neurologically Controlled Hybrid Assistive Limb Exoskeleton: Is There a Difference in Outcome? *Front. Neurorobot.* **2021**, *15*, 728327. [CrossRef]

21. Cruciger, O.; Schildhauer, T.A.; Meindl, R.C.; Tegenthoff, M.; Schwenkreis, P.; Citak, M.; Aach, M. Impact of locomotion training with a neurologic controlled hybrid assistive limb (HAL) exoskeleton on neuropathic pain and health related quality of life (HRQoL) in chronic SCI: A case study. *Disabil. Rehabil. Assist. Technol.* **2016**, *11*, 529–534. [CrossRef]

22. Cruciger, O.; Tegenthoff, M.; Schwenkreis, P.; Schildhauer, T.A.; Aach, M. Locomotion training using voluntary driven exoskeleton (HAL) in acute incomplete SCI. *Neurology* **2014**, *83*, 474. [CrossRef]

23. Ikumi, A.; Kubota, S.; Shimizu, Y.; Kadone, H.; Marushima, A.; Ueno, T.; Kawamoto, H.; Hada, Y.; Matsumura, A.; Sankai, Y.; et al. Decrease of spasticity after hybrid assistive limb® training for a patient with C4 quadriplegia due to chronic SCI. *J. Spinal Cord Med.* **2017**, *40*, 573–578. [CrossRef]

24. Shimizu, Y.; Nakai, K.; Kadone, H.; Yamauchi, S.; Kubota, S.; Ueno, T.; Marushima, A.; Hiruta, K.; Endo, A.; Kawamoto, H.; et al. The Hybrid Assistive Limb® intervention for a postoperative patient with spinal dural arteriovenous fistula and chronic spinal cord injury: A case study. *J. Spinal Cord Med.* **2018**, *41*, 710–717. [CrossRef] [PubMed]

25. Yilmaz, E.; Schmidt, C.K.; Mayadev, A.; Tawfik, T.; Kobota, K.; Cambier, Z.; Norvell, D.D.; Chapman, J. Does treadmill training with Hybrid Assistive Limb (HAL) impact the quality of life? A first case series in the United States. *Disabil. Rehabil. Assist. Technol.* **2019**, *14*, 521–525. [CrossRef] [PubMed]

26. Watanabe, H.; Marushima, A.; Kawamoto, H.; Kadone, H.; Ueno, T.; Shimizu, Y.; Endo, A.; Hada, Y.; Saotome, K.; Abe, T.; et al. Intensive Gait Treatment Using a Robot Suit Hybrid Assistive Limb in Acute Spinal Cord Infarction: Report of Two Cases. *J. Spinal Cord Med.* **2019**, *42*, 395–401. [CrossRef] [PubMed]

27. Soma, Y.; Kubota, S.; Kadone, H.; Shimizu, Y.; Takahashi, H.; Hada, Y.; Koda, M.; Sankai, Y.; Yamazaki, M. Hybrid Assistive Limb Functional Treatment for a Patient with Chronic Incomplete Cervical Spinal Cord Injury. *Int. Med. Case Rep. J.* **2021**, *14*, 413–420. [CrossRef]

28. Baunsgaard, C.; Nissen, U.; Brust, A.; Frotzler, A.; Ribeill, C.; Kalke, Y.; León, N.; Gómez, B.; Samuelsson, K.; Antepohl, W.; et al. Exoskeleton gait training after spinal cord injury: An exploratory study on secondary health conditions. *J. Rehabil. Med.* **2018**, *50*, 806–813. [CrossRef]

29. Hubscher, C.H.; Herrity, A.N.; Williams, C.S.; Montgomery, L.R.; Willhite, A.M.; Angeli, C.A.; Harkema, S.J. Improvements in bladder, bowel and sexual outcomes following task-specific locomotor training in human spinal cord injury. *PLoS ONE* **2018**, *13*, e0190998. [CrossRef]

30. Juszczak, M.; Gallo, E.; Bushnik, T. Examining the Effects of a Powered Exoskeleton on Quality of Life and Secondary Impairments in People Living with Spinal Cord Injury. *Top. Spinal Cord Inj. Rehabil.* **2018**, *24*, 336–342. [CrossRef]

31. Chun, A.; Asselin, P.K.; Knezevic, S.; Kornfeld, S.; Bauman, W.A.; Korsten, M.A.; Harel, N.Y.; Huang, V.; Spungen, A.M. Changes in bowel function following exoskeletal-assisted walking in persons with spinal cord injury: An observational pilot study. *Spinal Cord* **2020**, *58*, 459–466. [CrossRef]

Article

Design Approaches of an Exoskeleton for Human Neuromotor Rehabilitation

Cristian Copilusi [1], Sorin Dumitru [2,*], Ionut Geonea [1], Leonard Gherghe Ciurezu [1] and Nicolae Dumitru [1]

[1] Faculty of Mechanics, University of Craiova, 200512 Dolj, Romania; cristian.copilusi@edu.ucv.ro (C.C.); ionut.geonea@edu.ucv.ro (I.G.); leonard.ciurezu@edu.ucv.ro (L.G.C.); nicolae.dumitru@edu.ucv.ro (N.D.)

[2] Faculty of Automation, Computers and Electronics, University of Craiova, 200776 Dolj, Romania

* Correspondence: sorin.dumitru@edu.ucv.ro; Tel.: +40-744152497

Abstract: This paper addresses a design for an exoskeleton used for human locomotion purposes in cases of people with neuromotor disorders. The reason for starting this research was given by the development of some intelligent systems for walking recovery involved in a new therapy called stationary walking therapy. This therapy type will be used in this research case, through a robotic system specially designed for functional walking recovery. Thus, the designed robotic system structure will have a patient lifting/positioning mechanism, a special exoskeleton equipped with sensors and actuators, a treadmill for walking, and a command and control unit. The exoskeleton's lower limbs will have six orthotic devices. Thus, the exoskeleton's lower limbs' motions and orthoses angle variations will be generated by healthy human subjects on the treadmill with the possibility of memorizing these specific motions for obtaining one complete gait cycle. After this, the memorized motions will be performed to a patient with neuromotor disorders for walking recovery programs. The design core is focused on two planar-parallel mechanisms implemented at the knee and ankle joints of each leg's exoskeleton. Thus, numerical simulations for the design process were carried out to validate the engineering feasibility of the proposed leg exoskeleton.

Keywords: exoskeleton; robotic system; dynamic analysis; neuromotor disorders; planar-parallel mechanisms

Citation: Copilusi, C.; Dumitru, S.; Geonea, I.; Ciurezu, L.G.; Dumitru, N. Design Approaches of an Exoskeleton for Human Neuromotor Rehabilitation. *Appl. Sci.* **2022**, *12*, 3952. https://doi.org/10.3390/app12083952

Academic Editor: Alessandro de Sire

Received: 10 March 2022
Accepted: 4 April 2022
Published: 13 April 2022

Publisher's Note: MDPI stays neutral with regard to jurisdictional claims in published maps and institutional affiliations.

1. Introduction

Nowadays, exoskeletons have become popular in medical rehabilitation programs and extended their applications as power augmentation systems used in different domains, such as industry or defense.

Considering exoskeletons specially designed for human locomotion system recovery, these can be divided into two major categories: one category consists of exoskeletons that serve as robotic systems in an assistive mode for restoring gait patterns for persons who have partial mobility; the other category is represented by exoskeletons specially designed to help elderly persons perform daily activities. The exoskeletons from the first category were specially designed for people with neuromotor disorders or stroke accidents [1–8]. Others exoskeletons [9–15] were specially designed to fulfill human walking temporal recovery programs dedicated to persons who were exposed to orthopedic surgical interventions. In both mentioned categories, there are fully actuated exoskeletons [3,6] for all human lower-limb joints under rehabilitation programs. Considering the exoskeletons used on temporal recovery programs [11,13], these can have actuated individual joints from the patient locomotion system (one joint actuated, a pair of joints or a group of human lower limb joints). In addition, these are characterized by personalized therapy protocols that can recover a single joint, namely the one that was involved in a surgical intervention.

Thus, most of the existing exoskeletons have imposed several design conditions, such comfort during rehabilitation programs, complex actuations, low- or high- cost principles, easy operation features, and user-friendly interfaces.

In order to accomplish the aforementioned conditions, some exoskeletons [3,6] can fulfill a few of them, while other solutions [11,14] can cover the rest. The latest ones [11,14] recharacterized by a complex structure, leading to expensive rehabilitation programs for patients.

Most of the designed exoskeletons require a qualified rehabilitation team made from a proper physician, a medical engineer, a technician and some medical rehabilitation assistants.

By referring to the beginning of an exoskeleton system design, and taking into account the physician-specific requirements, it can be remarked that human exoskeletons are characterized by a complex actuation system that involves mechanisms with a minimum number of actuators [2,8,9,15].

There are also complex rehabilitation systems especially designed for human walking recovery [16,17] and these have in their structure an exoskeleton with a simplified mechanical structure, but all equivalent human locomotion joints are fully actuated through complex command and control protocols.

The criteria imposed by physician requirements in cases of persons disabled due to neuromotor disorders during walking activities must be considered as a starting research point in designing an exoskeleton, as it can be remarked in [18], and it can be observed that there are persons who need to recover some joints from the entire locomotion system. In this case, it is necessary to have an exoskeleton that can perform isolated motions for the affected joints or to have a complete exoskeleton that involves all human locomotion joints for neuromotor rehabilitation processes. Thus, this represents a research challenge for designing an exoskeleton used in both cases.

Another argument is the one that the existing exoskeletons [1,3–5] need to have a database characterized by locomotion motion law patterns acquired from persons with different anthropometric data, and the therapist needs to choose and adapt these motion patterns for the targeted patient. For accomplishing this argument, most of the exoskeletons were designed in a parameterized form and have segments with variable length and width. These segments can be adjusted through locking/unlocking mechanical devices. In some cases, the motion patterns cannot correspond to the patient's anthropometric data and, in this case, raise the idea of generating the motion pattern in a manual mode, namely an exoskeleton teaching mode function. This function will be accomplished by the therapist moving the exoskeleton segments manually while these segments are connected with the patient's legs. This function lets the therapist set up the motion limits during walking training for the analyzed patient. Moreover, this will allow operation with sensitive patients which have pains during movement. Thus, it is necessary to apply progressive therapy programs in their motion limits without pains.

In addition, the elaborated robotic system allows the physician to adapt alternative recovery therapies such as TENS (Transcutaneous Electrical Nerve Stimulation). The elaborated robotic system will permit manual motions of the exoskeleton limbs together with the patient's lower limbs in a controlled mode, and the robotic system command and control unit will memorize these motions under a special function called teaching mode. Similar procedures are mentioned in [19–21].

The research aim of this paper is to develop a new robotic system based on an exoskeleton structure. For this, it is necessary to elaborate a conceptual solution, experimental analyses with healthy persons and neuromotor-disabled persons, which are considered input data for the mathematical models specially developed for dynamic analysis and experimental tests. Thus, the experimental tests, which represent the purpose of this research, will be defined as preliminary ones, which do not involve any human volunteers and will be performed only for checking the exoskeleton's functionality.

In the proposed research, two major objectives can be identified for the proposed exoskeleton. One objective is to have a fully actuated test bed with a proper exoskeleton specially designed for walking activities and, in some particular cases, to permit motions on individual joints. The second objective is to have a teaching function application, as previously described, which can be applied to patients whose anthropometric data cannot be identified from a motion pattern database.

The mentioned objectives can be considered as starting point for a feasibility study in designing an exoskeleton for persons with neuromotor disorders. Thus, in the Section 1, several exoskeleton construction solutions which can partially fulfill the proposed objectives were analyzed.

The Section 2 is allocated to create a database by performing experimental analyses of human locomotion system motions during walking. This allows the generation of gait patterns implemented in command and control programs for controlling the designed exoskeleton.

Once we had an idea of the analyzed exoskeleton, in the third part of the research the conceptual exoskeleton design was elaborated. This will be eligible to perform an adequate dynamic analysis based on modern design techniques, which is presented in the fourth section for obtaining the reaction forces during ground contact and joints connection forces which will be further used as input data for virtual simulations. Through the performed dynamic analysis, the proposed exoskeleton conceptual solution is validated and in the last section, there is presented the elaborated prototype together with an experimental analysis performed in laboratory conditions.

2. Human Walking Database

In order to create the desired database used in exoskeleton design and programing and control protocols, it is necessary to perform experimental analyses on a set of 30 healthy persons during walking activity. This database consists of gait patterns and anthropometric data of the analyzed human healthy subjects.

The proposed group comprised 15 male and 15 female human subjects with ages between 30 to 58 years old, weights between 55 kg and 95 kg, heights from 1.55 m to 1.90 m.

These experimental analyses were performed in laboratory conditions with the aid of VICON equipment from the University of Craiova Research Centre—Biomechanical laboratories [22,23]. In the case of each analyzed human subject, a protocol based on the workflow was applied, as schematized in Figure 1.

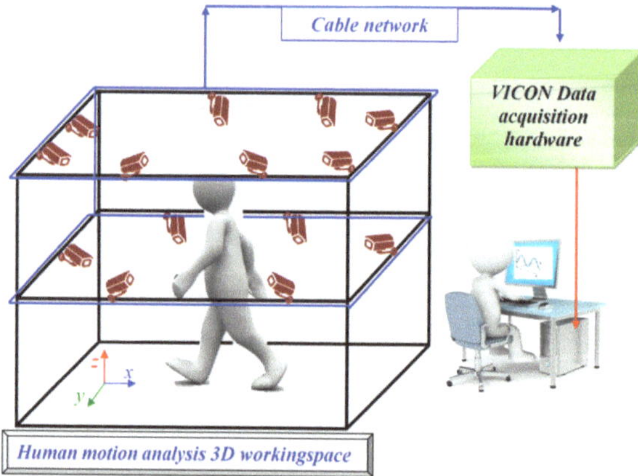

Figure 1. VICON equipment workflow for creating the proposed database.

Each protocol consists of preparing the person for experimental tests, motion analysis equipment setup, camera calibration and global reference system definition, numerical data processing, and obtaining the filtered angular data and exporting these as a report of each experimental analysis.

Each person's experimental analysis was developed in about 30 min and the acquired filtered angular data were obtained for a complete gait. A complete gait of the analyzed persons was developed in a time limit interval between 1.40 to 1.82 s.

This motion video analysis equipment has 14 high-speed cameras that can simultaneously track and record the attached markers, and these cameras work in IR-mode characterized by a frame rate of 1000 frames/s.

A snapshot of modeling the human walking during an experimental test is shown in Figure 2. In this case, a female human subject was analyzed, as an example, with known anthropometric data (age 31 years, dimensional leg parameters known, 61 kg weight, 1.75 m height). This person performed a complete gait in a time interval between 0 to 1.45 s. This time interval is essential for dynamic analyses of the proposed exoskeleton, but also virtual simulations of this. The acquired data were considered for a complete gait when the human subject's heel enters in contact with the ground and finalizes this act at the next contact with the same heel. Moreover, during a complete gait, the ankle joint performs two motions as referring to a horizontal line, namely plantar and dorsal flexion. Thus, dorsal flexion is the one when the ankle moves above the horizontal line and the other motion is below the horizontal line [23,24].

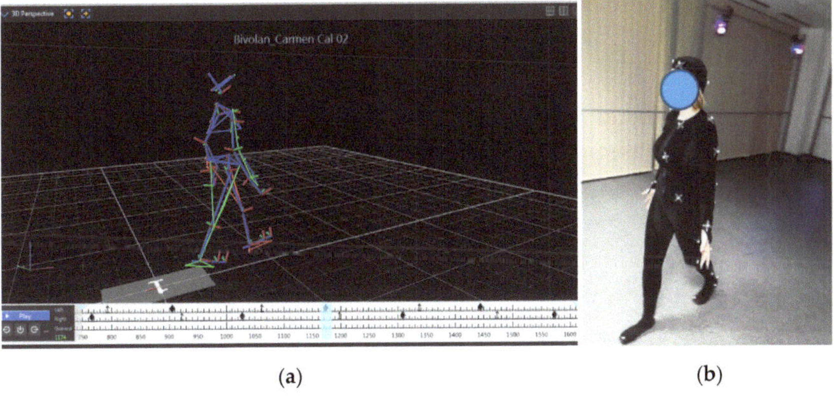

(a)   (b)

Figure 2. Video capture during an experimental test: (**a**) print screen during video motion analysis; (**b**) snapshot during experimental analysis.

The important processed data were represented by each joint angular variation from the human subject locomotion system. The analyzed human joints were hips, knees, and ankles, according to angular data variations presented in Figures 3–6.

From Figure 3, it can be remarked that the angular variation for human hip joints has an interval between −23.6575 degrees to 19.537 degrees. In particular, from Figure 4, the angular variation of a human knee for a complete gait was developed between 0 to 51.261 degrees. In the case of Figure 5, the angular variation for analyzed human ankle joints during a complete gait is situated in an interval of −33.237 degrees to 32.428 degrees. Moreover, these numerical values are very important due to the use of these as reference data and input data for numerical simulations, as well as implementing them as input data for planar-parallel mechanisms on knee and ankle actuation.

The obtained result will be useful for accomplishing comparative analyses between the numerical computation of the designed exoskeleton and also to validate the proposed prototype. These results were validated by considering data from the specialty literature [23–27] and at some points during this research, especially in the case of virtual simulations, these were converted in radians as measuring units.

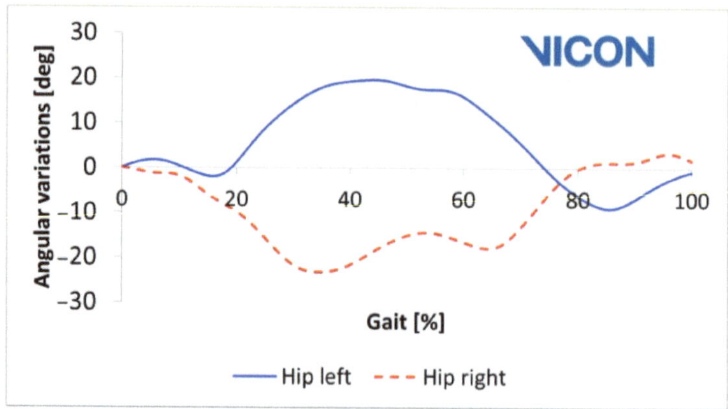

Figure 3. Hip joints angular variations during one gait vs. time.

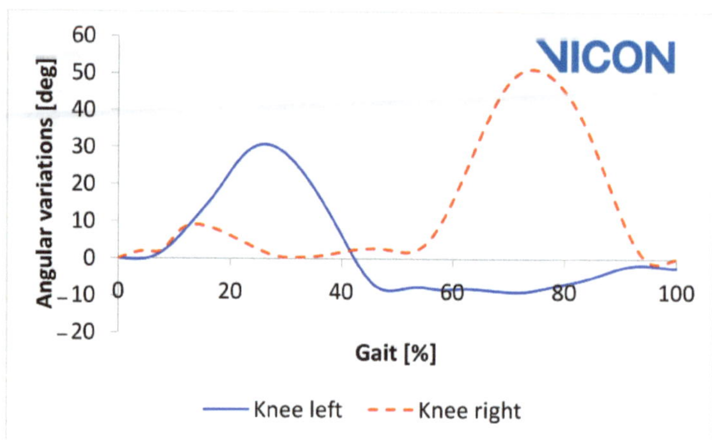

Figure 4. Knee joints angular variations during one gait vs. time.

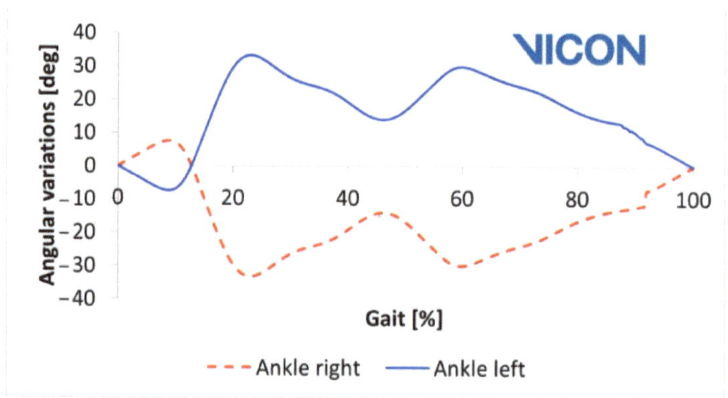

Figure 5. Ankle joints angular variations during one gait vs. time.

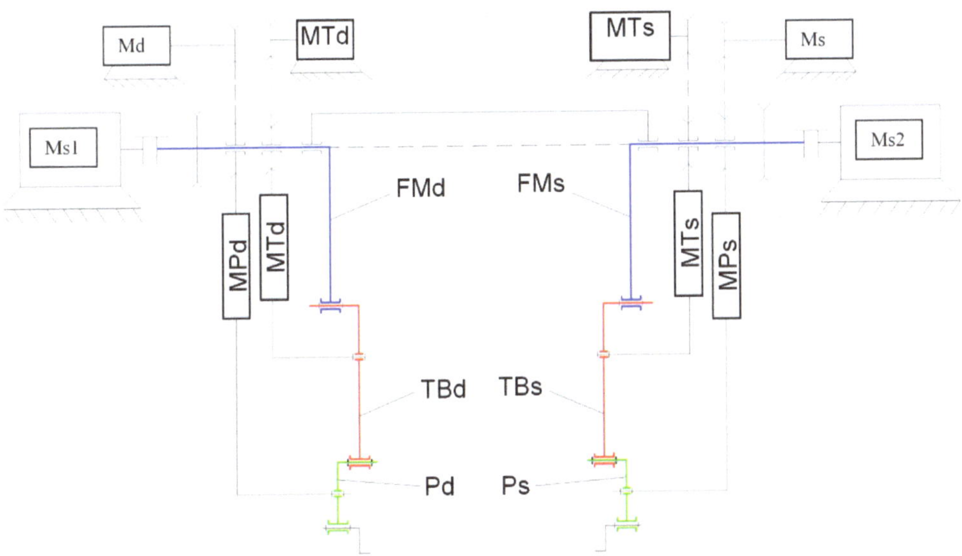

Figure 6. Exoskeleton functional principle design.

3. Exoskeleton Functional Principle

Considering similar existing state-of-the-art exoskeleton models, the designed prototype should have 6 actuators for the main joints of the human locomotion system. Moreover, these actuators must be placed outside the exoskeleton structure, on a special frame, by assuring them an optimal functionality. In order to transmit energy parameters (power, torque, angular speed), mechanical transmissions and planar mechanisms must be used.

The proposed conceptual solution is shown in Figure 6 and it has, in its own structure, the following components: **Md**—servomotor for right ankle and foot segment actuation; **MTd**—servomotor for right tibia equivalent segment and right knee actuation; **Ms1**—servomotor for direct actuation of right hip; **MPd**—planar mechanism for right ankle joint actuation; **MTd**—planar mechanism for actuating the right tibia equivalent segment; **FMd**—right femur equivalent segment; **TBd**—right tibia equivalent segment; **Pd**—right foot equivalent segment; **Ms2**—servomotor for left ankle and foot segment actuation; **MTs**—servomotor for left tibia equivalent segment and left knee actuation; **Ms2**—servomotor for direct actuation of left hip; **Ms**—servomotor for left ankle and foot segment actuation; **MTs**—planar mechanism for actuating the left tibia equivalent segment; **MPs**—planar mechanism for left ankle joint actuation; **FMs**—left femur equivalent segment; **TBs**—left tibia equivalent segment; **Ps**—left foot equivalent segment.

Starting from this conceptual solution, the first model of the proposed exoskeleton prototype was developed, which is presented in Figure 7. This is structured by four planar-parallel mechanisms equivalent to the lower limbs of the human locomotion system and two major actuation groups equipped with servomotors for actuating hip, knee, and ankle joints. These servomotor groups are placed near the pelvic zone on a special frame specially designed with size adjustment mechanical devices. All four planar-parallel mechanisms are deployed in a mirror form, i.e., a pair for the left lower limb and a pair for the right one. These mechanisms are actuated through chain transmissions which assure the energy parameters transfer from the servomotors to the planar-parallel mechanisms drive links.

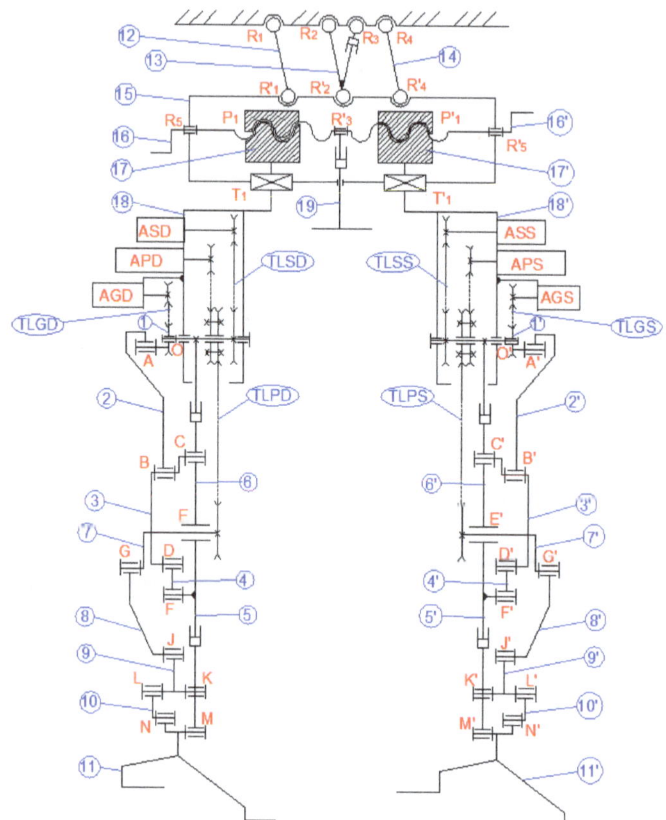

Figure 7. Exoskeleton structural scheme.

Considering the structural scheme from Figure 7, the whole designed robotic system has the following components: **ASD**—right hip servomotor; **APD**—right ankle servomotor; **AGD**—right knee servomotor; **ASS**—left hip servomotor; **APS**—left ankle servomotor; **AGS**—left knee servomotor.

The mechanical system for exoskeleton width adjustment situated near the patient pelvic zone was designed with power screws, which can be actuated manually and move inside or outside the exoskeleton lower limbs, depending on the patient pelvic zone width. This can be performed by actuating the elements **16**, respectively **16'**. The power screw was designed with two thread sectors made in opposite directions, namely **P1** and **P1'**. These will move two nuts **17** and **17'** which are connected with the exoskeleton lower limbs through translational joints **T1** and **T1'**, respectively. The power screws are articulated without the frame **15** through ball bearings **R5**, **R3**, and **R5'**. The role of component **19** is to maintain the patient body in an orthostatic position during walking recovery procedures.

The whole assembly is articulated onto main support by connection elements **12**, **13**, and **14** through proper articulations (revolute joints **R1**, **R2**, **R3**, **R4**, and **R1'**, **R2'**, **R4'**) which allows the whole structure to oscillate in a vertical plane during walking recovery activities. Component **13** has a dumping system for shock-absorbing during walking motion when the patient body will modify the mass center in a vertical direction due to the foot and ground contact.

In the case of exoskeleton right lower limb, the following revolute joints can be seen in Figure 7: **O**—right hip revolute joint; **E**—right knee revolute joint; **M**—right ankle and foot revolute joint. The right link equivalent to femur was labeled with **6** and for the right tibia can be found the link labeled with **5**. For the right foot equivalent link, this can be found labeled with **11**. Thus, these links are actuated through two planar-parallel mechanisms, namely one for the right knee joint **E** and the other for the right ankle joint and foot **M**. Right hip joint is actuated directly through a chain transmission, respectively **TLSD**, by the **ASD** servomotor unit. Regarding the right knee joint actuation, this can be done by **AGD** servomotor unit through a second chain transmission labeled with **TLGD**, towards planar-parallel drive link mechanism situated at the femur level **6**. Thus, the motion from the drive chain wheel connected with the drive link **1** is transmitted through a revolute joint **A** to connecting rod **2**. This connecting rod **2** is articulated through the **B** revolute joint and will transmit motion to the support plate **3**. The support plate **3** is articulated onto the right femur equivalent segment by revolute joint **C**. Practically support plate **3** will perform an oscillating motion imposed to the connection link **4**, which is articulated through revolute joints **D** and **F**. This connection link **4** will actuate the right tibia equivalent segment **5** for a proper flexion motion.

The second right planar-parallel mechanism is situated at the right tibia segment **5** and it will actuate the right ankle joint **M**. This mechanism will receive motion from the **APD** servomotor unit through a third chain transmission **TLPD** straight to drive link **7** situated at the right knee joint level. This drive link **7** motion will be imposed to another support plate **9** through connecting rod **8** articulated with revolute joints **G** and **J**. The support plate **9** will perform an oscillating motion towards the right tibia equivalent segment through revolute joint **K**. This motion is transmitted through connection link **10** to the right ankle and foot segment **11**, for performing the dorsal/plantar flexion during walking activity recovery. The connection link **10** is articulated by the whole planar-parallel mechanism through revolute joints **L** and **N**. In this mode the exoskeleton right lower limb will be actuated and the in case of the left limb, the motions are the same, only the difference is the exoskeleton architecture, where left lower limb is mirrored by a vertical plane from the right one. For this reason, the exoskeleton left lower limb functionality is similar to the right one, except for some small peculiarities, namely: **O′**—left hip revolute joint; **E′**—left knee revolute joint; **M′**—left ankle and foot revolute joint. The left link is equivalent to the femur was labeled with **6′** and for the left tibia can be found the link labeled with **5′**. For the left foot equivalent link, this can be found labeled with **11′**. These links are actuated by another pair of planar-parallel mechanisms, one for the left knee joint **E′** and the other for the left ankle joint **M′**.

The left hip joint is actuated directly through a chain transmission, **TLSS**, by the **ASS** servomotor unit. Regarding the left knee joint actuation, this can be done by **AGS** servomotor unit through a second chain transmission labeled with **TLGS**, towards the planar-parallel drive link mechanism situated at the femur level **6′**. Thus, the motion from the drive chain wheel connected with the drive link **1′** is transmitted through a revolute joint **A′** to connecting rod **2′**. This connecting rod **2′** is articulated through **B′** revolute joint and will transmit motion to the support plate **3′**. The support plate **3′** is articulated onto the left femur equivalent segment by revolute joint **C′**. Practically the support plate **3′** will perform an oscillating motion imposed to the connection link **4′**, which is articulated through revolute joints **D′** and **F′**. This connection link **4′** will actuate the left tibia equivalent segment **5′** for a proper flexion motion.

The second right planar-parallel mechanism is situated at the left tibia segment **5′** and will actuate the left ankle joint **M′**. This mechanism will receive motion from the **APS** servomotor unit through a third chain transmission **TLPS** straight to drive link **7′** situated at the left knee joint level. This drive link **7′** motion will be imposed to another support plate **9′** through connecting rod **8′** articulated with revolute joints **G′** and **J′**. The support plate **9′** will perform an oscillating motion towards the left tibia equivalent segment through revolute joint **K′**. This motion is transmitted through connection link **10′** to the left ankle and foot segment **11′**, for performing the dorsal/plantar flexion during walking activity recovery. The connection link **10′** is articulated by the whole planar-parallel mechanism through revolute joints **L′** and **N′**.

At the end, an exoskeleton lower limb has on its base two planar-parallel mechanisms, three servomotors for actuating the main joints, namely hip, knee, and ankle joints (revolute joints O, **E**, and **M**).

4. Dynamic Analysis of the Proposed Exoskeleton

4.1. Exoskeleton Mathematical Modeling for Dynamic Analysis

To accomplish a dynamic analysis of the proposed exoskeleton concept, a kinematic scheme for both mechanisms that were to actuate knee and ankle joints was elaborated, as can be seen in Figure 8. This was performed in accordance with [28–30]. From this kinematic scheme, the following components can be identified: **O**—hip revolute joint, a direct motor joint; **E**—knee revolute joint which also assures support for the knee and foot actuated mechanism; **M**—knee revolute joint; **A, B, C, D, F**—revolute joints for knee mechanism actuation; **G, J, L, N, K**—revolute joints for ankle mechanism actuation. Link no. 1 represents the drive link for knee mechanism actuation and this mechanism is made from the connecting rod no. 2, support link no. 3 articulated in **CBD** revolute joints, and connection link no. 4 which actuates the tibia equivalent link no. 5 through revolute joint F. The equivalent link of the femur will assure the entire support for the knee actuation mechanism and this can be identified as link no. 6.

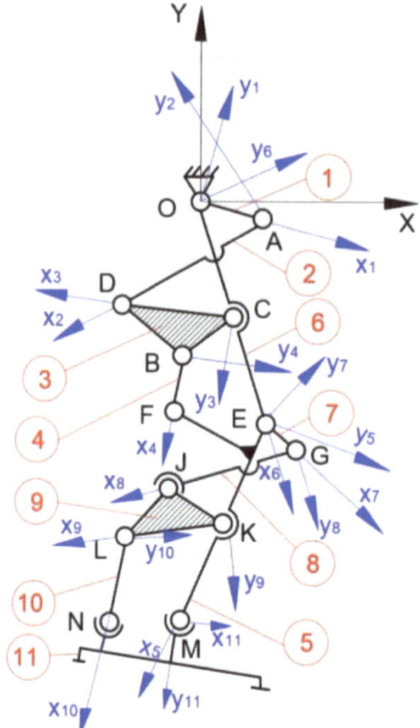

Figure 8. Kinematic scheme of the exoskeleton lower limb.

By having insight into the experimental analysis database, we considered a case subject with the proper motion laws of hip, knee, and joints which will serve as input data through ϕ_{06}, ϕ_{65}, ϕ_{57} variation angles. This will be used also as input data for an inverse kinematic analysis presented in [31] and the aim of this is to identify the drive joints' proper angular speeds.

For this dynamic analysis, the Newton–Euler method completed with Lagrange multipliers will be used. Thus, the kinematic constraints equations for a mobile mechanical system are:

$$\varphi(q,t) = 0 \tag{1}$$

where: t is time and q represents the generalized coordinates.

By differentiating the Equation (1) depending on time, we will obtain:

$$\overset{\bullet}{\varphi}(q,t) = J_q\overset{\bullet}{q} + \frac{\partial \varphi_i}{\partial t} \tag{2}$$

$$\overset{\bullet\bullet}{\varphi}(q,t) = J_q\overset{\bullet\bullet}{q} + \left(J_q\overset{\bullet}{q}\right)_q\overset{\bullet}{q} + 2\frac{\partial^2 \varphi_i}{\partial t\,\partial q_j}\overset{\bullet}{q} + \frac{\partial^2 \varphi_i}{\partial t^2} \tag{3}$$

$$\left(J_q\overset{\bullet}{q}\right)_q = \sum_{k=1}^{n_c} \frac{\partial^2 \varphi_i}{\partial q_j\,\partial q_k}\overset{\bullet}{q}_k \tag{4}$$

where: $i = \overline{1, n_h}$ represents the kinematic constraints equations number and $j = \overline{1, n_c}$ represents the generalized coordinates number.

The following notation will be performed:

$$J_q\overset{\bullet}{q} = \nu = -\frac{\partial \varphi}{\partial t} \tag{5}$$

$$J_q\overset{\bullet\bullet}{q} = -\left(J_q\overset{\bullet}{q}\right)_q\overset{\bullet}{q} - 2\frac{\partial^2 \varphi_i}{\partial t\,\partial q_j}\overset{\bullet}{q} - \frac{\partial^2 \varphi_i}{\partial t^2} = a \tag{6}$$

The Newton–Euler formalism motion equations, completed with Lagrange multipliers method, are:

$$\begin{bmatrix} M & J_q^T \\ J_q & O \end{bmatrix}\begin{bmatrix} \overset{\bullet}{q} \\ a \end{bmatrix} = \begin{bmatrix} Q^a \\ a \end{bmatrix} \tag{7}$$

where: M is the mass matrix, λ—Lagrange multipliers, $\ddot{q}$—accelerations, Q^a—active generalized forces vector, a—is a term from Equation (6).

From Equation (7), it can be written:

$$M\overset{\bullet\bullet}{q} + J_q^T \cdot \lambda = Q^a \tag{8}$$

The Lagrange multipliers will be determined as:

$$\lambda = J_q^{-1}\left[Q^a - M\overset{\bullet\bullet}{q}\right] \tag{9}$$

The connection forces from the k kinematic joint, depending on the axis global coordinates system can be determined with:

$$F_k = \lambda_k \tag{10}$$

Thus, the knee actuating mechanism is made from 1 to 6 kinematic links according to Figure 8. For these links, the proper generalized coordinates are:

$$q_1 = \{X_E, Y_E, \phi_{06}, X_F, Y_F, \phi_{65}, X_B, Y_B, \phi_{63}, X_A, Y_A, \phi_{01}, \phi_{32}\}^T \tag{11}$$

For an inverse dynamic analysis, the kinematic constraint equations in the case of knee actuating mechanism are:

$$
\phi_1(q_1, t) =
\begin{bmatrix}
X_E - X_E^6 \cos \phi_{06} + Y_E^6 \sin \phi_{06} - X_0 \\[4pt]
Y_E - X_E^6 \sin \phi_{06} - Y_E^6 \cos \phi_{06} - Y_0 \\[4pt]
X_F - X_F^5 \cos \phi_{65} + Y_F^5 \sin \phi_{65} - X_E \\[4pt]
Y_F - X_F^5 \sin \phi_{65} - Y_F^5 \cos \phi_{65} - Y_E \\[4pt]
X_D - X_D^4 \cos \phi_{54} + Y_D^4 \sin \phi_{54} - X_F^5 \cos \phi_{65} + Y_F^5 \sin \phi_{65} - X_E^6 \cos \phi_{06} + Y_E^6 \sin \phi_{06} - X_0 \\[4pt]
Y_D - X_D^4 c \sin \phi_{54} - Y_D^4 \cos \phi_{54} - X_F^5 \sin \phi_{65} - Y_F^5 \cos \phi_{65} - X_E^6 \sin \phi_{06} - Y_E^6 \cos \phi_{06} - Y_0 \\[4pt]
X_B - X_B^3 \cos \phi_{63} + Y_B^3 \sin \phi_{63} - X_C \\[4pt]
Y_B - X_B^3 \sin \phi_{63} + Y_B^3 \cos \phi_{63} - Y_C \\[4pt]
X_A - X_A^1 \cos \phi_{01} + Y_A^1 \sin \phi_{01} - X_0 \\[4pt]
Y_A - X_A^1 \sin \phi_{01} - Y_A^1 \cos \phi_{01} - Y_0 \\[4pt]
X_D^4 \cos \phi_{54} - Y_D^4 \sin \phi_{54} + X_F - X_D^3 \cos \phi_{63} + Y_D^3 \sin \phi_{65} - X_C \\[4pt]
X_D^4 \sin \phi_{54} + Y_D^4 \cos \phi_{54} + Y_F - X_D^3 \sin \phi_{63} - Y_D^3 \cos \phi_{65} - Y_C \\[4pt]
X_A^2 \cos \phi_{32} - Y_A^2 \sin \phi_{32} + X_B - X_A^1 \cos \phi_{01} + Y_A^1 \sin \phi_{01} - X_0 \\[4pt]
X_A^2 \sin \phi_{32} + Y_A^2 \cos \phi_{32} + y_B - X_A^1 \sin \phi_{01} - Y_A^1 \cos \phi_{01} - Y_0 \\[4pt]
\phi_{06} - 0.05436169159 + 1.049144646 \cdot t - 3.528717373 \cdot t^2 - 4.459178549 \cdot t^3 + \\
16.81669021 \cdot t^4 - 12.95743095 \cdot t^5 + 3.037974275 \cdot t^6 \\[4pt]
\phi_{65} - 0.05572857950 + 3.917643736 \cdot t - 46.09421696 \cdot t^2 + 142.2026539 \cdot t^3 - \\
183.2410652 \cdot t^4 + 106.6799384 \cdot t^5 - 23.24793765 \cdot t^6
\end{bmatrix}
\tag{12}
$$

The corresponding Jacoby for q_1 generalized coordinates, according with Equation (9) is:

$$
J_{q1} = \frac{\partial \varphi_{1i}}{\partial q_{1j}}; \quad i, j = \overline{1, 16}
\tag{13}
$$

The proper mass matrix is:

$$
M_1 = diag \begin{bmatrix} m_6 & m_6 & J_6 & m_5 & m_5 & J_5 & m_4 & m_4 & J_4 & m_3 & m_3 & J_3 & m_1 & m_1 & J_1 & J_2 \end{bmatrix}
\tag{14}
$$

where: m_i—link mass i; J_i—mechanical inertia momentum corresponding to i link.

The generalized forces matrix:

$$
Q_1^a = \begin{bmatrix} 0 & -m_6 g & 0 & 0 & -m_5 g & 0 & 0 & -m_4 g & 0 & 0 & -m_3 g & 0 & 0 & -m_1 g & 0 & -\tfrac{1}{2} m_2 g \cdot l_{AB} \end{bmatrix}^T
\tag{15}
$$

The matrix made from linear and angular accelerations, corresponding to generalized coordinates from Equation (11) is:

$$
a_{q1} = \ddot{q}_1 = \begin{bmatrix} aX_E & aY_E & \varepsilon_6 & aX_F & aY_F & \varepsilon_5 & aX_D & aY_D & \varepsilon_4 & aX_B & aY_B & \varepsilon_3 & aX_A & aY_A & \varepsilon_1 & \varepsilon_2 \end{bmatrix}^T
\tag{16}
$$

The Lagrange multipliers matrix is:

$$
\lambda_l^1 = J_{q_1}^{-1} \left[Q_1{}^a - M_1 \ddot{a}_{q_1} \right]
\tag{17}
$$

Thus, the kinematic joints connection forces are:

$$F_{l1}^k = \lambda_l^{1k} \tag{18}$$

In the case of the ankle actuation mechanism, by having insight to Figure 8, this consists of 7 to 11 links. Thus, the proper generalized coordinate vector is:

$$q_2 = \left\{ X_M, Y_M, \phi_{65}, X_N, Y_N, \phi_{511}, X_L, Y_L, \phi_{1110}, X_J, Y_J, \phi_{59}, X_G, Y_G, \phi_{57}, \phi_{98} \right\}^T \tag{19}$$

For this, the kinematic constraint equations in the case of ankle actuation mechanism are:

$$\varphi_2(q_2, t) = \begin{bmatrix}
X_M - \cos \phi_{65} X_M^5 + \sin \phi_{65} Y_M^5 - X_E \\[4pt]
Y_M - \sin \phi_{65} X_M^5 - \cos \phi_{65} Y_M^5 - Y_E \\[4pt]
X_N - \cos \phi_{5\,11} X_N^{11} + \sin \phi_{5\,11} Y_N^{11} - X_M \\[4pt]
Y_N - \sin \phi_{5\,11} X_N^{11} - \cos \phi_{5\,11} Y_N^{11} - Y_M \\[4pt]
X_L - \cos \phi_{11\,10} X_L^{10} + \sin \phi_{11\,10} Y_L^{10} - X_N \\[4pt]
Y_L - \sin \phi_{11\,10} X_L^{10} - \cos \phi_{11\,10} Y_L^{10} - Y_N \\[4pt]
X_J - \cos \phi_{59} X_J^9 + \sin \phi_{59} Y_J^9 - X_K \\[4pt]
Y_J - \sin \phi_{59} X_J^9 - \cos \phi_{59} Y_J^9 - Y_K \\[4pt]
X_G \quad \cos \phi_{57} X_G^7 + \sin \phi_{57} Y_G^7 - X_E \\[4pt]
Y_G - \cos \phi_{57} X_G^7 + \sin \phi_{57} Y_g^7 - X_E \\[4pt]
\cos \phi_{11\,10} X_L^{10} - \sin \phi_{11\,10} Y_L^{10} + X_N - \cos \phi_{59} X_L^9 + \sin \phi_{59} Y_L^9 - X_K \\[4pt]
\sin \phi_{11\,10} X_L^{10} + \cos \phi_{11\,10} Y_L^{10} + Y_N - \sin \phi_{59} X_L^9 - \cos \phi_{59} Y_L^9 - Y_K \\[4pt]
\cos \phi_{98} X_G^8 - \sin \phi_{98} Y_G^8 + X_J - \cos \phi_{57} X_G^7 + \sin \phi_{57} Y_G^7 - \cos \phi_{06} X_E^6 - X_0 \\[4pt]
\cos \phi_{98} X_G^8 - \sin \phi_{98} Y_G^8 + X_J - \cos \phi_{57} X_G^7 + \sin \phi_{57} Y_G^7 - \cos \phi_{06} X_E^6 - X_0 \\[4pt]
\phi_{65} - 0.05572857950 + 3.917643736 \cdot t - 46.09421696 \cdot t^2 + 142.2026539 \cdot t^3 - \\ 183.2410652 \cdot t^4 + 106.6799384 \cdot t^5 - 23.24793765 \cdot t^6 \\[4pt]
\phi_{5\,11} - 0.07468922928 - 4.125887507 \cdot t + 46.18350162 \cdot t^2 - 139.7502390 \cdot t^3 + \\ 185.4096885 \cdot t^4 - 113.2636093 \cdot t^5 + 25.99845833 \cdot t^6
\end{bmatrix} \tag{20}$$

The corresponding Jacoby for generalized coordinates that satisfy Equation (20) is:

$$J_{q2} = \frac{\partial \varphi_{2i}}{\partial q_{2j}}; \quad i, j = \overline{1, 16} \tag{21}$$

The mass matrix, which corresponds to the ankle actuation mechanism, is:

$$M_2 = diag \begin{bmatrix} m_5 & m_5 & J_5 & m_{11} & m_{11} & J_{11} & m_{10} & m_{10} & J_{10} & m_9 & m_9 & J_9 & m_7 & m_7 & J_7 & J_8 \end{bmatrix} \tag{22}$$

For this, the linear and angular accelerations matrix corresponding to generalized coordinates given by Equation (19):

$$a_{q2} = \ddot{q}_2 = \begin{bmatrix} aX_M & aY_M & \varepsilon_5 & aX_N & aY_N & \varepsilon_{11} & aX_L & aY_L & \varepsilon_{10} & aX_J & aY_J & \varepsilon_9 & aX_G & aY_G & \varepsilon_7 & \varepsilon_8 \end{bmatrix}^T \tag{23}$$

$$Q_2^a = \begin{bmatrix} 0 & -m_5g & 0 & 0 & -m_{11}g & 0 & 0 & -m_{10}g & 0 & 0 & -m_9g & 0 & 0 & -m_7g & 0 & -\tfrac{1}{2}m_8g \cdot l_{AB} \end{bmatrix}^T \tag{24}$$

The Lagrange multipliers matrix is:

$$\lambda_l^2 = J_{q2}^{-1}\left[Q_2{}^a - M_2\ddot{a}_{q2} \right] \tag{25}$$

The kinematic joints connection forces, in a global coordinates system, will be:

$$F_{l2}^k = \lambda_l^{2k} \tag{26}$$

4.2. Numerical Processing

For the numerical processing of the presented mathematical models numerical processing, a program under MAPLE environment was created, which allows computing a kinematic and dynamic calculus through a direct and inverse method. This program was elaborated for the entire mechanism presented in Figure 8. The computing sequence was processed for a complete predefined gait. This program has on its base the Newton–Raphson method.

The input data for numerical processing, were represented by the positions, speeds, and accelerations of some characteristic points of the kinematic linkage shown in Figure 8. Moreover, the generalized coordinates' variation laws depending on time [31], links mass and inertia properties were considered, and the contact force between foot and ground is represented by the following equation:

$$\begin{aligned} F_t = &\, 16.1241335150856 + 856.883813681185 \cdot t - 407.327793973914 \cdot t^2 + 10389.4221994136 \cdot t^3 - \\ &\, 31172.2612478359 \cdot t^4 + 29652.5101430907 \cdot t^5 - 9228.95270543672 \cdot t^6 \end{aligned} \tag{27}$$

For the output data, the kinematic joints connection forces will be obtained. These forces helped to choose the proper bearings for the exoskeleton structure prototype.

4.3. Exoskeleton Virtual Prototyping with the Aid of MSC Adams Software

By considering the kinematic scheme from Figure 8, an exoskeleton virtual prototype was built with the aid of SolidWorks program. This was a parameterized model and it was imported into the MSC Adams software environment. For this model were defined the mass properties and material characteristics (longitudinal elasticity module: $E = 7.1 \times 10^{10}$ Pa, density: $2770\ \text{kg/m}^3$ and Poisson ratio: 0.33), respectively, as well as kinematic links mobile reference systems and 15 kinematic joints. The virtual prototyping was performed in accordance with specific instructions provided by [32]. Moreover, 6 motor joints for hips, knees, and ankles were identified. By knowing the motion angle variations for hips, knees, and ankle joints, the drive joints angle variation laws φ_{01} and φ_{07} were obtained, according to Figure 8. These joints, as in the case of the mathematical model, actuate these two actuation mechanisms for the ankle and for knee joints, respectively.

The angle variations for hips, knees, and ankle joints were established for a case study, through experimental analysis, as was shown in Figures 3–5. The imported exoskeleton prototype with proper revolute joints is shown in Figure 9. These drive joints, as in the case of the mathematical model, will actuate two mechanisms, one for the knee joint and other for ankle joint. Additionally, the contact force between ground and foot and the patient weight were introduced through the parameterized mannequin weighing 61.7 kg.

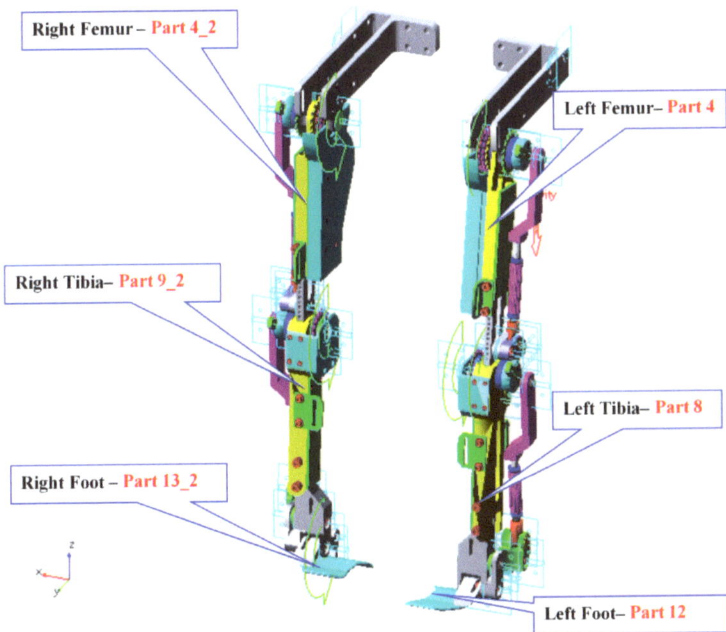

Figure 9. Exoskeleton parametrized virtual model with proper joints prepared for virtual simulations in a dynamic mode.

Thus, an inverse dynamic analysis was processed with the aid of MSC Adams software, in order to obtain the proper connection forces variation over time of the analyzed kinematic joints. The obtained results were materialized through diagrams presented in a comparative analysis between computed mathematical models and the ones obtained through numerical simulations with MSC Adams software.

For the proposed exoskeleton prototype, a parameterized mannequin was introduced, with anthropometric data similar to the healthy human subject used in the experimental analysis section. Thus, four sequences during virtual simulations processed under the MSC Adams environment are presented in Figure 10.

In Figure 11, the connection force component in X-axis of the knee joint is represented, which corresponds to a complete gait sequence. This component has a high value when the analyzed models start walking and this reaches a value of 23. This component has a high value when the analyzed models start walking and this reaches a value of 23.567 Newton. This value was obtained when the contact between ground and foot occurs, for both analysed models (mathematical model and Adams model).

It can be observed that the obtained trajectories for both models are appropriate, recording values between −10 Newton and 11.875 Newton. The time interval also was appropriate, namely 0 to 1.425 s, respectively. An error of 2.25% was obtained, which is quite small.

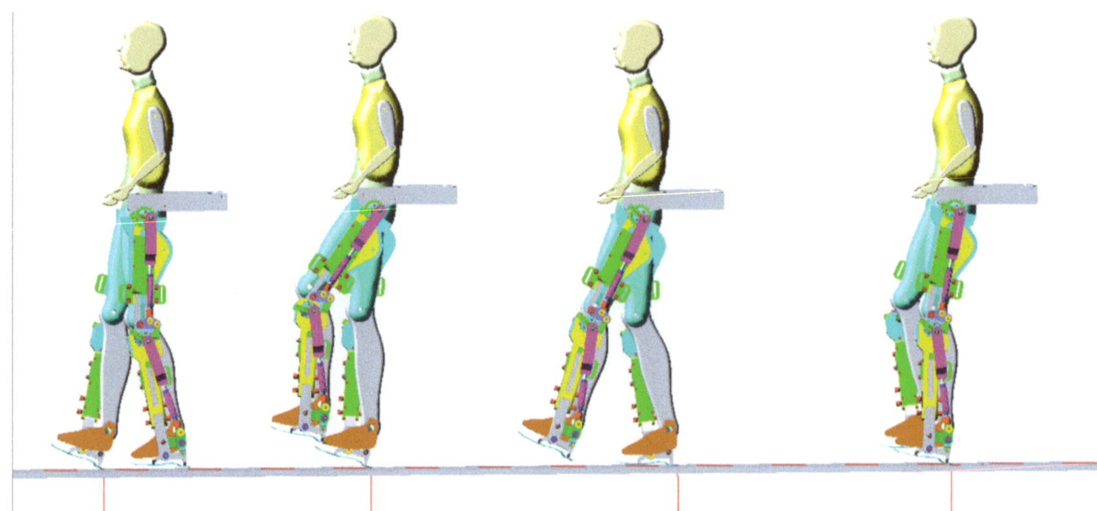

Figure 10. A snapshot during virtual simulations in dynamic mode for a complete gait.

The desired results, in order to validate this model, were given by the connection forces' variations over time; a comparative analysis can be seen in Figures 11–16.

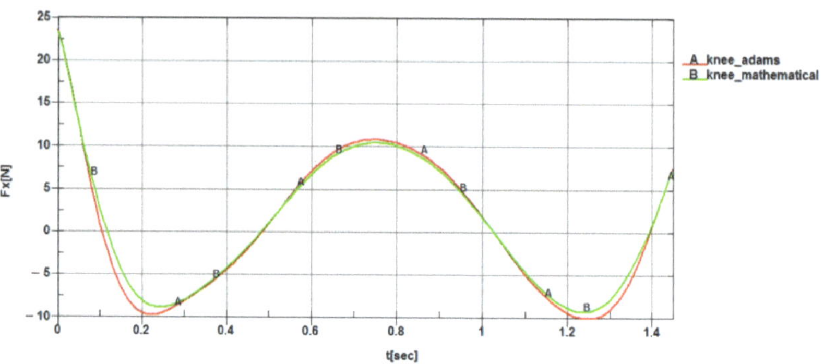

Figure 11. Connection forces component on x-axis for knee joint vs. time (without ground contact).

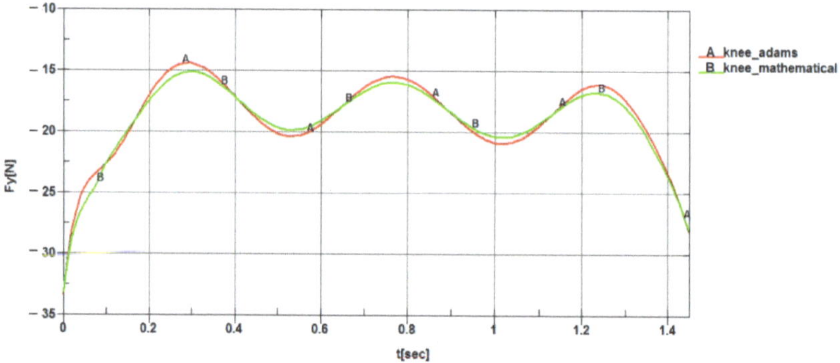

Figure 12. Connection forces component on y-axis for knee joint vs. time (without ground contact).

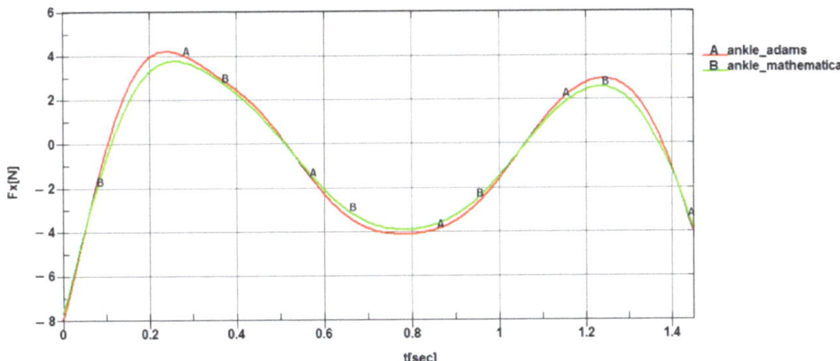

Figure 13. Connection forces component on x-axis for ankle joint vs. time (without ground contact).

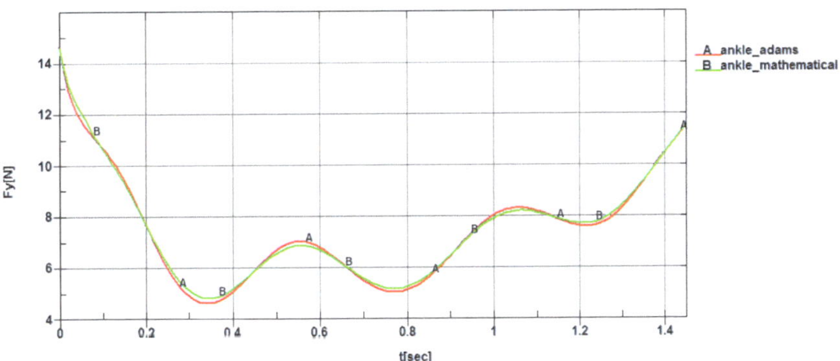

Figure 14. Connection forces component on y-axis for ankle joint vs. time (without ground contact).

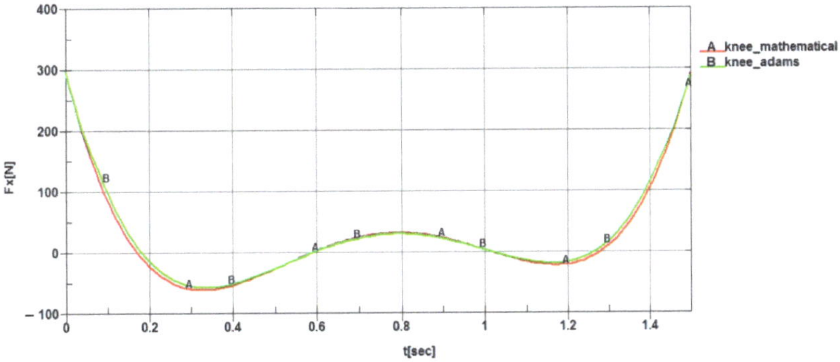

Figure 15. Connection forces component on x-axis for knee joint vs. time (with ground contact).

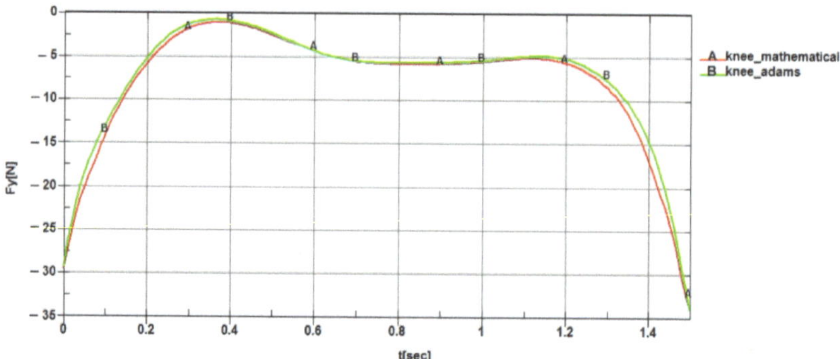

Figure 16. Connection forces component on y-axis for knee joint vs. time (with ground contact).

In the case of the connection force component on y-axis, for the same joint, high values can be observed in Figure 12 when gait starts and at the end of a complete gait. These values are recorded between an interval of -22.5 Newton and -16.74 Newton. Moreover, this analyzed diagram has appropriate trajectory paths with an error of approximately 1.87%.

Analyzing the ankle joint diagram, the connection forces are shown in Figures 13 and 14. For these, values between -4.27 Newton and 4.37 Newton were recorded during a complete gait for the x-axis component. Both analyzed models have similar trajectories and the error between the analyzed paths records a value of 2.58%. Similar to the knee joint, high values were recorded when gait starts and when gait ends. By having insight into the y-axis connection force component, the analyzed trajectories are almost identical and an error of 0.86% was obtained. Thus, forces between 4.75 Newton and 14.68 Newton were noted. Another important argument was given by the fact that the processed models are valid only when the contact between foot and ground was neglected.

By taking into account the contact force between foot and ground, the connection forces' values will differ. Thus, Figures 15 and 16 present the connection force components in the case of the knee joint. Thus, the connection force component on x-axis in the case of the knee joint records values between -65 Newton and 300 Newton. These values were obtained for a complete gait cycle and the presented trajectories shown in Figure 15 are similar for both analyzed models, and the obtained error for this case reaches 0.92%.

Regarding the connection force component on the y-axis represented in the diagram from Figure 16, the recorded values are between -34.46 Newton and -2.57 Newton. Furthermore, in this case, the recorded trajectories for both models have smooth paths with an error of 1.68%.

5. Exoskeleton 3D Modeling

Considering the structural scheme presented in Figure 7, a virtual model was created for the entire exoskeleton in a constructive form with the aid of SolidWorks. This is shown in Figures 17 and 18. In these figures, the correspondence between virtual model, structural scheme, and kinematic scheme shown in Figures 7 and 8 can be identified.

Thus, the exoskeleton's lower limb has, in its structure, 15 kinematic joints and 11 kinematic elements (links). Each lower limb will be actuated through three servomotor groups and the energetic parameters will be transmitted to planar-parallel mechanism drive links through chain transmissions with 1:1 ratio.

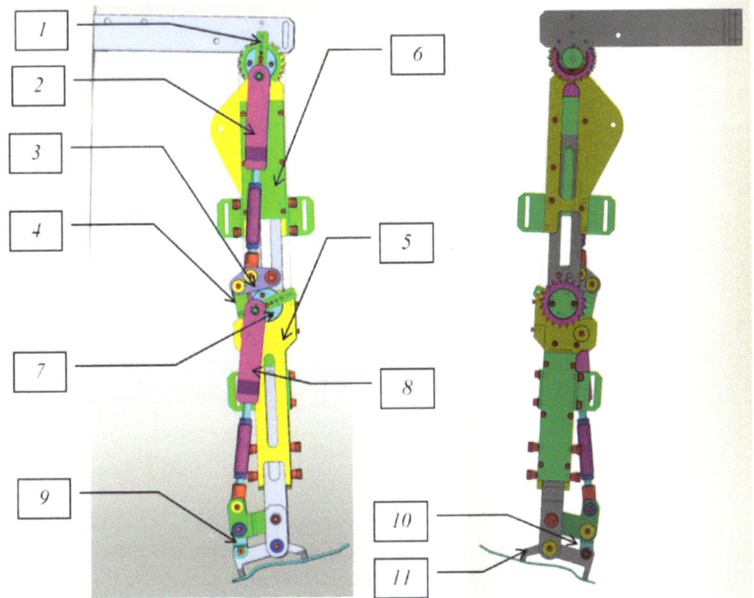

Figure 17. Left leg exoskeleton virtual model.

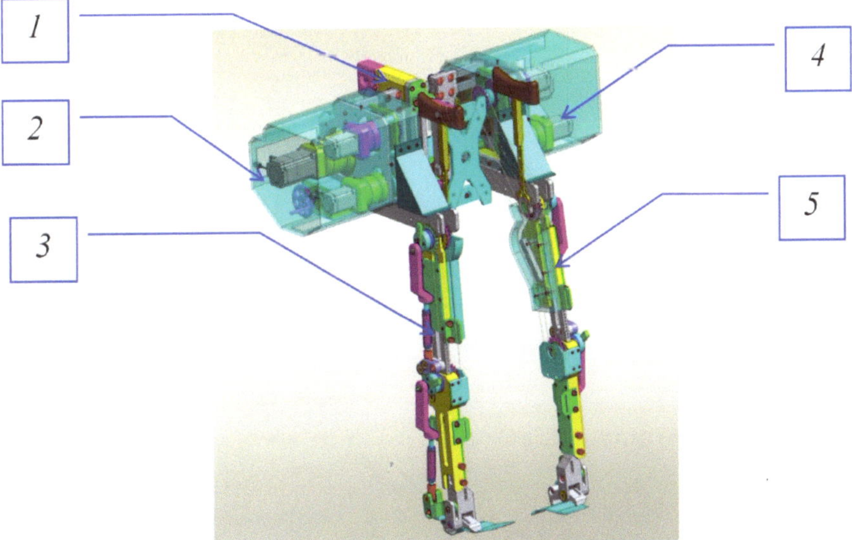

Figure 18. 3D model of the NeuRob exoskeleton with actuation unit.

From Figure 18, the following components are identified: **1**—frame; **2**—servomotors unit for left exoskeleton lower limb; **3**—exoskeleton left lower limb with proper segments equivalent to femur, tibia, and foot; **4**—servomotors unit for right exoskeleton lower limb; **5**—exoskeleton right lower limb with proper segments equivalent to femur, tibia, and foot. Each exoskeleton lower limb was designed with proper modular mechanical systems for dimensional parameters adjustments, as can be seen in Figure 19.

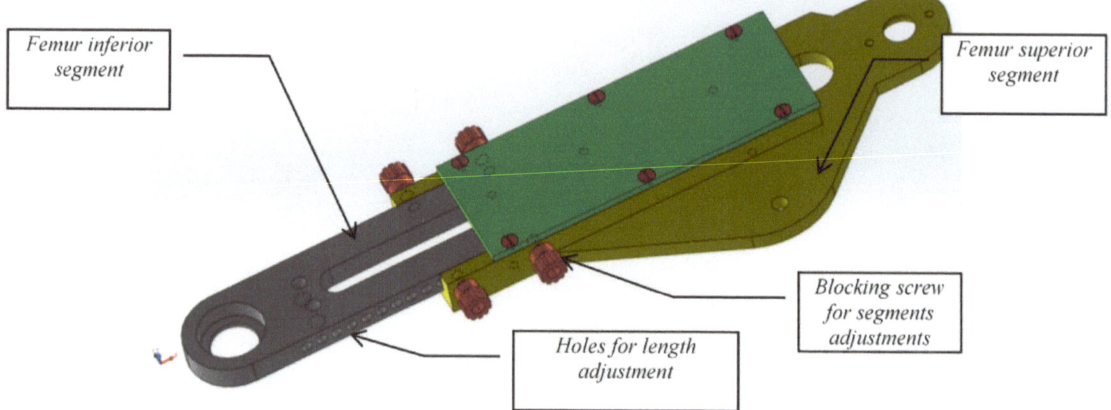

Figure 19. Femur segment with variable length.

As regarding the exoskeleton limbs' adjustment, for adapting them to different anthropometrics, both lower limbs are fixed in a console type through a specific frame specially designed with a mechanical transmission nut-screw type as it can be seen in Figure 20.

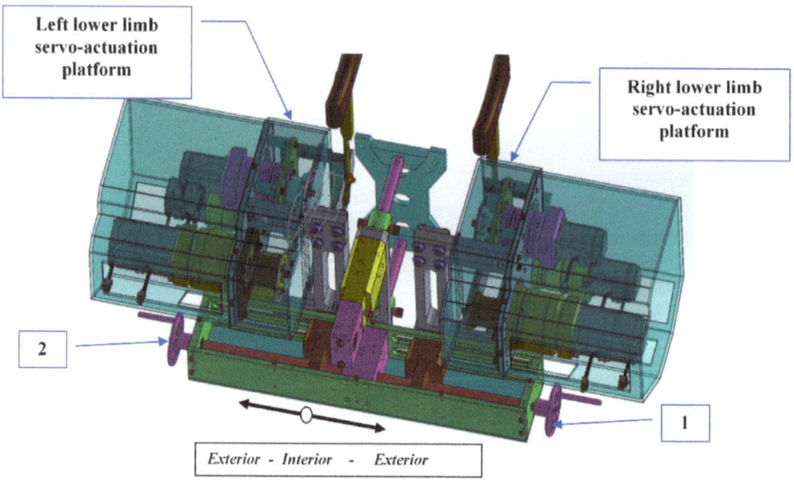

Figure 20. Detailed view of the exoskeleton actuation system.

This mechanical transmission can be actuated manually by the user through 1 and 2 rotational elements. These will rotate the powered screw designed with two threaded sectors (left and right thread sense). In this manner, the actuation platforms will move closer to or further from each other, to adapt the width of the exoskeleton pelvic segment. Regarding the patient lifting and positioning system on the specific treadmill, the following design data were considered:

-   a patient weight should be between 50 to 120 kg;
-   a patient height will be between 1.55 to 1.85 m.

The entire robotic system design was made by having insight into specific kinetotherapy protocols adapted to this system, which need to assure the schematized functions represented in Figure 21.

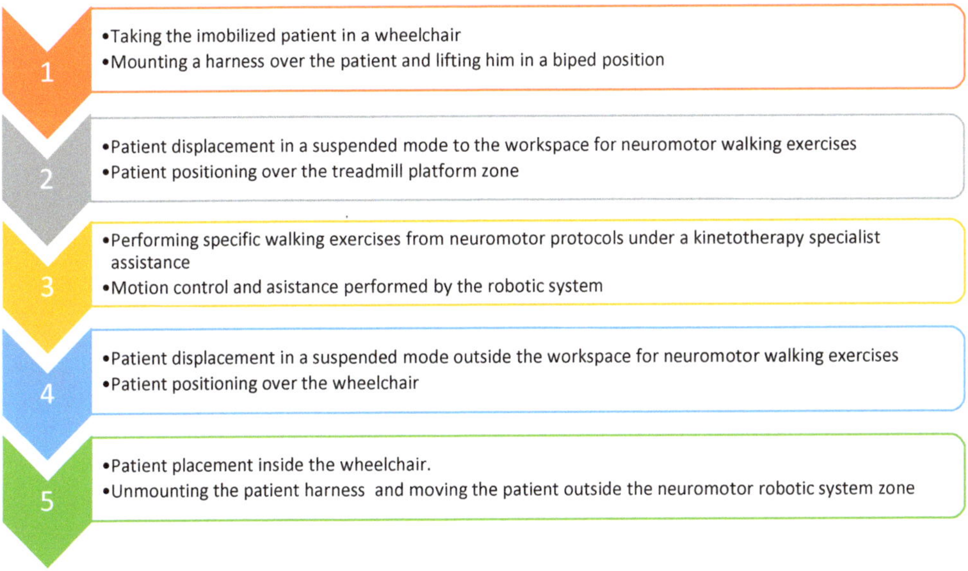

1
•Taking the imobilized patient in a wheelchair
•Mounting a harness over the patient and lifting him in a biped position
2
•Patient displacement in a suspended mode to the workspace for neuromotor walking exercises
•Patient positioning over the treadmill platform zone
3
•Performing specific walking exercises from neuromotor protocols under a kinetotherapy specialist assistance
•Motion control and asistance performed by the robotic system
4
•Patient displacement in a suspended mode outside the workspace for neuromotor walking exercises
•Patient positioning over the wheelchair
5
•Patient placement inside the wheelchair.
•Unmounting the patient harness and moving the patient outside the neuromotor robotic system zone

Figure 21. Desired functions procedure which will be assured by the desired exoskeleton prototype.

By having insight into the mentioned functions in Figure 21, and keeping in the background previous research developed for this robotic system, the entire robotic system was designed, as shown in Figure 22. The designed robotic system presented in Figure 22 is characterized by the following main parts: (**A**)—command and control system; (**B**)—patient lifting/positioning on the treadmill zone; (**C, C′**)—proper exoskeleton which models, under mechanical viewpoints, the left and right lower limbs; (**D**)—mechanical system for adjusting the exoskeleton width to patient anthropometric data.

B
Winches support
Aluminium frame
Winches 1 and 2
Patient flexible support
A
C'
Treadmill

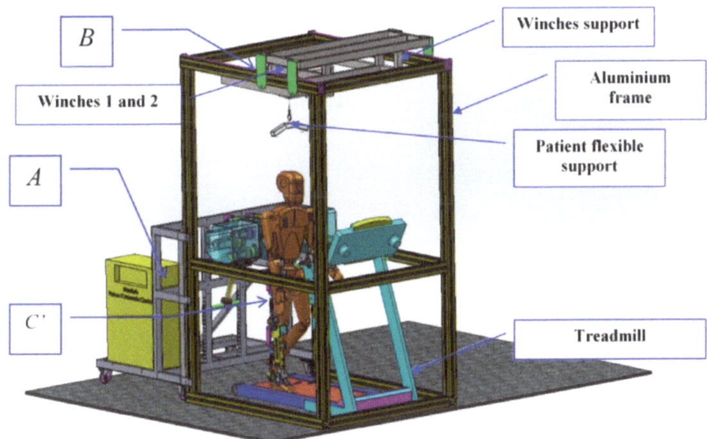

Figure 22. Robotic system virtual prototype with proper parts.

To accomplish the mentioned functions schematized in Figure 21, (**B**)—patient lifting/positioning on the treadmill zone—needs to be activated. The entire structure is placed on an aluminum frame in a cubic shape with L × l × h = 1700 × 1400 × 2940 mm.

On this frame, the mobile subassembly (**B**) is placed, made of a metallic cassette, where two electric winches can be found. These winches are controlled by the human operator in order to place the patient into a neuromotor recovery workspace. Both winches are the same and can lift a maximum weight of 250 kg. One electric winch will lift the patient up from the wheelchair and the other will move the suspended patient into the neurmotor recovery workspace (under the treadmill).

The mobile subassembly consists of a metallic cassette and the mentioned electric winches can move on a specific rail with a V shape through four metallic rollers. This mobile subassembly was designed with mechanical limitations and electronic sensors in order to move precisely the patient in or out of the neuromotor recovery workspace in safety conditions.

6. Exoskeleton Experimental Tests

The performed experimental tests were done in laboratory conditions, using a similar protocol to the one presented at the human experimental analysis for creating the desired database. Thus, the time interval for performing a complete gait cycle was considered as equal with 1.45 s. This time interval is the same as the one for mathematical models processing and virtual simulations environment. Also this interval was setup in a closed-loop for achieving four complete steps in a total time of 5.8 s.

For the experimental evaluations, motion video analysis equipment called CONTEM-PLAS was used, which is similar to VICON Equipment, except for the fact that it uses only two cameras and has the advantages of being mobile equipment. Thus, by having a similar working principle, a total of 10 reflexive markers were attached to the exoskeleton prototype limbs, respectively in the main joint centers of the desired characteristic points (hip, knee, and ankle joints for each lower limb), as can be seen in Figure 23.

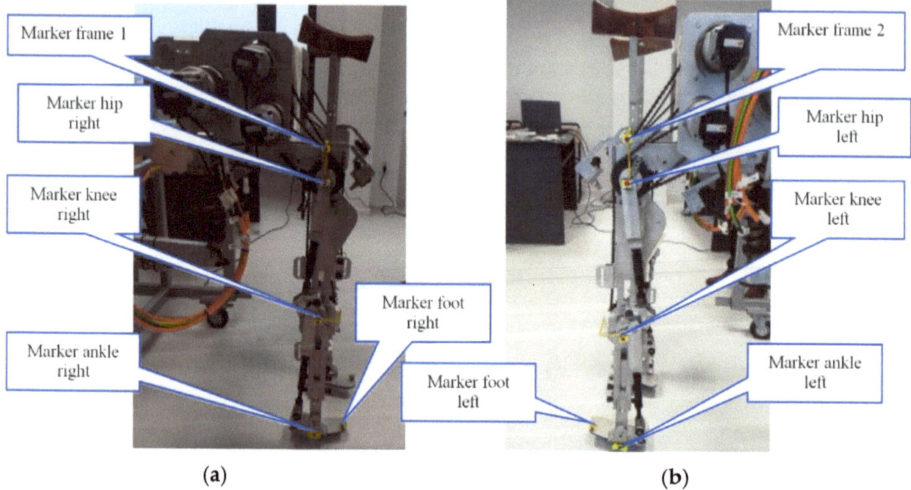

Figure 23. Markers' positions for the exoskeleton prototype experimental tests: (**a**)—right lower limb; (**b**)—left lower limb.

Thus, several snapshots during these tests were presented in Figures 24–26. Informed consent was obtained from all subjects involved in the study.

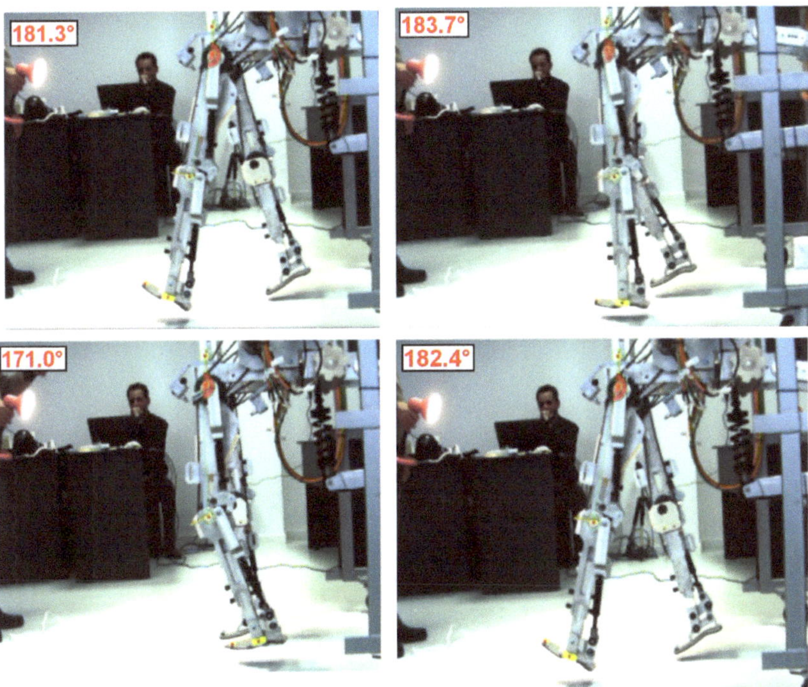

Figure 24. Snapshots during exoskeleton left ankle joints without foot and ground contact.

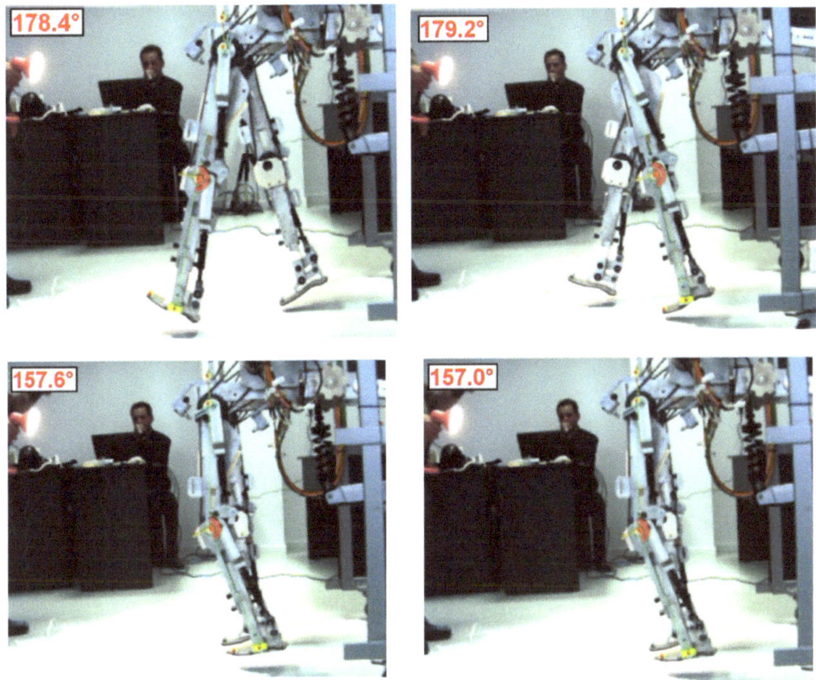

Figure 25. Snapshots during exoskeleton left knee joint without foot and ground contact.

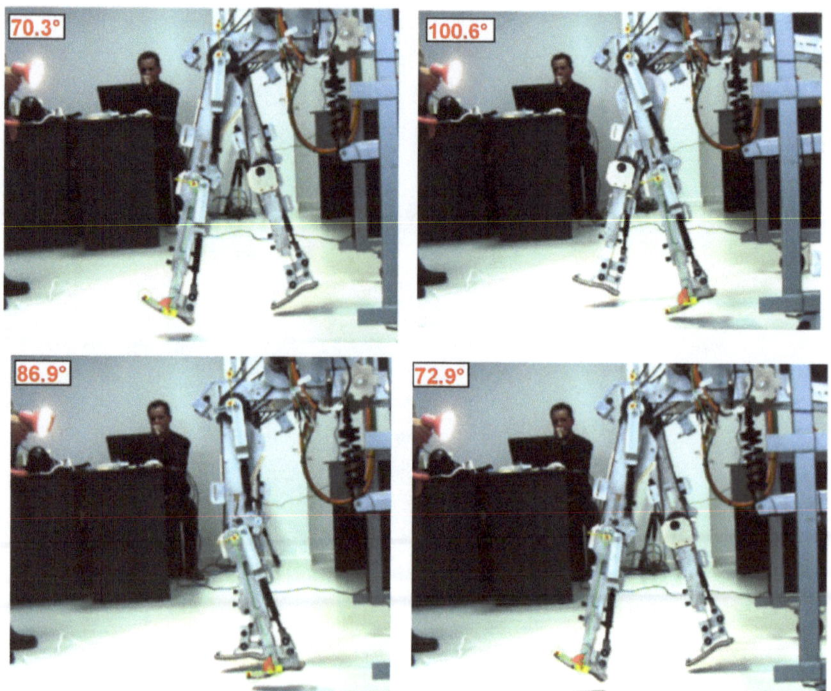

Figure 26. Snapshots during exoskeleton left ankle joint without foot and ground contact.

These tests were performed without foot and ground contact to observe the exoskeleton behavior during walking exercises. This experimental evaluation process was not used by any human volunteers to test this and other experimental tests will be presented in future research.

After processing the video sequences, the desired results were obtained for both exoskeleton lower limbs. These are shown in Figures 27–29.

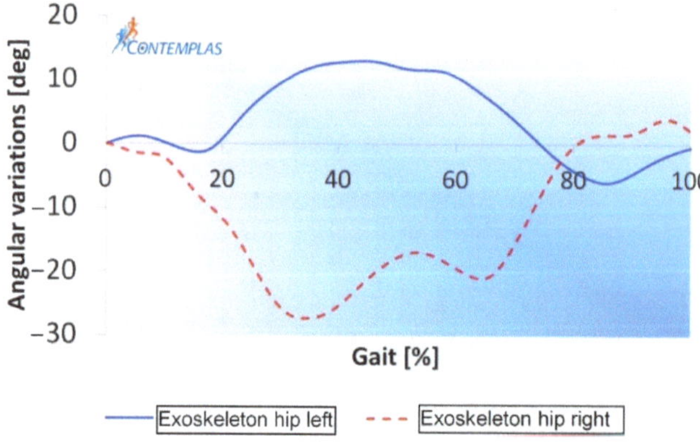

Figure 27. Exoskeleton hip joints angular variation for a complete gait without foot and ground contact.

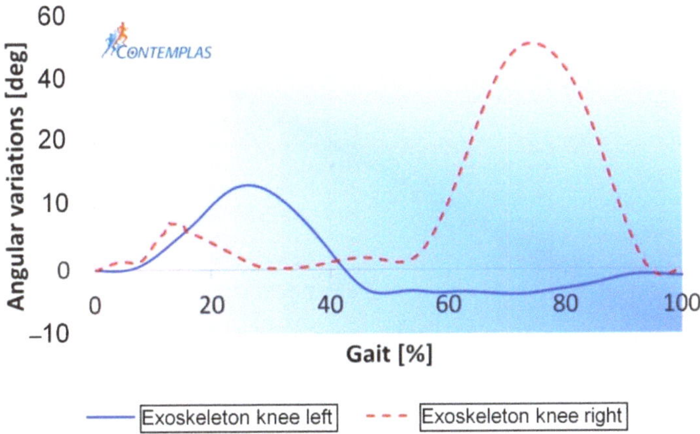

Figure 28. Exoskeleton knee joints angular variation for a complete gait without foot and ground contact.

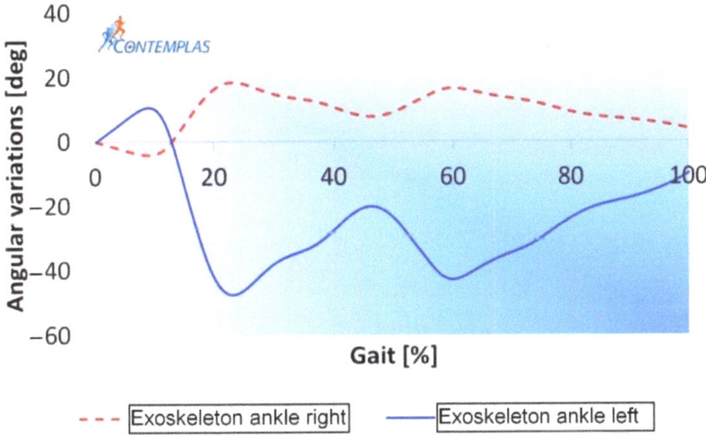

Figure 29. Exoskeleton ankle joints angular variation for a complete gait without foot and ground contact.

Analyzing the presented diagrams in Figures 27–29, it can be seen that the obtained angular variations have similar trajectories with the ones obtained in the case of mathematical models' processing and virtual simulations, but also in the case of the experimental analysis of a healthy human subject. Thus, in the case of the hip joints, values between −26.785 and 16.238 degrees were recorded, and the angular amplitude in the case of the analyzed exoskeleton's hips was 44.023 degrees.

By having insight into the previous experimental database and also the existent specialty literature data [23,27], this angular amplitude is correct from a numerical viewpoint and it corresponds to a normal gait in the case of a healthy subject. On the other hand, the introduced data inside of the robotic system's command and control program contained data acquired in the previous experimental analysis of a healthy human subject case study.

Regarding the exoskeleton knee joints, it can be observed in Figure 28 that an angular amplitude of 47.753 degrees was recorded, which is higher than a normal angular amplitude obtained in the case of a healthy subject. This high value was obtained due to the high inertia momentums of the exoskeleton segments corroborated with high angular speeds. In future experimental test protocols, this will not affect the patient recovery behaviour due to fact that these high values were recorded in knee flexion moments during the swing phase of a complete gait. It is possible that the use of a patient together with the designed robotic system prototype will increase the stiffness and this angular amplitude will record lower values.

In the case of ankle joints' experimental evaluations, from the presented results in Figure 29, it can be remarked that the angular amplitude records a value of 48.589 degrees and this value is the correct one, by having insight the specialty literature data limits and the previous experimental analysis.

7. Results Analysis and Discussions

We developed a flexible program that allows the processing of the created mathematical models, according to the virtual prototyping of the elaborated robotic system under an Adams environment correlated with experimental analyses in an integrated system.

Thus, a comparative analysis was performed between the experimental results of the analysed human subject and the ones obtained during experimental tests of the elaborated prototype. Through this analysis, the error of the acquired results will be identified.

To be able to compare the results of the analysed human subject during experimental analysis characterized by numerical processing with the ones obtained from the exoskeleton prototype experimental analysis, the kinematic response of the desired exoskeleton joints versus human subject joints was monitored, namely hips, knees, and ankles. The obtained results are represented by diagrams reported in Figures 30–35.

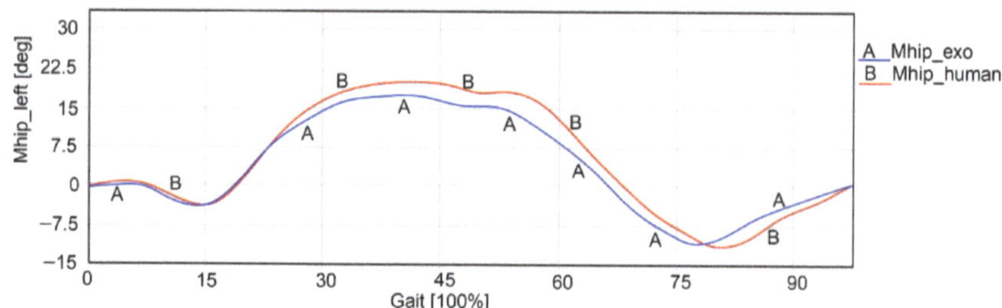

Figure 30. Left hip angular variation exoskeleton vs. human subject during a complete gait.

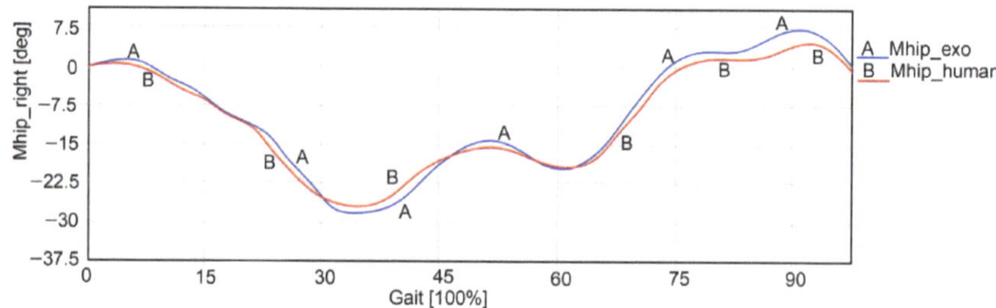

Figure 31. Right hip angular variation exoskeleton vs. human subject during a complete gait.

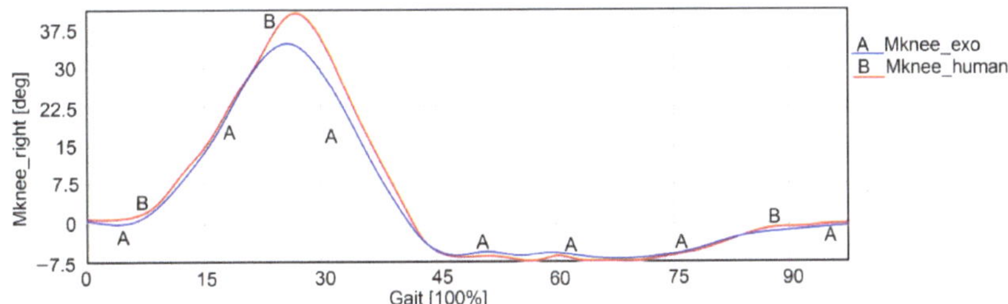

Figure 32. Right knee angular variation exoskeleton vs. human subject during a complete gait.

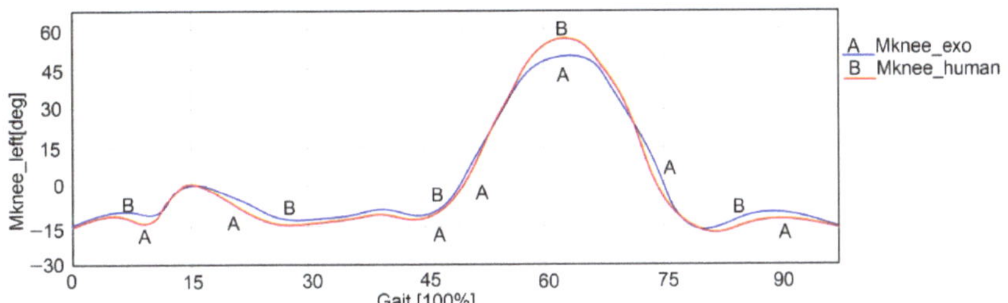

Figure 33. Left knee angular variation exoskeleton vs. human subject during a complete gait.

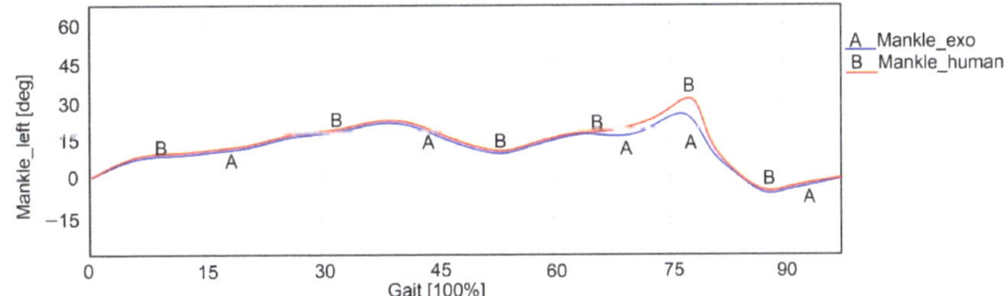

Figure 34. Left ankle angular variation exoskeleton vs. human subject during a complete gait.

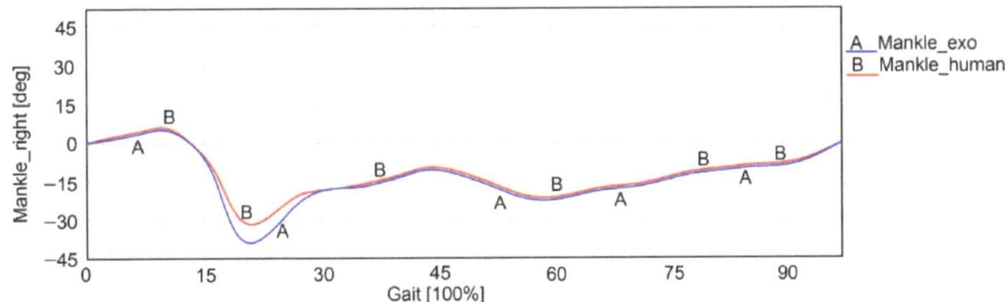

Figure 35. Right ankle angular variation exoskeleton vs. human subject during a complete gait.

Considering the comparative results reported in the diagram from Figure 30, it can be seen that the left hip angular variation has an appropriate path, and the maximum value was obtained in the case of the left human hip, namely 19.537 degrees. In the case of the exoskeleton's left hip, the maximum value was equal to 16.238 degrees. Thus, the error of the reported results presented in this graph was equal to 2.943%.

By having insight into the right hip, it can be remarked a maximum value of 23.6775 degrees was reported for the healthy human subject and in the case of exoskeleton's right hip, this records a value of 26.785. Both reported curves from Figure 31 have similar paths and the error was equal to 2.875%.

Regarding the knee joints, the reported results as a comparative analysis can be found in Figures 32 and 33. In fact, for the left knee joint in the case of the healthy human subject, a value of 38.273 degrees was recorded, and for the left knee joint of the analyzed exoskeleton this variation reached a value of 33.583 degrees. In this case, the error increases and reaches a value of 4.993%. In the case of right knee joints for both models, it reached high values, namely 51.261 degrees for healthy human subject analysis and 47.753 degrees in the case of the knee exoskeleton joint. Thus, an error of 4.211% was observed, and the paths of the reported curves were similar.

In the case of the ankle joints, the obtained results through the reported comparative analysis are shown in Figures 34 and 35. For the left ankle joint, a maximum angular value of 32.428 degrees in the case of the human subject was recorded, and for the exoskeleton left ankle joint this reached a value of 27.105 degrees. The obtained error was around 4.878% and the paths of both analyzed cases have appropriate trajectories.

Similar results were obtained also for the right ankle joints analysis and, as can be seen in Figure 35, for the human subject a maximum value of 33.237 degrees was obtained. In the case of the exoskeleton ankle joint, this reached a value of 37.584 degrees. Thus, the acquired error was around 4.34%.

This comparative analysis validates the exoskeleton functionality. As a major remark, these experimental tests in the case of exoskeleton prototype experimental analysis were performed in laboratory conditions without the involvement of any humans. It can be remarked that the obtained curves reported in Figures 30–35 have similar paths and the computed error does not exceed 5%. This error depends mostly on the video analysis equipment's accuracy but also, in the case of identifying the joint center positions of the analyzed human subject. In the experimental setup of human subject motion evaluation during walking, a crucial role in the analyzed joints' angular variation is represented by identifying the joint center position, which was performed by hand palpation.

8. Conclusions

In this study, the motions variations depending on time for a complete gait in case of the hips, knees, and ankles, for different interest points from lower limbs structure were obtained, which were considered as a mandatory group for walking activity.

Through this research, the following objective was accomplished: obtaining a database after experimental analyses of 30 healthy human subjects and 5 subjects with neuromotor deficiencies. This analysis consists of angle variations motion laws of the main joints of human locomotion (hips, knees, and ankles).

By having insight into the kinematic functions established through motion experimental analysis, an exoskeleton mechanism made from kinematic linkages that allows the actuating of the desired joints from human lower limb structure, namely hip, knee, and ankle joints, was conceived.

The developed structural analysis allows us to identify the motor joints for different sequences from a complete gait cycle and to design the kinematic schemes compatible with kinematic functions established experimentally with the aid of video analysis equipment.

We performed a dynamic analysis of the exoskeleton mechanism where the input data for motor joints were established through an inverse kinematic analysis based on the experimental motion laws of hips, knees, and ankles.

The dynamic analysis consists of elaborating the mathematical models for direct and inverse methods. The direct method allows the generalized coordinate variation laws to be established by having insight, as input data, into the geometrical characteristics, mass, inertia, and contact force with the ground. The inverse method (dynamic synthesis), allows the connection forces variation (depending on time) from kinematic joints to be established, having the generalized coordinates variations, mechanism geometry, mass and inertia characteristics, and generalized forces as inputs.

We created a program under the MAPLE software environment, which allowed us to perform a numerical process of the elaborated mathematical models for both methods used in dynamic analysis. The obtained numerical results were compared via two methods, computed mathematical modeling and numerical simulation, with the aid of MSC Adams software.

Thus, connection forces variation laws that occur at the exoskeleton joints during motions were established, and these were used for design and manufacturing phases.

There were performed experimental tests of the entire robotic system and, for these tests, servomotor motions were considered as input data. These motions were verified analytically and numerically with MSC Adams software.

We presented numerical results, namely, connection forces' variations over time in the case of knee and ankle joints, obtained through two methods: analytical method and numerical simulations, respectively, both with MSC Adams. The obtained results were considered as input data for a comparative analysis developed with LS Dyna software. The obtained results through this comparative analysis present small accuracy errors under 3% which confirms that the mathematical models and numerical simulation developed in the virtual environment are correct.

We performed preliminary experimental tests, in laboratory conditions, and the obtained results validated the exoskeleton prototype.

The developed software, which was implemented at the level of the exoskeleton command and control unit, allowed us to implement motion laws for the drive joints to actuate 6 servomotors from the exoskeleton prototype structure.

Moreover, the obtained robotic system is characterized by a function called "teaching mode" which permits users to manually move the hip, knee, and ankle joints with the possibility to record each position and establish the maximum and minimum angular variation limits. Thus, the entire robotic system experimental tests, performed in laboratory conditions without a human subject, permit this research to verify the exoskeleton's kinematic functions and compare these with the ones obtained experimentally on a healthy human subject.

An exoskeleton robotic system with a compact structure in a modular concept was obtained, with a high adaptability level due to its structural–kinematic adjustments. This was done by having insight into the kinematic linkages equivalent to human lower limbs, with four planar-parallel mechanisms for actuating the hips, knees, and ankle joints. This is also characterized by its necessary support to sustain 6 servomotors, and its specific adjustment possibilities in two directions (horizontal and vertical).

A comparative analysis between experimental motion laws versus experimental motions developed by the exoskeleton for each main joint (hips, knees, and ankles) was accomplished.

The experimental tests performed with the entire robotic system were done without a patient and the evaluated motions were not entered in contact with the ground.

We consider our objectives to have been accomplished by: the comparative results through mathematical models processing; numerical and virtual simulations with MSC Adams program; experimental analysis onto human subjects; and experimental tests of the entire robotic system.

Through this research, we achieved all the proposed investigation objectives and the conceptual solution was validated through a comparative analysis which certifies the use of the prototype in future research.

Author Contributions: Conceptualization, S.D., methodology, N.D.; software, S.D.; validation, N.D., S.D. and C.C.; formal analysis, I.G.; investigation, L.G.C., resources, S.D.; data curation, I.G.; writing—original draft preparation, S.D.; writing—review and editing, C.C.; visualization, S.D., C.C. and N.D.; supervision, N.D.; project administration, S.D., C.C. and N.D.; funding acquisition, S.D., C.C. and N.D. All authors have read and agreed to the published version of the manuscript.

Funding: This research was funded by Romanian National Authority for Scientific Research and Innovation, CNCS/CCCDI – UEFISCDI, grant number PNIII-P2-2.1-PED-2016-0934, within PNCDI III.

Institutional Review Board Statement: Not applicable.

Informed Consent Statement: Not applicable.

Data Availability Statement: Not applicable.

Conflicts of Interest: The authors declare no conflict of interest.

References

1. Esquenazi, A.; Talaty, M.; Jayaraman, A. Powered exoskeletons for walking assistance in persons with central nervous system injuries: A narrative review. *PM&R* **2017**, *9*, 46–62.
2. Louie, D.R.; Eng, J.J. Powered robotic exoskeletons in post-stroke rehabilitation of gait: A scoping review. *J. Neuroeng. Rehabil.* **2016**, *13*, 1–10. [CrossRef] [PubMed]
3. Bortole, M.; Venkatakrishnan, A.; Zhu, F.; Moreno, J.C.; Francisco, G.E.; Pons, J.L.; Contreras-Vidal, J.L. The H2 robotic exoskeleton for gait rehabilitation after stroke: Early findings from a clinical study. *J. Neuroeng. Rehabil.* **2015**, *12*, 1–14. [CrossRef] [PubMed]
4. Federici, S.; Meloni, F.; Bracalenti, M.; de Filippis, M.L. The effectiveness of powered, active lower limb exoskeletons in neurorehabilitation: A systematic review. *J. Neuro-Rehabil.* **2015**, *37*, 321–340. [CrossRef] [PubMed]
5. Pons, J.L. *Wearable Robots: Biomechatronic Exoskeletons*; John Wiley & Sons, Ltd.: Chichester, UK, 2008.
6. Banala, S.K.; Agrawal, S.K.; Scholz, J.P. Active leg exoskeleton (ALEX) for gait rehabilitation of motor-impaired patients. In Proceedings of the IEEE 10th International Conference on Rehabilitation Robotics, Noordwijk, The Netherlands, 13–15 June 2007; pp. 401–407. [CrossRef]
7. Sergi, F. *Biomechatronic Design of Wearable and Operational Robots for Rehabilitation and Assistive Applications Terza Linea*; Universita' Campus Bio-Medico di Roma School of Engineering: Roma, Italy, 2011.
8. Dumitru, N.; Didu, A.; Copilusi, C. Mechatronic system for locomotion rehabilitation. In *Advances in Engineering and Management*; TransTech Publications: Zurich, Switzerland, 2016; pp. 53–54. ISSN 2537-4443.
9. Dumitru, N.; Copilusi, C.; Geonea, I.; Tarnita, D.; Dumitrache, I. Dynamic Analysis of an Exoskeleton New Ankle Joint Mechanism. In *New Trends in Mechanism and Machine Science*; Flores, P., Viadero, F., Eds.; Mechanisms and Machine Science; Springer: Cham, Switzerland, 2015; Volume 24.
10. Beil, J.; Perner, G.; Asfour, T. Design and Control of the Lower Limb Exoskeleton KIT-EXO-1. In Proceedings of the IEEE International Conference on Rehabilitation Robotics (ICORR), Singapore, 11–14 August 2015; pp. 119–124.
11. Yu, S.; Han, C.; Cho, I. Design considerations of a lower limb exoskeleton system to assist walking and load-carrying of infantry soldiers. *Appl. Bionics Biomech.* **2014**, *11*, 119–134. [CrossRef]
12. Lopez, R.; Aguilar, H.; Salazar, S.; Lozano, R.; Torres, J.A. Modelado y Control de un Exoesqueleto para la Rehabilitacio'n de Extremidad Inferior con dos grados de libertad. *Rev. Iberoam. Autom. Inform. Ind. RIAI* **2014**, *11*, 304–314. [CrossRef]
13. Galle, S.; Malcolm, P.; Derave, W.; De Clercq, D. Adaptation to walking with an exoskeleton that assists ankle extension. *Gait Posture* **2013**, *38*, 495–499. [CrossRef] [PubMed]
14. Strausser, K.A.; Zoss, A.B.; Swift, T.A.; Zoss, A.B.; Kazerooni, H.; Bennett, B.C. Mobile exoskeleton for spinal cord injury: Development and testing. In Proceedings of the ASME 2011 Dynamic Systems and Control Conference, Arlington, VA, USA, 31 October–2 November 2011; pp. 419–425. [CrossRef]
15. Geonea, I.D.; Margine, A.; Dumitru, N.; Copiluși, C. Design and simulation of a mechanism for human leg motion assistance. In *Advanced Materials Research*; Trans Tech Publications Ltd.: Bäch, Switzerland, 2014; Volume 1036, pp. 811–816.
16. Ghobrial, G.M.; Wang, M.Y. The next generation of powered exoskeleton use in spinal cord injury. *Neurosurg. Focus* **2017**, *42*, E16. [CrossRef] [PubMed]
17. You, S.H. Concurrent Validity and Test-Retest Reliability of the Novel Walkbot-K System. *J. Mech. Med. Biol.* **2016**, *16*, 1640029.
18. Keijsers, N.L.W.; Stolwijk, N.M.; Renzenbrink, G.J.; Duysens, J. Prediction of walking speed using single stance force or pressure measurements in healthy subjects. *J. Gait Posture* **2016**, *43*, 93–95. [CrossRef] [PubMed]
19. D'Apuzzo, F.; Minervini, G.; Grassia, V.; Rotolo, R.P.; Perillo, L.; Nucci, L. Mandibular Coronoid Process Hypertrophy: Diagnosis and 20-Year Follow-Up with CBCT, MRI and EMG Evaluations. *Appl. Sci.* **2021**, *11*, 4504. [CrossRef]
20. Minervini, G.; Nucci, L.; Lanza, A.; Femiano, F.; Contaldo, M.; Grassia, V. Temporomandibular disc displacement with reduction treated with anterior repositioning splint: A 2-year clinical and magnetic resonance imaging (MRI) follow-up. *J. Biol. Regul. Homeost. Agents* **2020**, *34*, 151–160. [PubMed]
21. Minervini, G.; Lucchese, A.; Perillo, L.; Serpico, R.; Minervini, G. Unilateral superior condylar neck fracture with dislocation in a child treated with an acrylic splint in the upper arch for functional repositioning of the mandible. *Cranio* **2017**, *35*, 337–341. [CrossRef] [PubMed]
22. VICON. *NEXUS Product Guide for Motion Analysis Equipment*. Edited by VICON Industries; VICON: New York, NY, USA, 2016.
23. Steinicke, F.; Visell, Y.; Campos, J.; Lécuyer, A. *Human Walking in Virtual Environments*; Springer: New York, NY, USA, 2013.
24. Duffy, J. *Statics and Kinematics with Applications to Robotics*; Cambridge University Press: New York, NY, USA, 1996; p. 173.

25. Pirker, W.; Katzenschlager, R. Gait disorders in adults and the elderly: A clinical guide. *Wien. Klin. Wochenschr.* **2016**, *129*, 81–95. [CrossRef] [PubMed]
26. Muscolo, G.; Caldwell, D.; Cannella, F. Biomechanics of Human Locomotion with Constraints to Design Flexible-Wheeled Biped Robots. In Proceedings of the IEEE International Conference on Advanced Intelligent Mechatronics, Munich, Germany, 3–7 July 2017; pp. 1273–1278.
27. Bertomeu-Motos, A. Biomechanics of human walking and stability descriptive parameters. *Rev. Dr. UMH* **2015**, *1*, 4–8. [CrossRef]
28. Chiacchio, P.; Chiaverini, S.; Sciavicco, L.; Siciliano, B. Closed-loop inverse kinematics schemes for constrained redundant manipulators with task space augmentation and task priority strategy. *Int. J. Robot. Res.* **1991**, *10*, 410–425. [CrossRef]
29. Norton, R.L. *Design of Machinery*; McGraw-Hill: New York, NY, USA, 2008.
30. Wilson, C.E.; Sadler, J.P. *Kinematics and Dynamic of Machinery*; HarperCollins College Publishers: London, UK, 1993.
31. Copilusi, C.; Dumitru, S.; Geonea, I.; Colici, F.; Dumitru, N. Design and Numerical Characterization of a New Leg Exoskeleton for Human Neuromotor Rehabilitation. *Acta Napoc. Appl. Math. Mech. Eng.* **2021**, *64*, 2.
32. Basic ADAMS Full Simultion Training Guide. Version 11.0. Part Number: 110VIEWTR-03. MSC 2017. Adams Software User Manual. Available online: https://www.mscsoftware.com/product/adams (accessed on 20 February 2021).

Article

Energy-to-Mass Ratio—A Novel Selection Criterion of Pneumatic Motors Used for the Actuation of Wearable Assistive Devices

Andrea Deaconescu and Tudor Deaconescu *

Department of Industrial Engineering and Management, Transilvania University of Brasov, 500036 Brasov, Romania; deacon@unitbv.ro
* Correspondence: tdeacon@unitbv.ro; Tel.: +40-74-575-7850

Abstract: The requirements to be met by a wearable assistive device are compactness, lightweight and energy efficiency. While the literature discusses the construction and performance of such devices, no information is provided as to the criteria to be applied in selecting such an actuator, capable of satisfying the mentioned conditions. Ensuring the high autonomy of a wearable assistive device requires actuators that can store a large quantity of energy in a small as possible volume, for example, actuators with a high energy density. This paper presents a comparative study of the performance of two types of pneumatic actuators: single-acting cylinders and pneumatic muscles, respectively, and offers information that will enable users to select an optimum solution. The quality indicators considered in conducting the comparative study are size, mass, the developed force and the energy-to-mass ratio. A method is proposed to determine the energy developed by the motors over the entire stroke; based on that, the energy-to-mass ratio is subsequently calculated. This indicator is a valuable tool made available to designers of wearable assistive devices. The conclusion yielded by the study asserts that while pneumatic muscles have larger radial and axial dimensions, they present benefits as to the developed forces and the energy-to-mass ratios.

Keywords: pneumatic muscle; single-acting cylinder; wearable assistive device; energy-to-mass ratio

Citation: Deaconescu, A.; Deaconescu, T. Energy-to-Mass Ratio—A Novel Selection Criterion of Pneumatic Motors Used for the Actuation of Wearable Assistive Devices. *Appl. Sci.* **2022**, *12*, 6459. https://doi.org/10.3390/app12136459

Academic Editors: Ionuţ Daniel Geonea and Cristian Copilusi Petre

Received: 8 June 2022
Accepted: 22 June 2022
Published: 25 June 2022

Publisher's Note: MDPI stays neutral with regard to jurisdictional claims in published maps and institutional affiliations.

1. Introduction

Mobile robots are capable of moving in a predefined environment and can be autonomous (AMR—autonomous mobile robots) or nonautonomous. The latter deploy guiding devices allowing movement along a predefined navigation route. A significant characteristic of mobile robots is that their moving structure includes all the components that are required for achieving the proposed task, from the energy source to the effectors.

Due to its operational independence mobile robotics lends themselves to a practically unlimited number of applications. In addition to utilization in industry, agriculture or road traffic safety (self-driving motor vehicles), mobile robots can also perform tasks that are potentially dangerous to humans (e.g., disaster control or military applications). Frequently, mobile robots can be found in everyday household applications, such as, for example, semi-autonomous vacuum cleaners [1]. Rehabilitation medicine is another growing field of applicability, where mobile robotics is present in the form of wearable assistive devices.

Wearable assistive devices are designed to facilitate the performance of tasks by persons who, for various reasons, have lost certain abilities. Such devices can be of passive type (not powered) or can be powered by electrical, pneumatic or hydraulic actuation systems.

Orthotic and prosthetic devices fall into the passive category and are externally applied wearable devices: the former being used to modify the structural and functional characteristics of the neuromuscular and skeletal system, while the second is used to replace wholly, or in part, an absent or deficient limb segment [1]. According to the International Society for Prosthetics and Orthotics (ISPO), a prosthetic or orthotic device allows a person with motor deficiencies to remain active and independent, and to live a normal life without the need for formal support services [2]. Orthotic and prosthetic devices are complex technical systems that, depending on the patient's needs, can be custom-made to exact measurements.

Powered exoskeletons are portable devices of higher complexity. They were developed as augmentative devices to improve the bearer's physical performance or as orthotic devices for the rehabilitation of walking or locomotory assistance. Compared to stationary systems, powered exoskeletons must be compact, lightweight, safe, reliable, intelligent and portable, and can be potentially used at home [3,4].

The mechanical power needed for an exoskeleton is obtained by pneumatic, hydraulic, and electrical actuators. Pneumatic and hydraulic actuation systems, while benefitting from a good power-to-weight ratio, are unattractive from the viewpoint of positioning precision. Electric motors, on the other hand, can be controlled with high precision, while having an unfavorable power-to-weight ratio [5,6]. In [7], Asbeck et al. compare two wearable assistive devices with different actuation systems, pneumatic and electro-mechanical, respectively. The masses of the two systems are 7.1 kg for the pneumatic and 10.1 kg for the electro-mechanical one. The conclusion reached by this study was to choose the pneumatic actuation for wearable assistive suits.

Actuator positioning poses a challenge to designers of powered exoskeletons. Placing the motors near the joints entails a simpler construction, but inertial forces are generated that render the entire system more difficult to control. If the actuators are placed at a distance from the joints, the weight of the mobile components is reduced, but power transmission mechanisms are required that complicate the mechanical structure of the entire assembly.

The growing use of compressed air in the construction of powered exoskeletons is determined by its benefits, such as easy generation and storage, non-flammability, minimum risk of explosion, minimal maintenance requirements of the pneumatic systems, compliance, etc. By their generated motion, pneumatic motors are linear or rotary, and, as to their construction, they are equipped with a piston, blades or membrane [8]. The literature reveals the predominant deployment of pneumatic cylinders and pneumatic muscles in the construction of powered exoskeletons. The main applications of such systems are the rehabilitation and exercising of the upper and lower limbs.

Bae et al. present in [9] the construction of an exoskeleton device for active wrist rehabilitation based on pneumatic cylinders. The flexion and extension of the wrist is conducted by means of a bar system actuated by a pneumatic cylinder. In [10], Goergen et al. put forward a simple and low-cost pneumatic robotic mechanism for lower limb rehabilitation. Operational safety and force control are achieved by adequately adjusting the pressures in the chambers of the pneumatic cylinders that drive the rehabilitation robot. Other exoskeleton devices actuated by pneumatic cylinders are described in papers [11–14].

The utilization of pneumatic muscles is gaining terrain in the construction of wearable assistive devices. This is due to the advantages offered by pneumatic muscles, such as higher force-to-weight ratio and the capacity to generate larger forces than other pneumatic motors. The low weight and flexibility of pneumatic muscles renders them easily integrated into an exoskeleton device [15]. In [16], Bogue presents an overview of recent developments in exoskeletons and robotic prosthetics that highlights the growing presence of pneumatic muscles in this field.

Portable medical systems actuated by pneumatic muscles devised for the rehabilitation of the upper and lower limb joints are presented in a number of publications. Balasub-ramanian et al. propose in [17] a wearable upper extremity therapy robot, actuated by pneumatic muscles. Abe et al. also use pneumatic muscles to build a soft power support suit for the upper limb [18]. Other examples of deploying pneumatic muscles for the actuation of exoskeletons designed for mobilizing upper and lower limb joints are discussed in papers [19–24]. In [25,26], the authors present two patented applications of pneumatic muscles in the field of medical rehabilitation equipment, designed for the wrist, and hip and knee and ankle, respectively.

In addition to characteristics such as compactness, low mass and portability, the published studies revealed a further significant requirement to be met by a powered exoskeleton, namely the highest possible energy efficiency. While all analyzed papers available in the literature describe the construction and performance of powered exoskeletons, the criteria for selecting a certain type of pneumatic motor (pneumatic cylinder or pneumatic muscle) are not explained.

As a novelty, this paper introduces an energy-related approach to selecting a certain pneumatic actuator. In consideration of the global concern related to energy consumption, the paper presents a method for determining the energy-to-mass ratio (gravimetric energy density) of pneumatic actuators and puts forward this indicator to be used for their selection. Gravimetric energy density is the amount of energy stored in a given system or region of space per unit mass. The higher the gravimetric energy density of the air, the more energy can be stored or transported for the same amount of mass [27].

The reviewed literature discusses this energy-related indicator as a criterion for selecting pneumatic motors only to a rather slight extent. Only in [28], Plettenburg proposes the energy-to-mass ratio as a criterion for selecting a certain type of pneumatic actuator, his calculations yielding the conclusion that the highly praised advantages of pneumatic muscles are exaggerated. Other papers refer to the concept of power-to-weight ratio, without, however, offering a corresponding calculation course [29–31]. Paper [32] presents a comparison of a hydraulic and an electric motor as to their energy efficiency, based on the power-to-mass ratio. The conclusion asserts a higher power density of hydraulic motors compared to that of electric motors.

From all analyzed articles, it followed that regardless of the field of applicability, a certain type of pneumatic actuator is selected generally strictly on economic or technical grounds (dimensions, masses, efficiency, stroke, work pressures, etc.). Starting from these considerations, the paper proposes a methodology for computing the energy-to-mass ratio for two types of pneumatic motors (pneumatic cylinders and muscles, respectively). The presented methodology is new and can be applied to other categories of actuators as well. The comparative study conducted by the authors is also significant for analyzing the possibility of using either type of pneumatic actuator in the construction of powered exoskeletons. It is known that high energy density systems store a larger quantity of energy that in, its turn, ensures the increased autonomy of the rehabilitation system.

For practical purposes pneumatic muscles and pneumatic cylinders of the same manufacturer were selected for the comparative study. Actuators made by other manufacturers have a behavior (performance) similar to those of the studied motors, slight differences existing only in relation to the exterior design.

The aim of the study was to provide the necessary information for a correct selection based on quantifiable quality indicators of the optimum type of pneumatic actuator. The quality indicators considered in the comparative study are size, mass, developed force and the energy-to-mass ratio, meaning the specific energy or gravimetric energy density of an actuator. All of these quantities are easily measured or calculated, thus yielding reliable results.

The main contributions of this paper consist in (i) introducing a computational method of the energy developed by the studied pneumatic actuators and (ii) conducting an energy-to-mass ratio based comparative study of the performance of pneumatic cylinders and pneumatic muscles, respectively.

The paper continues with a second section that describes the studied pneumatic actuators. In the third section, the energy-to-mass ratio is defined, and its values determined for the studied motors. The third section of the paper concludes with two concrete examples of rehabilitation equipment for which the energy-to-mass ratios are calculated. The paper is completed by the conclusions and recommendations resulting from the study. Depending on the comparison criteria that are of interest to a particular user, one or the other variant of pneumatic actuator can be selected.

2. Pneumatic Muscles vs. Pneumatic Cylinders

The proposed working method consists of going through a number of steps with the aim of obtaining the energy-to-mass ratio. The sequence of steps is as follows:

- Computation of the forces developed by the analyzed pneumatic actuators;
- Determining the required dimensional specifications of the analyzed actuators;
- Establishing the dependency of the actuators mass on the length of the stoke;
- Determining the dependency equations of the forces developed by the actuators versus stroke.

Based on these data, subsequently, the energy-to-mass ratio is calculated.

2.1. Computation of the Forces Developed by the Analyzed Pneumatic Actuators

A number of papers on applications of pneumatic muscle assert that this type of actuator offers more advantages than pneumatic cylinders, without, however, following up with concrete arguments in this respect. Some of the advantages of pneumatic muscles enumerated in [33] are:

- The developed initial force is significantly larger at the same diameter;
- Small mass per unit force;
- Smaller consumption of compressed air;
- Superior energy density, etc.

In order to prove these assertions, further discussion on the authors present and a comparison between single-acting pneumatic cylinders and pneumatic muscles manufactured by Festo are presented (Figure 1). These specific actuators were selected for the study because the forces developed by them have similar evolutions over the length of the stroke. The maximum force is developed at the beginning and reaches its minimum by the end of the stroke.

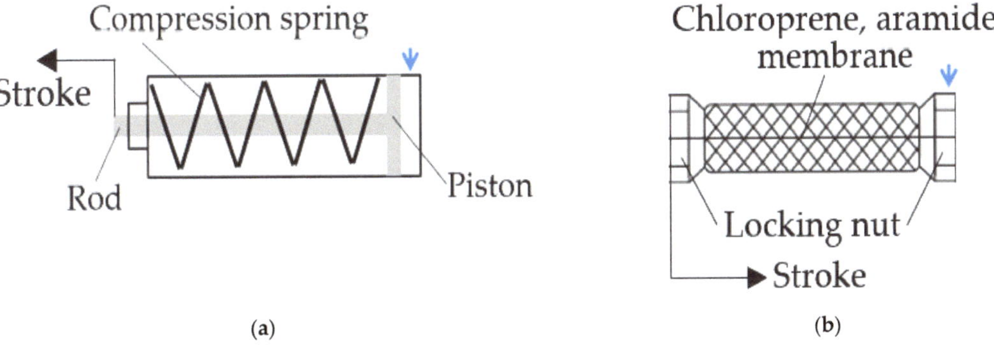

Figure 1. Types of studied pneumatic motors: (**a**) Single-acting cylinder; (**b**) Pneumatic muscle.

In the case of single-acting cylinders, compressed air is applied to a single face of the piston and, consequently, motion is generated in a single direction. The piston is then reversed automatically by means of a built-in spring. The force developed by a cylinder of this type is calculated by Equation (1):

$$F = p \times \frac{\pi \cdot D^2}{4} - F_{FR} - F_{Spr} \tag{1}$$

where p is the pressure of the compressed air, D—the piston diameter, F_{FR}—the friction forces, F_{Spr}—the spring force.

The pneumatic muscle generates displacement by changing its geometrical form when fed compressed air. As the air pressure increases, a traction force acting along the muscle axis causes it to shorten. Such axial contraction represents the stroke of the motor. The maximum axial contraction is of about 20% of the muscle's initial length. The axial force is maximum at the onset of contraction and falls to zero as the end of the stroke is reached. The force developed by a pneumatic muscle is calculated by Equation (2) [30]:

$$F = p \times \frac{\pi}{4} \times d^2 \times \left[\frac{3 \times \cos^2 \alpha - 1}{1 - \cos^2 \alpha} \right] \tag{2}$$

where p is the working pressure, d the diameter of the pneumatic muscle, and α is the winding angle of the synthetic fiber tissue the envelopes the inner tube of the muscle.

The following models of the two types of pneumatic motors were selected for the purpose of this comparative study:

- Round cylinders, single-acting, pushing, based on ISO 6432, manufactured by FESTO, Esslingen, Germany, shown in Figure 2a [34];
- Pneumatic muscles, models DMSP-xx-N, manufactured by FESTO, Esslingen, Germany, shown in Figure 2b [35].

(a)   (b)

Figure 2. Studied pneumatic motors: (**a**) Single-acting cylinder (ESNU) [34]; (**b**) Pneumatic muscle (DMSP) [35].

2.2. Determining the Required Dimensional Specifications of the Analyzed Actuators

Three sizes of each motor type were selected and paired for comparison. For each pair of compared motors, the active diameters and the strokes were the same. Table 1 presents the characteristics of these pneumatic motors [34,35]:

Table 1. Characteristics of the studied pneumatic motors.

Motor Type	Motor Code	Interior Diameter [mm]	Stroke Length [mm]
Round cylinders	ESNU-10-50-P-A	10	50
	ESNU-20-50-P-A	20	50
	ESNU-40-50-P-A	40	50
Pneumatic muscles	DMSP-10-250N-RM-CM	10	50
	DMSP-20-250N- RM-CM	20	50
	DMSP-40-250N- RM-CM	40	50

A comparison of the two types of pneumatic motors used for the evaluation criteria their dimensions, masses, developed forces and the energy-to-force ratio.

From the viewpoint of sizes, Table 2 features the main dimensions of the two types of motors, as indicated in their technical specifications [34,35].

Notably, single-acting cylinders are of smaller axial and radial dimensions, which may represent an advantage for building compact powered exoskeletons.

Table 2. Dimensions of the studied pneumatic motors.

Motor Type	Motor Code	Maximum Exterior Diameter [mm]	Total Stroke Length [mm]
Round cylinders	ESNU-10-50-P-A	16	86
	ESNU-20-50-P-A	30	132
	ESNU-40-50-P-A	50	177.6
Pneumatic muscles	DMSP-10-250N-RM-CM	30.1	318.2
	DMSP-20-250N-RM-CM	45.6	335
	DMSP-40-250N-RM-CM	65.5	361

2.3. Establishing the Dependency of the Actuators Mass on the Length of the Stoke

The masses of the pneumatic motors depend on the lengths of their strokes and their axial and radial dimensions. In the case of single-acting cylinders based on their technical specifications, the equations of mass versus stroke variations can be determined:

$$m_{\mathrm{ESNU\_10}}(s) = 0.27 \times s + 0.0373 \tag{3}$$

$$m_{\mathrm{ESNU\_20}}(s) = 0.72 \times s + 0.1868 \tag{4}$$

$$m_{\mathrm{ESNU\_40}}(s) = 2.4 \times s + 0.661 \tag{5}$$

In the three equations, the masses m are calculated in kg and the stroke length s is entered in meters.

Equations (6)–(8) describe the variation in the mass of a pneumatic muscle (in kg) versus the length l (in meters) of its active part (excluding the end connections). The equations resulted from the information included in the technical specifications of the pneumatic muscles.

$$m_{\mathrm{DMSP\_10}}(l) = 0.094 \times l + 0.077 \tag{6}$$

$$m_{\mathrm{DMSP\_20}}(l) = 0.178 \times l + 0.238 \tag{7}$$

$$m_{\mathrm{DMSP\_40}}(l) = 0.34 \times l + 0.675 \tag{8}$$

It is known that upon being fed compressed air, the pneumatic muscle contracts by up to 20% of its initial length. The maximum stroke carried out by the free end of the muscle is described by Equation (9):

$$s = \Delta l = 0.2 \times l \tag{9}$$

Thus, the equations describing the variation in the mass of the pneumatic muscles versus the stroke s becomes:

$$m_{\mathrm{DMSP\_10}}(s) = 0.47 \times s + 0.077 \tag{10}$$

$$m_{\mathrm{DMSP\_20}}(s) = 0.89 \times s + 0.238 \tag{11}$$

$$m_{\mathrm{DMSP\_40}}(s) = 1.7 \times s + 0.675 \tag{12}$$

Table 3 includes the masses of the six studied pneumatic motors, and Figure 3 shows the variation lines of the masses of the two types of motors versus their strokes.

Table 3. Masses of the studied pneumatic motors.

Motor Dimension [mm]	Mass Calculated for a 50 mm Stroke [kg]	
	ESNU	DMSP
10	0.0508	0.1005
20	0.2228	0.2825
40	0.795	0.76

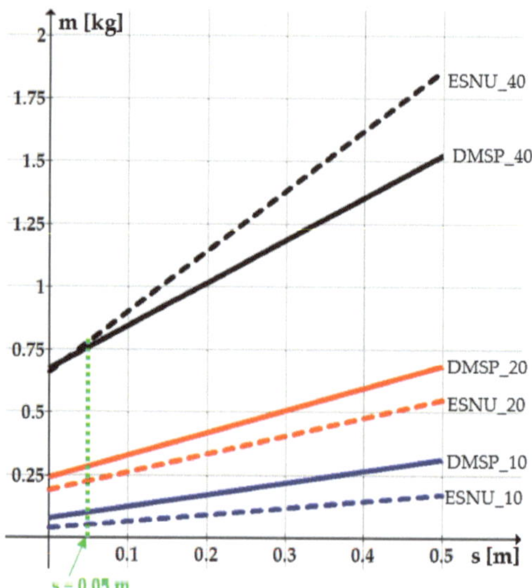

Figure 3. Dependency of the masses of the studied pneumatic motors on the stroke (m = f(s)).

The values in Table 3 and the lines in Figure 3 suggest that for the same diameter of the motors, their masses do not differ significantly. For diameters of 10 mm and 20 mm, single-acting cylinders have a slightly smaller mass, while at 40 mm diameter it is the pneumatic muscle that has the smaller mass. In the case of 40 mm diameter actuators, both cylinders and muscles, as the strokes required to be carried out and implicitly their lengths increase, the difference between cylinder and muscle mass increases, i.e., the pneumatic muscles become increasingly lighter than the cylinders.

2.4. Determining the Dependency Equations of the Forces Developed by the Actuators Versus Stroke

Tables 4 and 5 feature the variability intervals of the forces developed by the six pneumatic motors. All presented values were determined for a maximum pressure of 6 bar, and the data were taken from the technical specifications of the pneumatic motors.

Table 4. Forces developed by the ESNU pneumatic cylinders.

Motor Code	Maximum Force [N]	Minimum Force [N]
ESNU-10-50-P-A	41	4.8
ESNU-20-50-P-A	169	13.6
ESNU-40-50-P-A	688	30

Table 5. Forces developed by the DMSP pneumatic muscles.

	Specific Axial Contraction ε = s/l					
	0	0.04	0.08	0.12	0.16	0.20
Motor Code	Stroke [m]					
	0	0.01	0.02	0.03	0.04	0.05
	Developed Force [N]					
DMSP-10-250N- RM-CM	397.4	318.8	226.6	154.4	90.9	32.1
DMSP-20-250N- RM-CM	1199.1	1175.5	910.8	691.5	498.1	310
DMSP-40-250N- RM-CM	3991.7	3852.3	3712.3	2852.3	2126.3	1489.6

Based on the data in Tables 4 and 5, the lines and curves shown in Figures 4 and 5 were plotted that describe the dependency of the forces developed by the pneumatic motors on the stroke. The dependencies are expressed by Equations (13)–(18).

$$F_{\text{ESNU\_10}}(s) = -724 \times s + 41 \tag{13}$$

$$F_{\text{ESNU\_20}}(s) = -3108 \times s + 169 \tag{14}$$

$$F_{\text{ESNU\_40}}(s) = -97656 \times s + 6372.4 \tag{15}$$

$$F_{\text{DMSP\_10}}(s) = 5.3333 \times 10^5 \times s^3 - 1821.4286 \times s^2 - 8556.5476 \times s + 398.95 \tag{16}$$

$$F_{\text{DMSP\_20}}(s) = 1.0866 \times 10^7 \times s^3 - 9.1088 \times 10^5 \times s^2 + 549.1005 \times s + 1211.2778 \tag{17}$$

$$F_{\text{DMSP\_40}}(s) = 2.7745 \times 10^7 \times s^3 - 2.9408 \times 10^6 \times s^2 + 2.7968 \times 10^4 \times s + 3960.6746 \tag{18}$$

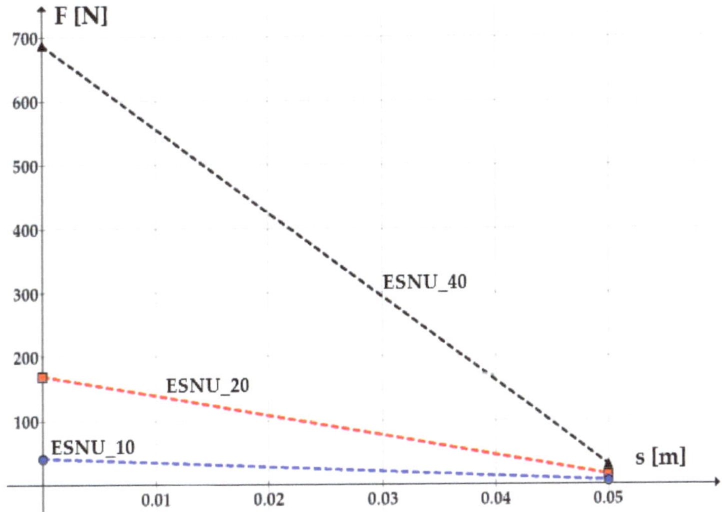

Figure 4. Variation of the forces developed by the studied pneumatic cylinders.

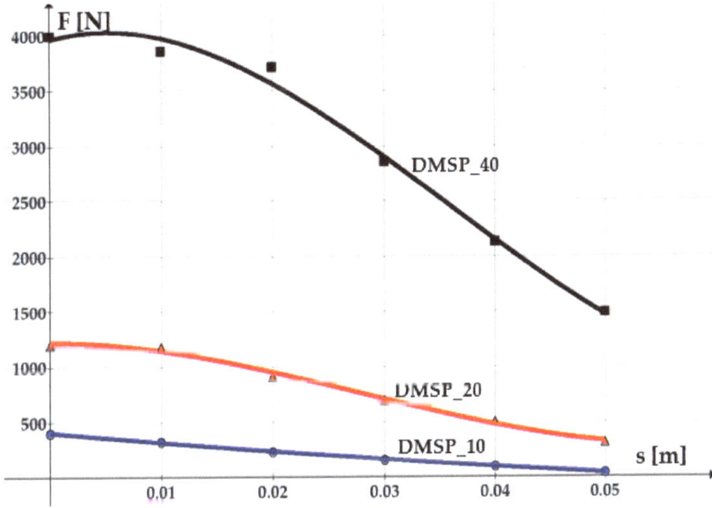

Figure 5. Variation of the forces developed by the studied pneumatic muscles.

Figures 4 and 5 show that the forces developed by pneumatic muscles are about 5 to 10 times greater than those developed by single-acting cylinders.

3. Determination of the Energy-to-Mass Ratio

The energy-to-mass ($R_{E/m}$) ratio is a quantity devised for comparing the energy efficiency of different types of actuators. In [26], Plettenburg proposes the computational relationship of this indicator:

$$R_{\frac{E}{m}} = \frac{\int F(s) \cdot ds}{m(s)} \tag{19}$$

where $F(s)$ is the force developed by the pneumatic motor, s is the stroke and m is the mass of the motor.

For pneumatic muscles, Equation (19) is adapted by taking into account the specific axial contraction $\varepsilon = s/l$:

$$R_{\frac{E}{m}} = \frac{l \cdot \int F(\varepsilon) \cdot d\varepsilon}{m(s)} = \frac{\int F(s) \cdot ds}{m(s)} \tag{20}$$

The numerator in Equations (19) and (20) represents the energy developed by the motors over the entire stroke and is obtained by the integration of the force Equations (13)–(18). The integrals in the numerator represent the surfaces between the variation curves of the forces, the abscissa of the graphs and the maximum and minimum limits of the motors' displacements. Figure 6 highlights these surfaces for the ESNU cylinders and the DMSP pneumatic muscles.

Equations (21)–(23) represent the calculation formulae of the energy-to-mass ratios for the three studied pneumatic cylinders:

$$R_{\frac{E}{m}, \text{ESNU\_10}} = \frac{1.145}{0.27 \times s + 0.0373} \tag{21}$$

$$R_{\frac{E}{m}, \text{ESNU\_20}} = \frac{4.565}{0.72 \times s + 0.1868} \tag{22}$$

$$R_{\frac{E}{m}, \text{ESNU\_40}} = \frac{17.95}{2.4 \times s + 0.661} \tag{23}$$

For the pneumatic muscles, first the equations of the energies developed over the working interval are determined. Figure 6b shows the surface that materializes the integrals in Equation (20).

$$E_{\text{DMSP\_10}}(s) = \int_{0}^{0.05} F_{\text{DMSP\_10}}(s) \cdot ds = 10.0093 \tag{24}$$

$$E_{\text{DMSP\_20}}(s) = \int_{0}^{0.05} F_{\text{DMSP\_20}}(s) \cdot ds = 40.2748 \tag{25}$$

$$E_{\text{DMSP\_40}}(s) = \int_{0}^{0.05} F_{\text{DMSP\_40}}(s) \cdot ds = 153.8121 \tag{26}$$

The energy-to-mass ratios of the three pneumatic muscles will be:

$$R_{\frac{E}{m}, \text{DMSP\_10}} = \frac{E_{\text{DMSP\_10}}(s)}{m_{\text{DMSP\_10}}(s)} - \frac{10.0093}{0.47 \times s + 0.077} \tag{27}$$

$$R_{\frac{E}{m}, \text{DMSP\_20}} = \frac{E_{\text{DMSP\_20}}(s)}{m_{\text{DMSP\_20}}(s)} = \frac{40.2748}{0.89 \times s + 0.238} \tag{28}$$

$$R_{\frac{E}{m}, \text{DMSP\_10}} = \frac{E_{\text{DMSP\_40}}(s)}{m_{\text{DMSP\_40}}(s)} \quad \frac{153.8121}{1.7 \times s + 0.675} \tag{29}$$

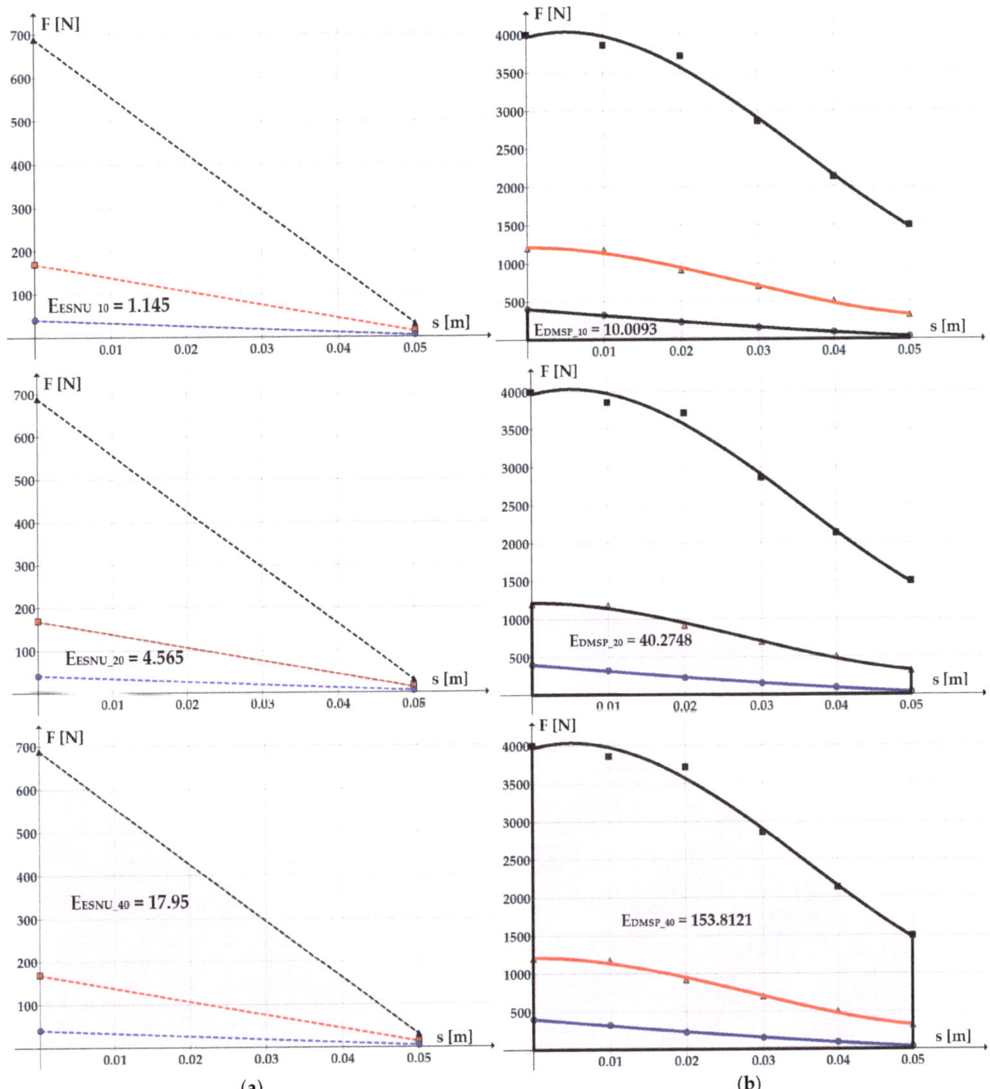

Figure 6. Energy developed by: (**a**) ESNU pneumatic cylinders; (**b**) DMSP pneumatic muscles.

Figure 7 shows the variation in the energy-to-mass ratios versus stroke for the six pneumatic motors and Table 6 presents the values of these ratios for a 50 mm stroke.

Table 6. Energy-to-mass ratios of the studied pneumatic motors.

Motor Dimension [mm]	Energy-to-Mass Ratio for a 50 mm Stroke [Nm/kg]	
	ESNU	**DMSP**
10	22.5394	99.595
20	20.4892	142.5657
40	22.9834	202.3843

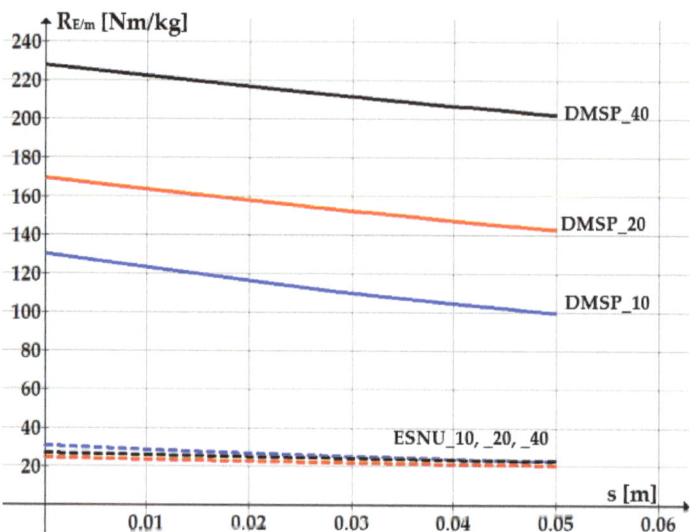

Figure 7. Variation of the energy-to-mass ratios versus strokes of the pneumatic motors.

Figure 7 reveals that in the case of single-acting cylinders, the values of the energy-to-mass ratios are approximately the same regardless of their diameters. This observation does not apply in the case of pneumatic muscles, where the energy-to-mass ratio in a 40 mm diameter muscle is about two-times larger than that in a 10 mm diameter one.

The above table shows that for the studied pneumatic actuators with a 50 mm stroke, the energy-to-mass ratios of the pneumatic muscles are about 4- to 9-times larger than those of single-acting cylinders. Consequently, from the viewpoint of energy efficiency, pneumatic muscles are recommended for applications that require the development of a large force while ensuring maximum energy efficiency.

Table 7 presents an overview of the results obtained by the comparative study of the two types of linear pneumatic motors. The (+) sign indicates a favorable characteristic while (−) indicates a disadvantage.

Table 7. Overview of the results obtained by the comparative study of the two types of pneumatic motors.

Motor Type	Maximum Exterior Diameter	Total Length	Mass	Force	Energy-to-Mass Ratio
Round cylinders ESNU	+	+	=	−	−
Pneumatic muscles DMSP	−	−	=	+	+

In articles [25,26], the authors have presented and discussed two pieces of medical rehabilitation equipment, one designed for the recovery of the lower limb (Figure 8), and the other for the recovery of the wrist (Figure 9). Both devices are actuated by pneumatic muscles, a technical solution selected due to the favorable energy-to-mass ratio.

For these devices, Table 8 features a comparison of the computational results obtained by the method proposed in this paper. The values of the energy-to-mass ratios of the utilized pneumatic muscles were compared to those of the single-acting cylinders that could have been used alternatively as actuators of the two devices. Table 8 includes the dimensions of the compared actuators.

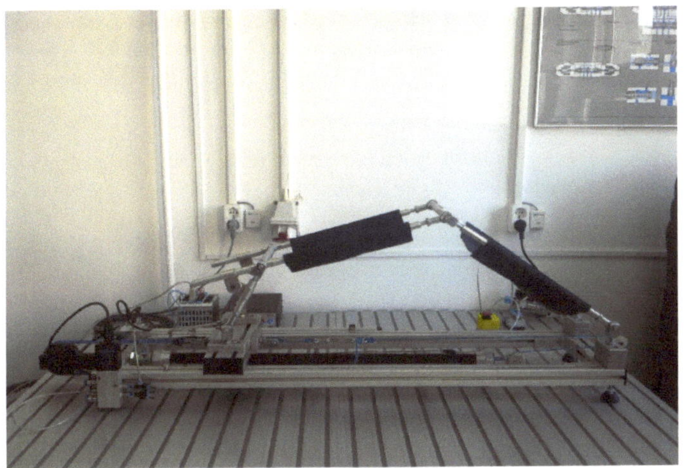

Figure 8. Assistive rehabilitation device for the joints of the lower limb.

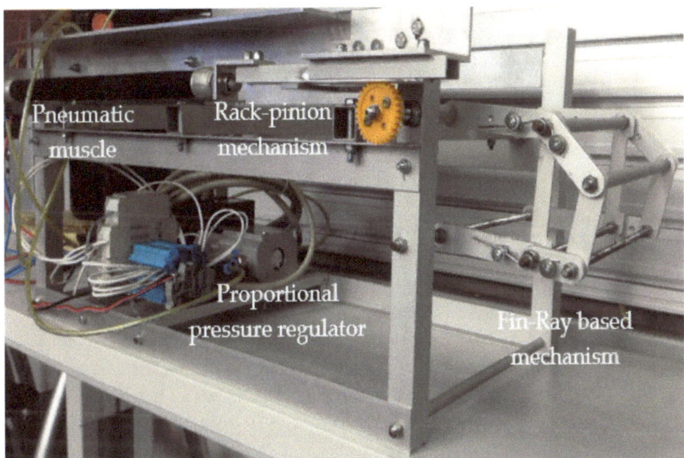

Figure 9. Pneumatic muscle actuated wrist rehabilitation equipment.

Table 8. Comparison of the energy-to-mass ratios computed for the two types of actuators.

	Energy-to-Mass Ratio [Nm/kg]
Assistive rehabilitation device for the joints of the lower limb	
Pneumatic muscle L = 750 mm; d = 20 mm; s = 150 mm	108.4113
Single-acting cylinder D = 20 mm; s = 150 mm	15.4851
Wrist rehabilitation equipment	
Pneumatic muscle L − 300 mm; d = 10 mm; s = 60 mm	95.1454
Single-acting cylinder D = 10 mm; s = 60 mm	21.4019

It needs be pointed out that in the case of the assistive rehabilitation device for the joints of the lower limb, the value of the energy-to-mass ratio a of the pneumatic muscle is

7-times greater than that of the equivalent single-acting cylinder, while in the case of the wrist rehabilitation equipment, the proportion is of 4.45. These two examples confirm the superior behavior of pneumatic muscles from the energy point of view.

4. Conclusions

Thus, the methodology for determining the energy-to-mass ratio entails a sequence of steps consisting of (i) computing the actuator's mass; (ii) determining the evolution of the force developed over the entire working stroke; (iii) computing the developed energy and (iv) computing the energy-to-mass ratio. Given its short running time, the computational methodology of the energy-to-mass ratio put forward in the paper also lends itself to other types of actuators.

The following conclusions resulted in relation to the quality characteristics used and analyzed for the comparison of the two types of pneumatic actuators:

- From the viewpoint of axial and radial dimensions, single-acting cylinders are more compact than pneumatic muscles;
- The masses of the two types of pneumatic motors of the same diameter do not differ significantly. Nevertheless, in the case of 40 mm diameter actuators, as the strokes required to be carried out and implicitly their lengths increase, the difference between the mass of the cylinder and that of the muscle increases, in the sense that the pneumatic muscles become increasingly lighter than the cylinders.
- The forces developed by pneumatic muscles are about 5- to 10-times greater than those developed by single-acting cylinders;
- The energy-to-mass ratios of the pneumatic muscles are about 4- to 9-times larger than those of single-acting cylinders.

The results of the study prove that despite their radial and axial larger sizes, pneumatic muscles offer benefits from the viewpoint of developed forces and energy-to-mass ratios. In applications that require large forces and high energy efficiency, replacing single-acting cylinders with pneumatic muscles has become a necessity.

The analysis of the energy-to-mass ratios confirm the assertions found in the literature related to pneumatic muscles that motors have an energy efficiency superior to that of single-acting pneumatic cylinders.

The flexibility of pneumatic muscles, the large, developed forces and a high energy-to-mass ratio render these actuators an optimum solution for manufacturing exoskeleton devices. In addition, their light weight, portability, safe operation, reliability, compliance and easy integration in wearable assistive devices render pneumatic muscles a viable actuation alternative to such rehabilitation systems.

Author Contributions: Conceptualization, A.D. and T.D.; methodology, A.D. and T.D.; resources, A.D.; data curation, T.D.; writing—original draft preparation, A.D.; writing—review and editing, A.D. and T.D.; supervision, A.D. All authors have read and agreed to the published version of the manuscript.

Funding: This research received no external funding.

Institutional Review Board Statement: Not applicable.

Informed Consent Statement: Not applicable.

Data Availability Statement: Not applicable.

Conflicts of Interest: The authors declare no conflict of interest.

References

1. Harmonic Drive SE. Mobile Robotics. Available online: https://harmonicdrive.de/en/glossary/mobile-robotics (accessed on 17 January 2022).
2. International Society for Prosthetics and Orthotics. Prosthetics and orthotics services. Available online: https://www.ispoint.org/page/POservices (accessed on 18 January 2022).

3.   Sirlantzis, K.; Larsen, L.B.; Kanumuru, L.K.; Oprea, P. Robotics. In *Handbook of Electronic Assistive Technology*; Cowan, D., Najafi, L., Eds.; Elsevier Ltd.: London, UK, 2019; pp. 311–345.

4.   Martinez-Hernandez, U.; Metcalfe, B.; Assaf, T.; Jabban, L.; Male, J.; Zhang, D. Wearable Assistive Robotics: A Perspective on Current Challenges and Future Trends. *Sensors* **2021**, *21*, 6751. [CrossRef] [PubMed]

5.   Gopura, R.A.R.C.; Bandara, D.S.V.; Kiguchi, K.; Mann, G.K.I. Developments in hardware systems of active upper-limb exoskeleton robots: A review. *Robot. Auton. Syst.* **2016**, *75*, 203–220. [CrossRef]

6.   Knaepen, K.; Beyl, P.; Duerinck, S.; Hagman, F.; Lefeber, D.; Meeusen, R. Human-robot interaction: Kinematics and muscle activity inside a powered compliant knee exoskeleton. *IEEE Trans. Neural. Syst. Rehabil. Eng.* **2014**, *22*, 1128–1137. [CrossRef] [PubMed]

7.   Asbeck, A.T.; De Rossi, S.M.M.; Galiana, I.; Ding, Y.; Walsh, C.J. Stronger, Smarter, Softer: Next-Generation Wearable Robots. *IEEE Robot. Autom. Mag.* **2014**, *21*, 22–23. [CrossRef]

8.   Deaconescu, T.; Deaconescu, A. Linear pneumatic motors—A comparative study. *MATEC Web Conf.* **2017**, *112*, 5007. [CrossRef]

9.   Bae, J.; Moon, I. Design and control of an exoskeleton device for active wrist rehabilitation. In Proceedings of the 12th International Conference on Control, Automation and Systems, Jeju, Korea, 17–21 October 2012; pp. 1577–1580.

10.   Goergen, R.; Valdiero, A.C.; Rasia, L.A.; Oberdörfer, M.; de Souza, J.P.; Gonçalves, R.S. Development of a Pneumatic Exoskeleton Robot for Lower Limb Rehabilitation. In Proceedings of the IEEE 16th International Conference on Rehabilitation Robotics (ICORR), Toronto, ON, Canada, 24–28 June 2019; pp. 187–192.

11.   Pavithrana, A.; Skariaa, E.; Rajana, R.; Joseb, J. Design and fabrication of a Pneumatic Exoskeleton. *Int. J. of Adv. Res. Trends Eng. Technol.* **2017**, *4*, 36–42.

12.   Krishna, G.; Hosmutt, P.; Nyamagoud, B.M.; Patil, M.V.; Hunnur, O. Design and Fabrication of Pneumatic Powered Exoskeleton Suit for Arms. *Int. Res. J. Eng. Tech.* **2018**, *5*, 4415–4419.

13.   Henderson, G.C. Pneumatically-Powered Robotic Exoskeleton to Exercise Specific Lower Extremity Muscle Groups in Humans. Master's Thesis, Georgia Institute of Technology, Atlanta, GA, USA, 20 March 2012.

14.   Pavana, K.B.; Krisantha, A.P.; Pereira, R.M.; Sahil, M.A.; Varun, S.; Mallya, V.S. Design and Fabrication of a Pneumatically Powered Human Exoskeleton Arm. *J. Mech. Eng. Autom.* **2017**, *7*, 85–88.

15.   Thalman, C.; Artemiadis, P. A review of soft wearable robots that provide active assistance: Trends, common actuation methods, fabrication, and applications. *Wearable Technol.* **2020**, *1*, E3. [CrossRef]

16.   Bogue, R. Exoskeletons and robotic prosthetics: A review of recent developments. *Ind. Robot.* **2009**, *36*, 421–427. [CrossRef]

17.   Balasubramanian, S.; Wei, H.R.; Perez, M.; Shepard, B.; Koeneman, E.; Koeneman, J.; He, J. Rupert: An exoskeleton robot for assisting rehabilitation of arm functions. In Proceedings of the 2008 Virtual Rehabilitatio, Vancouver, BC, Canada, 25–27 August 2008; pp. 163–167.

18.   Abe, T.; Koizumi, S.; Nabae, H.; Endo, G.; Suzumori, K.; Sato, N.; Adachi, M.; Takamizawa, F. Fabrication of "18 Weave" muscles and their application to soft power support suit for upper limbs using thin McKibben muscle. *IEEE Robot. Autom. Lett.* **2019**, *4*, 2532–2538. [CrossRef]

19.   Andrikopoulos, G.; Nikolakopoulos, G.; Manesis, S. Motion Control of a Novel Robotic Wrist Exoskeleton via Pneumatic Muscle Actuators. In Proceedings of the IEEE 20th Conference on Emerging Technologies & Factory Automation (ETFA), Luxembourg, 8–11 September 2015.

20.   Zhang, J.F.; Yang, C.J.; Chen, Y.; Zhang, Y.; Dong, Y.M. Modeling and control of a curved pneumatic muscle actuator for wearable elbow exoskeleton. *Mechatronics* **2008**, *18*, 448–457. [CrossRef]

21.   Tsagarakis, N.; Caldwell, D.G.; Medrano-Cerda, G.A. A 7 DOF pneumatic muscle actuator (pMA) powered exoskeleton. In Proceedings of the 8th IEEE International Workshop on Robot and Human Interaction, Pisa, Italy, 27–29 September 1999.

22.   Nassour, J.; Zhao, G.; Grimmer, M. Soft pneumatic elbow exoskeleton reduces the muscle activity, metabolic cost and fatigue during holding and carrying of loads. *Sci. Rep.* **2021**, *11*, 12556. [CrossRef] [PubMed]

23.   Chen, C.-T.; Lien, W.-Y.; Chen, C.-T.; Wu, Y.-C. Implementation of an Upper-Limb Exoskeleton Robot Driven by Pneumatic Muscle Actuators for Rehabilitation. *Actuators* **2020**, *9*, 106. [CrossRef]

24.   Irshaidat, M.; Soufian, M.; Al-Ibadi, A.; Nefti-Meziani, S. A Novel Elbow Pneumatic Muscle Actuator for Exoskeleton Arm in Post-Stroke Rehabilitation. In Proceedings of the 2nd IEEE International Conference on Soft Robotics (RoboSoft), Seoul, Korea, 14–18 April 2019; pp. 630–635.

25.   Deaconescu, T.; Deaconescu, A. Pneumatic Muscle Actuated Equipment for Continuous Passive Motion. In *IAENG Transactions on Engineering Technologies*; Ao, S.L., Chan, A.H.S., Katagiri, H., Xu, L., Eds.; Springer: Dordrecht, The Netherlands, 2009; Volume 3, pp. 311–321.

26.   Petre, I.; Deaconescu, A.; Sârbu, F.; Deaconescu, T. Pneumatic Muscle Actuated Wrist Rehabilitation Equipment Based on the Fin Ray Principle. *Stroj. Vestn. J. Mech. Eng.* **2018**, *64*, 383–392.

27.   Energy Density. Available online: https://en.wikipedia.org/wiki/Energy_density (accessed on 20 June 2022).

28.   Plettenburg, D.H. Pneumatic Actuators: A Comparison of Energy-to-Mass Ratio's. In Proceedings of the 2005 IEEE 9th International Conference on Rehabilitation Robotics, Chicago, IL, USA, 28 June–1 July 2005; pp. 545–549.

29.   Tsagarakis, N.; Caldwell, D.G. Improved modelling and assessment of pneumatic muscle actuators. In Proceedings of the IEEE International Conference on Robotics & Automation, San Francisco, CA, USA, 24–28 April 2000; pp. 3641–3646.

30. Medrano-Cerda, G.A.; Bowler, C.J.; Caldwell, D.G. Adaptive position control of antagonistic pneumatic muscle actuators. In Proceedings of the IEEE/RSJ International Conference on Intelligent Robots and Systems, Pittsburgh, PA, USA; 1995; pp. 378–383.
31. Nakamura, N.; Sekiguchi, M.; Kawashima, K.; Fujita, T.; Kagawa, T. Developing a robot arm using pneumatic artificial rubber muscles. In Proceedings of the Bath Workshop on Power Transmission & Motion Control, Bath, UK, 15 September 2002; pp. 365–375.
32. Sakama, S.; Tanaka, Y.; Kamimura, A. Characteristics of Hydraulic and Electric Servo Motors. *Actuators* **2022**, *11*, 11. [CrossRef]
33. Hesse, S. *The Fluidic Muscle in Application*; Blue Digest on Automation: Esslingen, Germany, 2003; pp. 27–28.
34. Round Cylinders ESNU. Available online: https://www.festo.com/cat/en-gb_gb/data/doc_ENGB/PDF/EN/ESNU_EN.PDF (accessed on 12 December 2021).
35. Fluidic Muscle DMSP/MAS. Available online: https://www.festo.com/rep/en_corp/assets/pdf/info_501_en.pdf (accessed on 12 December 2021).

applied sciences

Article

The Design of a Smart Lower-Limb Prosthesis Supporting People with Transtibial Amputation—A Data Acquisition System

Cristina Floriana Pană [1], Liviu Florin Manta [1,*], Ionel Cristian Vladu [2,*], Ştefan Irinel Cismaru [1], Florina Luminiţa Petcu (Besnea) [1], Dorian Cojocaru [1] and Nicu Bîzdoacă [1]

[1] Department of Mechatronics and Robotics, University of Craiova, RO-200440 Craiova, Romania; cristina.pana@edu.ucv.ro (C.F.P.); stefan.cismaru@edu.ucv.ro (Ş.I.C.); florina.petcu@edu.ucv.ro (F.L.P.); dorian.cojocaru@edu.ucv.ro (D.C.); nicu.bizdoaca@edu.ucv.ro (N.B.)

[2] Department of Electromechanical, Environmentaland Computer Science Applied in the Electrical Engineering, University of Craiova, RO-200440 Craiova, Romania

* Correspondence: florin.manta@edu.ucv.ro (L.F.M.); cristian.vladu@edu.ucv.ro (I.C.V.); Tel.: +40-740-292334 (L.F.M.); +40-728-148258 (I.C.V.)

Abstract: For people with amputated lower limbs, it is imperative to make high-performance prostheses that reproduce, as accurately as possible, the functions of the amputated limb. In this case, a preliminary study of the lower limbs from a kinematic and dynamic point of view is necessary. This paper proposes a prosthesis design and a system for acquiring the information needed to determine the stepping phase kinematic and dynamic parameters of the legs. This system consists of a sensory system attached to the legs and a acquisition data unit built around a microcontroller. The sensory system is based on a sensory system for determining the weight distribution on the sole, made of resistive pressure sensors. The sensory system will be subjected to measurement repeatability and homogeneity tests to evaluate and validate the accuracy and error of the proposed solution. The data obtained by the sensory system is transmitted in real-time, via wi-fi, to a computer system for interpretation. After processing and interpreting the data using standard data sets for comparison, the position of the legs, the type of gait and the phase of movement can be determined. Constructively, the system is configurable and can be adapted to any person, male or female, regardless of shoe size.

Keywords: prosthesis; resistive pressure sensors; acquisition system

Citation: Pană, C.F.; Manta, L.F.; Vladu, I.C.; Cismaru, Ş.I.; Petcu (Besnea), F.L.; Cojocaru, D.; Bîzdoacă, N. The Design of a Smart Lower-Limb Prosthesis Supporting People with Transtibial Amputation—A Data Acquisition System. *Appl. Sci.* **2022**, *12*, 6722. https://doi.org/10.3390/app12136722

Academic Editors: Ionuţ Daniel Geonea and Cristian Copilusi Petre

Received: 18 March 2022
Accepted: 28 June 2022
Published: 2 July 2022

Publisher's Note: MDPI stays neutral with regard to jurisdictional claims in published maps and institutional affiliations.

1. Introduction

Amputation of the limb or extremity of a limb, either from the upper extremity amputation (UEA) or lower extremity amputation (LEA) of the human body, affects people's quality of life. Globally in 2017, the highest number of trauma amputations was in East and South Asia, followed by Western Europe, North Africa and the Middle East, increasing in North America and Eastern Europe [1].

In a 28-year study (1990–2017) on the incidence of lower extremity amputations in 19 countries (EU15+) by Hughes et al. [2], significant geographic and temporal variability was observed. According to this study, Australia had the highest incidence of LEA in women and men during the study period. However, this incidence decreased steadily in the Netherlands and the USA, and a higher percentage reduction is found in the USA. Another recent study examined Germany's LEA incidence from 2015 to 2019 [3]. According to this study, in 2019, compared to 2015, the incidence of major amputations decreased by 7.3%, while the incidence of minor amputations increased by 11.8%, and there was also a decrease in the incidence of women to men.

In terms of the causes that led to the amputation of the lower limbs, demographically, there are many differences between developed and developing countries. For example, compared to developed countries where most amputations are due to disease processes such as diabetes, the causes of amputations are more related to environmental factors, living conditions, or wars in developing countries [4].

Individuals who have had lower limb amputations face many other physical challenges that can compromise their health and mobility. Hence the need for a technical replacement (prosthesis) that restores the biomechanical function of the amputated element and the body's integrity. Management of lower limb prostheses for these individuals is a complicated issue. Unfortunately, the candidates who want to benefit from the usefulness of a prosthesis are a heterogeneous group with distinct ages and needs. In addition, the choice of the perfect candidate depends on the following factors: the aetiology of limb loss, the level of amputation, comorbidities and health, the postoperative stage, and the state of rehabilitation. Lower limb prostheses can be classified into three types of devices: passive, semi-powered or powered. For example, individuals with transtibial amputation (the term transtibial indicates that the amputation occurred between the knee and the ankle) are usually prescribed a prosthesis for storage and return of passive elastic energy, made of carbon fibre and works as an arc without the ability to generate energy again or to articulate. Most ankle and foot prostheses available on the market until the early 2000s were completely passive. The simplest solution for transtibial amputations is a solid prosthesis of the foot without an ankle joint [5]. As a result, the mechanical properties did not adapt to the user's walking speed and terrain type. Individuals with transtibial amputation often adopt compensatory gait strategies. These can lead to significant changes in gait dynamics, the joints loading and working, and the muscular activity of the affected and unaffected leg [6]. In the last three decades, approaches in prosthetic technology have led to significant advances, especially regarding biomechanical and user comfort [7] and in walking symmetry and energy cost [8]. An excellent example of a marketable passive prosthetic foot is the C-Walk [9], equipped only with passive components but combined suitably. As a result, it is more effective from an energy point of view compared to other prostheses in this category.

The first electrically powered ankle-foot prosthesis was built in 1998 and was pneumatically operated [10]. Subsequently, from 2005 until now, studies have focused on designing and developing autonomous energy supply systems [11–18]. The main factors considered for the design of prostheses are both the mechanical properties [19,20] and the length of the prosthesis [21]. Another factor that should not be overlooked is the weight of the prosthetic components [22]. A higher weight also increases the stress on the socket—the residual connection of the limbs, which is one of the most critical elements in the prosthesis [23]. In addition, most studies in the literature have been conducted on the evaluation of kinematic and kinetic gait [24] and foot plantar pressure [25,26].

Magnetorheological fluids (MR) were successfully introduced into prosthetic devices after 2000. In 2001, a patent was published [27] for a variable torque magnetorheological knee prosthesis produced by Ossur Inc, Los Angeles, CA, USA [28]. Herr and Wilkenfeld [29], in 2003, presented a magnetorheological knee prosthesis that automatically adjusts the cushioning of the knee to walking amputated using only local detection of the knee, torque and strength position. In 2006, a study related to [27] was carried out, which was intended to be part of a project to create models with finite elements of the knee [30]. Another invention patent [31] was published in 2012 generally relating to powered human augmentation devices, such as lower-extremity prosthetic, orthotic, or exoskeleton apparatus, and/or humanoid robotic devices designed to emulate human biomechanics. Among the most current studies is [32], which analyzes the energy consumption of a magnetorheological active knee actuator that has been designed for transfemoral prostheses. The system was developed as an operational motor unit consisting of an EC motor, a harmonic drive, and a magnetorheological clutch (MR) parallel with an MR brake.

By analyzing the above literature, it can be said that significant advances have been made in the research and development of prostheses for the lower limbs, which has led to an increase in the function and quality of life for many people with amputations of the lower limbs living in developed countries. However, one downside of this new research and development is that many potential users live in developing countries and cannot benefit from this new technology. This fact is due to multiple causes: cost, durability, maintenance or access to these prostheses. Under these conditions, research needs to focus on designing and developing cost-effective foot prostheses that meet economic, environmental and physical standards to cope with unfavourable climates and working conditions. So far, many cheap prosthesis projects have been done to support the lower limbs, such as [33–35]. Our team aims to design and implement a low-cost prosthesis to support people with lower limb amputation (especially transtibial amputation). In the first phase, two goals were set: to design the smart ankle prosthesis and implement a solution to determine the weight distribution on the sole. The smart ankle prosthesis is intended to be developed as a passive one (from the articulation actuation point of view) and controllable (by using a magnetorheological fluid and controlling its properties to obtain a controllable damping effect inside the articulation). For the weight distribution on the sole, in this stage of our research, we will propose an alone sandal with eight pressure sensors. The first objective we set for the current study is to propose a smart ankle prosthesis design; the second objective is to design, develop, and validate a solution to determine the weight distribution on the sole; the third objective is to conduct a series of measurements which allow us to test the data acquisition system firstly and to compare the acquired data against other systems presented in the literature, and secondly to create a dataset which will be used to design a future algorithm which to detect the gait phases. With these parameters determined, it is possible to control the optimal position of the prosthesis joints, depending on the phase and the type of movement. The determinations were made for people without amputated limbs. The appropriate values for an amputated limb can be determined by mediating their values. This simplified and optimized system can be integrated into a prosthesis for its configuration in real-time, regardless of the phase or type of movement.

Our Prosthesis Design

The research effort in this field focuses on improving the characteristics of the artificial ankle to closely simulate the human ankle's functionality. The design of an artificial ankle involves many scientific and technical areas such as medicine, robotics and mechatronics, biomechanics, material science, mechanical engineering, electronics, and others. The challenge of the ankle prosthesis design is to find the means to achieve the functions of an intact ankle, especially the role of power generation.

This study presents an innovative solution for a smart ankle prosthesis based on smart fluids that will simulate the functionality of the human ankle for both walking and running activities.

The mechanical structure consists of the mechanical elements of a standard prosthesis (Figure 1). The innovative element will be a spherical joint based on smart fluids (class 4), which replaces the human ankle joint. The spherical joint allows two rotations, corresponding to the up and down movement and the lateral rotation of the foot. For each rotation, the spherical joint consists of two concentric hemispherical shells between which there is a magnetorheological fluid. The volume between the two spherical shells is divided in two by a fixed belt. Also, a spherical cap rotates between the two spheres. It is rigidly attached to the prosthesis elements, rotating with them and generating the rotational movement of the ankle. Angle 0 divides the fluid into two equal volumes for the relaxation position.

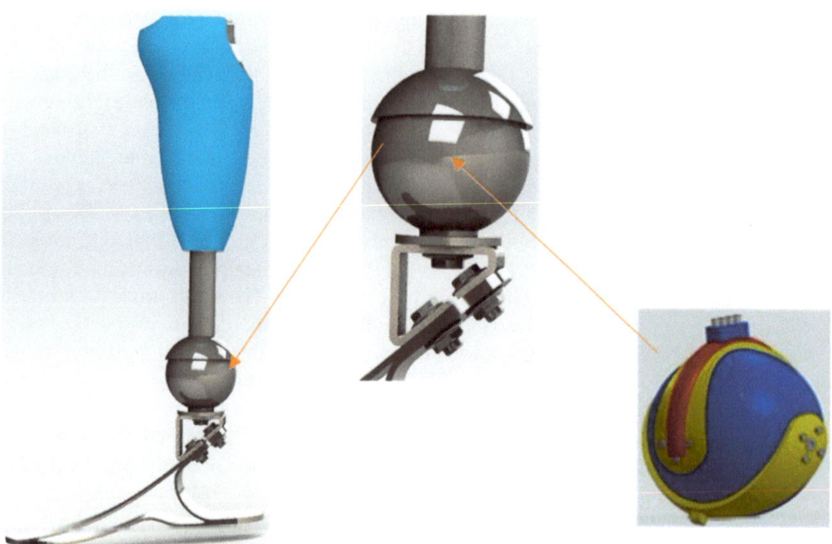

Figure 1. Ankle prosthesis with smart fluid joint.

At the rotation between the two hemispheres, the fluid is circulated from one space to another through a magnetorheological stop valve (outside the joint). Due to the incompressibility of the fluid through the control of the stop valve, the control of the rotation of the spherical cap between the two hemispheres is obtained, so implicitly, the control of the rotation of the joint. For the second axis of rotation, proceed similarly. The joint is shown in Figure 2. Its detailed description is presented in the works [36]. Also, the mechanical system contains the reconstruction of the leg components that have been amputated (ankle, leg, etc.).

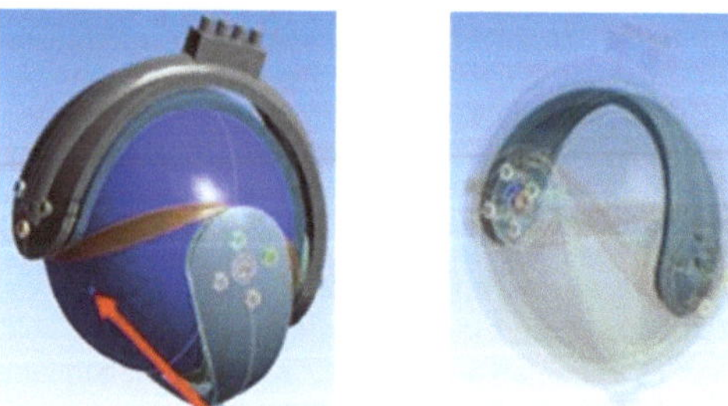

Figure 2. Spherical joint based on smart fluids (magnetorheological fluid).

The actuator system: The prosthesis uses elastic elements that maintain the foot's position in the relaxation position. Also, the ankle joint allows rotation (lateral and vertical) under the action of body weight (support on foot, walking, running). A stop-valve controls both movements of the joint with rheological fluid. After the cessation of the effort (due to the move), the elastic elements bring the paw to the position of relaxation.

The sensorial system: The sensory system provides information about the position of the articulation of the prosthesis and the force exerted on it due to the body's movement (size, direction). The information is provided to the control system. It consists of incremental rotation sensors (associated with the spherical joint) and force sensors (associated with the foot paw). In addition, the value of the working pressures for the stop valve is given by pressure sensors.

The control system: Prosthesis control systems must accomplish multiple tasks, such as recognizing the amputee's intended movements (high-level control), applying an appropriate control law based on the amputee's intent (mid-level control), and using local feedback to command the actuation systems within the prosthesis (low-level control). The control system will be capable of controlling the ankle joint across various ambulation modes (level-ground walking, ramp ascent/descent, stair ascent/descent, running); however, these control strategies are highly sophisticated.

There is a known ankle prosthesis that solves this problem through two constructive solutions, namely:

(1) The most used solution is the prosthetics leg for normal daily activities (upright position, walking, etc.). Generally, they are designed strictly for one person (weight, dimensional) and a spherical joint of class 5 that allows vertical movements of the foot. Most of these prostheses are passive, and the movement control is performed (strictly mechanical) by elastic elements or hydraulic/pneumatic cylinders. These have the disadvantage of the need for design strictly reported to a beneficiary. They also allow only one type of activity (e.g., walking) [37].

(2) Another solution is the prosthetic for particular activities (sports activities: running, jumping). They are generally built from a single elastic body without containing the rotating joint. Therefore, they are dedicated only to sports activities, designed only for certain types of requests, strictly for one person. To switch to daily activities, it is necessary for this person to change the prosthesis [38].

The classical spherical joints do not control the movement of interconnected elements, having only the role of a passive kinematic couple.

Our spherical joint based on smart fluids was proposed in a national patent application: Spherical joint based on intelligent fluids—A/00213/2019 [39] and was a gold medalist at Euro Invent 2019.

The rest of the paper is structured as follows: Section 2 presents the design of the prosthesis and the sensory system, as well as the data validation of the sensory system; Section 3 shows the results obtained from the simulations; Section 4 presents the challenges associated with the development of such systems and their potential solutions, as well as a discussion about the future research perspectives are given; finally, Section 5 is devoted to the conclusions.

2. Materials and Methods

2.1. Acquisition System for Determining the Phase of Passing and Walking Characteristics

The proposed sensorial system includes force sensors, which to provide a reliable plantar pressure measurement.

Plantar pressure measurement systems, including low-cost insole sensors, are trendy instruments used in diagnosing foot disorders [40] and for monitoring the rehabilitation process. We consider that the plantar pressure measurement can be used as an input for our active prosthesis control system. Currently, there are many commercial systems with a high density/large number of pressure sensors, having a limited charging time. From a transtibial active prosthesis control point of view, there is not a need for such a high density of plantar pressure measurement points. Also, the above-mentioned commercial solutions are designed to help determine the mode of complex information, i.e., the deviations of the spine and determinations made at rest. They are rigid systems of the type of pressure plates or flexible for the kind of insoles with insertion of pressure elements. Neither of the two systems satisfies the objective proposed in the paper—to provide reliable data about the plantar pressure to be used by the control system. There is also no integrated system that simultaneously reads the sole load and the kinematics or dynamics of the foot. Therefore, the first objective of the authors of this paper was to develop a low-cost insole prototype composed of an adequate, small number of sensors, which offer the optimum data about the plantar pressure.

A special shoe, like a sandal, with a rigid sole, was made for this. On its sole are mounted eight resistive pressure sensors in the three areas of interest. In addition, the foot straps are mounted on the top. The sandal is made modularly and, with minimal modifications, can be adapted to any size of shoes 36–41 women and 41–46 men. This implies the design of a highly customizable sandal sole, the manufacture of it using a 3D printer/ABS filament, identifying the optimal placement for the sensorial units, and the design and manufacture of an accurate and reliable data acquisition system.

2.1.1. Mechanical Component

This part consists of two main parts (heel and tip parts), joined by a rotating joint (Figure 3). The tip and the heel have two subcomponents: the common element (Figure 3, elements 1 and 3) and the frame (Figures 3 and 4, elements 2 and 4).

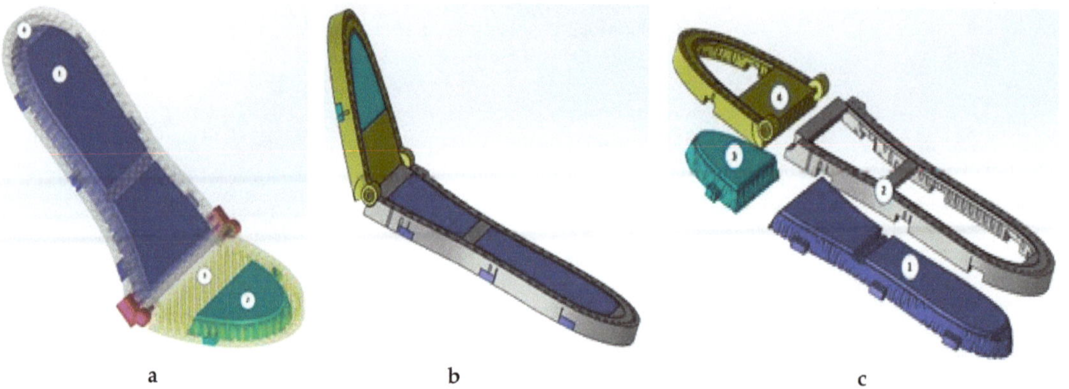

Figure 3. Sandal sole—Solidworks design: (**a**) inside view of the sole; (**b**) exterior view sole; (**c**) the elements of the sole: 1—heel sole; 2—heel sole frame; 3—toe sole; 4—toe sole frame.

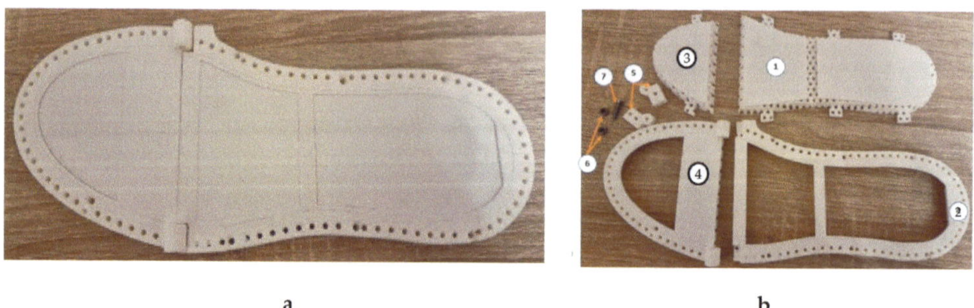

Figure 4. Sandal sole—3D printer prototyping: (**a**) inside view of the sole; (**b**) the elements of the sole: 1—heel sole; 2—heel sole frame; 3—toe sole; 4—toe sole frame; 5, 6 and 7: rotating joint components: 5—bearing support, 6—bearing, 7—screw.

The pressure sensors are mounted on the back of the sole, which supports the foot. The pressure sensors can be mounted anywhere and in any configuration on the sole because it is perforated (on the back) with a matrix network of threaded holes in which the sensors are mounted.

The system is obtained by assembling these elements, depending on the requirements. First, sensory elements and common elements are used for all models (woman/man, sizes 36–41). Then, the frame type elements are changed depending on the characteristics, and the sensors are repositioned according to the predetermined model. Our research team manufactured all the components forming the system using a 3D printer and ABS filament as raw material. The cost of a custom sandal sole, manufactured according to with user's particular specifications and needs, is around 5 Euro.

The components are assembled with screws. The fastening of the sandal straps is done on the sandal frame and allows any configuration of them due to a similar system with perforations. For different foot or lady/man sizes, change the frame, re-realizing the desired size of the shoes. This system is ergonomically minimizing the elements used. The straps have a VELCRO closure system, allowing a perfect fit.

2.1.2. Sensorial System

The sensory system comprises eight resistive pressure sensors mounted on the back of the sole. Sensors are positioned according to the pressure zones of the sole, having a proper positioning for each sandal size, made with a template. (Figure 5). The optimal positions were established by studying the existing insole models in trade, by studies performed on a group of people (male/female, different sizes) using pressure plates (for orthopaedical use), and by experimental positioning on the prototype obtained. As a result, an eight-sensor configuration was chosen to cover the sole's essential pressure areas.

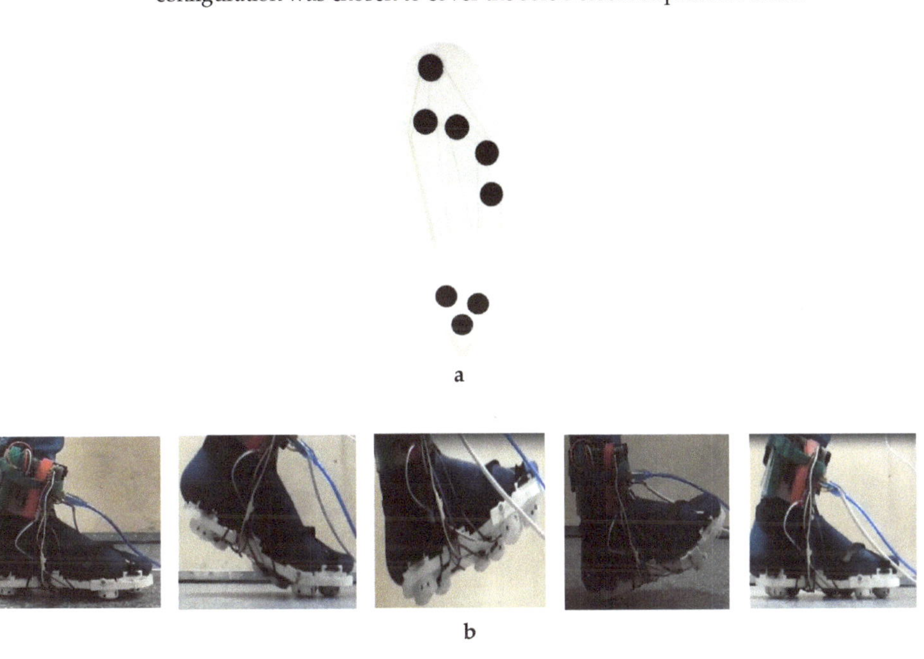

Figure 5. (**a**) Distribution of pressure sensors on the sole; (**b**) one-legged stepping cycle.

Each sensor is mounted in mechanical support fixed by a screw on the sandal's sole. This bracket consists of a cylindrical housing that supports the sensor. It is attached to the sole using an eccentric piece. When the sensor cannot be mounted precisely in the desired position, the eccentric part can reposition the cylindrical housing in the prescribed position. A sliding cylinder ending in a spherical cap represents the moving part and the contact with the ground. By pressing the sandal on the ground, this cap performs a translational movement in the cylindrical housing by pressing the pressure sensor. Thus the sensory network measures the distribution of body weight.

Figure 6 shows the conceptual photograph of the pressure sensor and an image of the sensor placed on the sole.

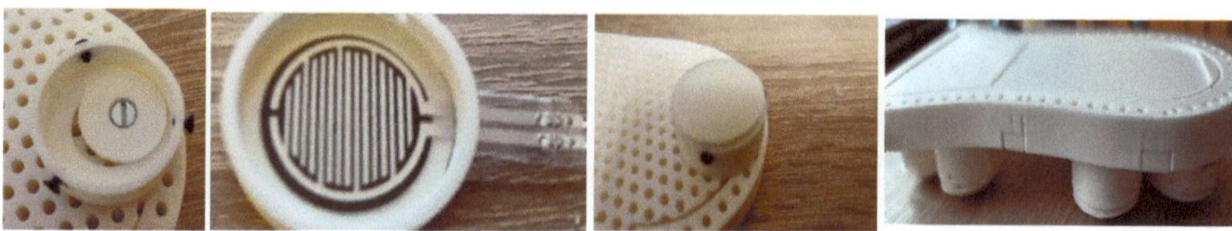

Figure 6. Sensor mounted on the sole of the sandal—prototype 3D printer, ABS—positioning detail using eccentric, sensor mounting detail in the housing.

All the components were designed using Solidworks and ANSYS, simulating accurate materials and stresses corresponding to the load with a body weight of 120 kg (Figure 7).

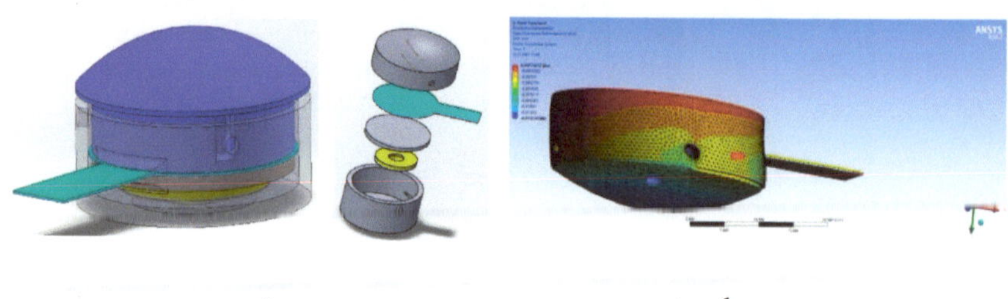

a   b

Figure 7. (**a**) Sensor support components—Solidworks design—cylindrical housing, eccentric support mounting, sensor support base, pressure sensor, ground contact cylinder (finished with spherical part); (**b**) Sensor simulation details using ANSYS.

The aim is to use pressure (weight) sensors to determine the main points where the pressure exerted by each step is high. Given many sensors, an ATMEGA2560 controller was adopted because 8 sensors will be used for each sole, and the number of inputs held by the controller must be significant.

An FSR sensor is made of a piezoresistive material that can modify its resistance depending on the weight it measures. These sensors are resistors that vary linearly, considering conductance vs. resistance under an applied weight. The proposed application's advantages include the operation with low energy consumption and the fact that the implementation, as a thin film, allows mounting in very narrow and tight places.

The sensors are read using the electronic current-to-voltage converter circuit to obtain precise values of the measured weight. Such circuitry is required because the evolution of the signal provided by the sensor, depending on the weight, is recorded as current values (mA). The electronic circuit in Figure 8 takes the current as an input, providing a voltage value as output that can be read with the microcontroller. The MCP6004 operational amplifier is part of this design because it was necessary to obtain a linear signal compared to a classic voltage divider, and the current bias input is minimal. In addition, compared to a simple divider, this design keeps the same voltage applied to the FSR sensor even if other resistors or other sensors are connected in parallel.

The cost of the data acquisition system (sensors, circuits, processing unit) is around 200 Euro.

In addition, the theoretical principle of having a virtual ground between the terminals of the FSR sensor is applied. Finally, the formula determines the output voltage of the circuit:

$$V_{OUT} = \frac{-V_{DRIVE}}{R_{FSR}} \times R_F$$

where V_{OUT} represents the output voltage, $-V_{DRIVE}$ represents the supply voltage, and R_F is the load resistance.

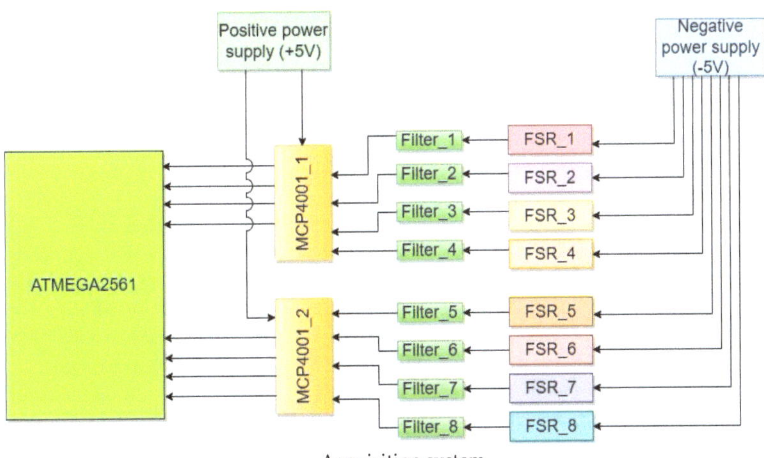

Figure 8. The hardware diagram.

It was decided to use the physical property. Namely, the conductance is obtained mathematically by the inverse of the resistance. The conductor's advantage is that the sensor has a linear evolution in weight detection based on this property. First, the conductor helps obtain the force, and then the force is transformed into mass.

2.2. Data Validation for the Sensorial System

Data validation is essential in any measurement process, mainly when the acquired data will be used as input for a control system—in our case, the gait control system for the lower limb prosthesis. We conducted a series of measurements to reveal the limits of the proposed sensorial systems under the terms of repeatability for each sensor, homogeneity among used sensors, and non-linearity. In this sense, the research team leased a calibration test weight kit, from an accredited metrological laboratory. The calibration test weight kit included several calibrated test weights, and the metrology laboratory presented the calibration certificate for them. The heaviest calibrated test weights included in the kit had the values of 1 kg, 5 kg and 10 kg, and they were used to test the repeatability of the sensorial element under various loads. The results obtained were correlated with determinations made with the pressure plate and were similar for several determinations.

2.2.1. General Testing Methodology

To obtain information as near as possible to the operation conditions, we conducted individual measurements for each of the eight sensorial element assemblies, (as shown in Figure 7, these are: the assembly formed by the cylindrical housing, eccentric support mounting, sensor support base, pressure sensor, and ground contact element). The methodology to record one set of data is:

- The team checks that the evaluated sensorial element's unique identifier is visible.
- The evaluated sensorial element is placed on a planar surface, with the cylindrical housing downwards and the ground contact element upwards.
- The evaluated sensorial element is electrically connected to the data acquisition system.
- All the electrical connections are verified visually for compliance with the wiring diagram.
- The calibrated test weights are placed for an easy and safe access.
- Preliminary readings are conducted to validate that the sensorial element, the data acquisition system and the data logging are working properly.
- The measurement protocol is conducted.
- The recorded data is verified for integrity.
- The recorded data is stored accordingly.

2.2.2. Repeatability Test Methodology

The repeatability test was designed to check the repeatability of the sensorial element, for three different calibrated test weights were used: 1 kg, 5 kg and 10 kg—thus also observing the (non)linearity. The measurement protocol for repeatability using a specific test weight is:

- The steps from the General methodology are conducted up until the conduction of the measurement protocol.
- On the measurement file, one research team member notes the evaluated sensorial element's unique identifier, the environmental conditions (temperature and humidity), and the value of the calibrated test weight used during the test.
- Another research team member places the calibrated test weight on the ground contact element of sensorial element (placed upwards, in this stage), assuring that the whole weight is supported only by the sensorial element.
- The weight is maintained on the sensorial element for 5 s.
- The research team member removes the weight from the sensorial element.
- The previous three steps are repeated ten times, respecting the repeatability conditions—within a short time interval and by the same research team member, without any recalibration or reinitialization of the data acquisition system. Each repetition is carried out as a standalone measurement—another research team member monitors the behaviour of the operator team member to be consistent and constant. The measurement is discarded if the monitoring team member observes deviations from those principles.

Following the data integrity check, the resulted dataset is subjected to validation and verification to find and eliminate possible random errors.

2.2.3. Dataset Processing for Repeatability

Since we expect to observe some oscillations of the values on the recorded datasets, we will determine the mean value for each repetition. In the computation of the mean value, we will consider only the values through stable measurements. Ten average values will be obtained for each repeatability dataset. Figure 9 shows an example of 5 recorded values. One can observe the loading phase of the sensorial element (noted with 1 on the figure), the stable measurement (noted 2) and the unloading stage (noted 3). The loading phase takes longer because the operator team member acts cautiously, not heavily, to hit the sensorial element. Unloading does not require such precautions.

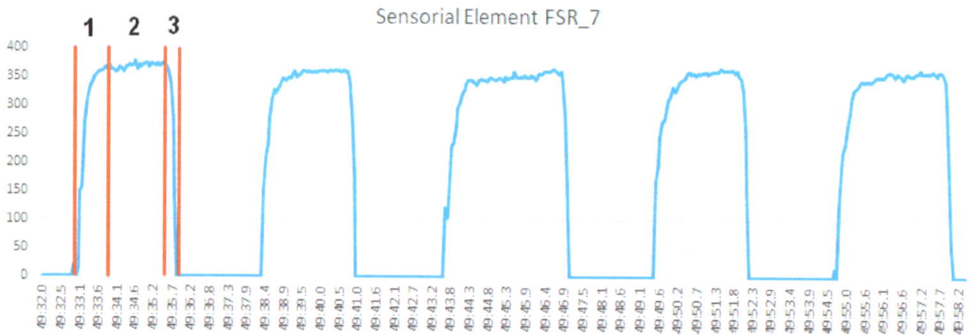

Figure 9. Sample from a repeatability dataset for FSR_7 sensor, using a 10 kg weight.

To conclude the sensorial element repeatability for a certain weight, we will compute the maximum recorded value, the minimum recorded value, the average and the standard deviation using the ten values, resulting in stable measurements. For example, suppose one sensorial element will exhibit a standard deviation higher than 10% of the mean value for any three recorded datasets (1 kg, 5 kg, 10 kg). In that case, the sensor will be rejected and replaced.

2.2.4. Homogeneity Test Measurement

After determining the repeatability of each sensor, which guarantees its stable behaviour, the research team will compare the results obtained for each sensorial element. If there will be differences among the mean values provided by different sensorial elements for the same loading weight, then we will determine the highest mean value among the eight sensorial elements.

For each of the other sensing elements, a gain factor will be determined as a proportional ratio between the highest determined mean value and the mean value for the current sensorial element.

The computed gain factor will be applied for all the future measurements.

Thus, all the sensorial element outputs will be normalized concerning the most responsive sensor.

2.3. Methodology for Measuring the Gait of Healthy Persons

The proposed lower limb prosthesis, on the one hand, will reproduce the gait of the person using it. On the other hand, it will be adapted and synchronized to the instantaneous gait of the person using it. To develop proper control algorithms for the lower limb prosthesis is necessary first to observe the gait of a regular person. To conduct those measurements, the following data recording methodology was established:

- The steps from the general methodology are completed up until the conduction of the measurement protocol.
- The measurement protocol is presented and explained to the subject.
- The subject is shod with the sandal soles.
- One research team member notes the subject's assigned ID (see Section 2.4).
- The subject is asked to walk several steps, gaiting as naturally as he can.
- The resulted dataset is subjected to a process of validation and verification to find and eliminate possible random errors.

2.4. Ethical Considerations

Since the measurements which will be presented in this paper involved human subjects, the proposed methodologies were designed, and the measures were conducted considering the applicable statements from the Ethical Principles for Medical Research Involving Human Subjects—Declaration of Helsinki developed by The World Medical Association (WMA), and all the applicable national legislation–Law 95/2006 regarding the health care system reform, Law 43/2003 regarding the patient rights, Ministry of Health Order No. 1502/2016, national guidelines for ethical principles in medical research on human subjects. Before conducting the measurements, the human subjects were informed about the purpose of the measurements, the goals it proposes, the expected duration for one measurement set, the procedures that will take place, the known risks and the possible inconveniences that it may cause, the payable benefits, the anonymization procedure of the data gathered during the measurements, how the recorded data it may be used it for research activities and the scientific research publication of the research results. All the human subjects were informed, and they understood and accepted that participation was voluntary, without being paid. Consent forms were presented to the participants, containing all the information described above. The research team ensured that the participants understood all the above information before signing the consent forms. Because the sensory system was designed to be worn over ordinary shoes and clothing without the need for medical drugs or invasive physical items, no requirements from the Ethics Commission for research involving human subjects were violated. Data anonymization is done using an alphanumeric code (ID) for each subject's dataset.

3. Results

Based on the methodologies presented in Section 2.3, the research team conducted the according to measurements. The obtained results are presented as follows.

3.1. Repeatability Measurements

All the sensorial elements were tested according to the methodology presented in Section 2.2.2 for repeatability. The research team conducted three repeatability tests—one repeatability test for each of the test weights—1 kg, 5 kg and 10 kg, and in the first stage, eight different sensorial elements were tested. The results are presented in Table 1.

Table 1. Repeatability test results.

Test Weight	Sensorial Element ID	Value 1	Value 2	Value 3	Value 4	Value 5	Value 6	Value 7	Value 8	Value 9	Value 10	Avg	SD (%)
1 kg	FSR_0	110	140	141	132	108	103	106	138	124	116	121.8	12.39
	FSR_1	119	118	96	121	73	64	100	59	110	84	94.4	24.46
	FSR_2	41	53	74	90	63	76	49	59	63	70	63.8	21.94
	FSR_3	102	90	99	101	98	99	108	94	102	91	98.4	5.56
	FSR_4	38	37	26	30	38	31	34	37	31	32	33.4	11.97
	FSR_5	45	52	43	57	37	52	50	52	44	48	48	12.10
	FSR_6	74	66	79	56	63	64	82	55	82	73	69.4	14.40
	FSR_7	102	67	98	97	73	50	79	68	65	51	75	25.40
5 kg	FSR_0	154	170	174	187	190	157	188	167	144	146	167	10.17
	FSR_1	204	186	157	147	165	197	189	199	193	228	186.5	12.93
	FSR_2	211	201	197	217	228	222	221	214	234	207	215.2	5.40
	FSR_3	200	210	204	196	201	181	198	197	214	193	199.4	4.56
	FSR_4	180	189	192	170	187	190	172	191	183	187	184.1	4.23
	FSR_5	55	60	52	72	60	60	68	56	61	61	60.5	9.73
	FSR_6	213	192	186	194	191	194	216	220	192	217	201.5	6.55
	FSR_7	214	222	227	219	207	220	196	222	212	211	215	4.20
10 kg	FSR_0	308	271	283	277	295	295	315	307	294	264	292	5.60
	FSR_1	287	248	368	317	366	437	345	351	345	270	331.2	15.91
	FSR_2	202	209	215	202	204	210	210	209	230	211	209.8	3.71
	FSR_3	275	336	260	267	284	279	322	316	340	293	298.5	9.36
	FSR_4	267	226	239	232	223	229	217	222	221	226	230.1	5.91
	FSR_5	144	136	114	147	175	105	116	102	101	100	123.6	19.51
	FSR_6	185	187	202	209	215	211	216	224	253	290	219.2	13.60
	FSR_7	325	316	317	323	310	311	312	311	294	304	312	2.73

The maximum load of the sensors used in the sensorial element is, according to their datasheet, 50 kg. The datasets were recorded as digital values, where 0 represents the absence of the weight load on the sensor, and the maximum digital value is 1024. Value 1 to Value 10 represents the mean value recorded during the stable measurement phase, presented in Section 2.2.2, Figure 10. Therefore, we considered it more beneficial to indicate the standard deviation as a percentage related to its average value computed over the ten measurements, under the SD (%) column.

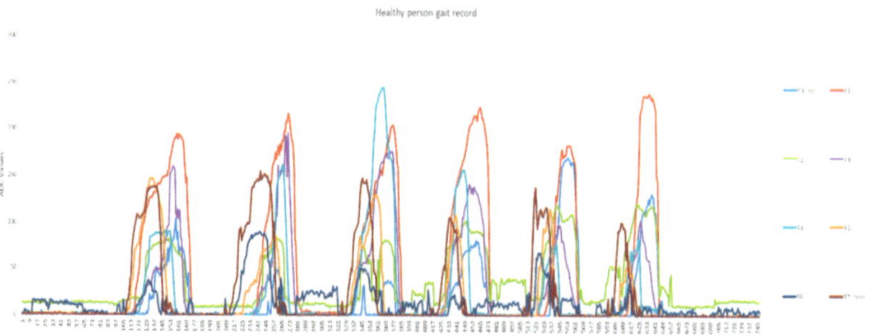

Figure 10. The gait record for a healthy person's left leg.

We analyzed the data from Table 1 following the criteria enounced in Section 2.2.2, especially the data obtained for measurements conducted for the 10 kg test weights. We noted that the FSR_1, FSR_5 and FSR_6 exhibit a standard deviation (SD%) higher than 10% of the mean value. Therefore, we concluded that the sensors used on the sensorial elements cannot provide reliable, repeatable data. Based on this analysis, the research team decided to replace them with similar sensors, which passed a similar repeatability test. Another conclusion regards the readings for the 1 kg load. Since the maximum load for a sensor is 50 kg, and 1 kg represents 2% of its full load, one can observe that the standard deviation exhibits high percentage values; thus, the repeatability is low. However, since such low loads appear only in short, transitory moments between gait phases, we can consider them acceptable.

Also, one can note that the output response of the sensors is powerful nonlinear. Thus, in future work, we will consider using more test weight values to determine the nonlinear characteristic.

3.2. Homogeneity Results

Another finding after concluding the repeatability tests is that the sensors' output response isn't homogenous—different sensorial elements offer different output values for the same load—still providing good repeatability for that output. Thus, the research team determined a proportional ratio between the highest value offered by one of the sensorial elements FSR_0 ... FSR_7 and the value for a specific sensorial element. Finally, the computed gain factors are applied for all the following presented measurements.

3.3. Records of the Gait of Healthy Persons

After the data validation for the sensorial system was concluded, the research team recorded the gait of healthy persons according to the methodology described in Section 2.4. The records were conducted for 10 healthy persons, chosen among the projects research team members We were looking to have representatives of both genders, covering as wide a range of age, weight and footwear size as possible. The measurements were conducted for 4 females and 6 males. We tested EU footwear sizes of 35, 36, 37 (for females) and 41, 42, 43 and 44 (for males). The body weight varied between 46 to 57 kg (for females) and 70 to 85 kg (for males). The height varied between 1.65m to 1.75m (for females) and 1.71 to 1.84 (for males). All the ten participants were evaluated, from an orthopedic point of view, by the specialist doctor also member the project—none exhibited orthopedic defficiencies. Also, their gait was evaluated.

Then, we analyzed the pressure distribution evolution over time, as revealed by the sensorial elements mounted on the sandal soles—Figure 11. Figure 11 represents the weight distribution on the sole, in the stepping cycle, for one foot: 1—foot in balance-contact with the heel, 2—contact with the heel plus the middle part of the sole, 3-4-5—firm contact with the sole, support on one foot, the other foot in the balance, 6—lifting the heel off the ground, 7—support only on the front of the sole, 8—support on the toe (toes), lifting the foot off the ground and entering the balance.

Each of the individual sensorial elements can sustain a load of up to 50 kg. The dataset illustrated in Figure 11. is recorded for a person weighing approx. 60 kg. Thus, the loading on each sensor does not reach high values, as one can observe.

The recorded data, displayed in Figures 10 and 11, are consistent with the plantar pressure measurements recorded using other traditional plantar measurement devices [11,41].

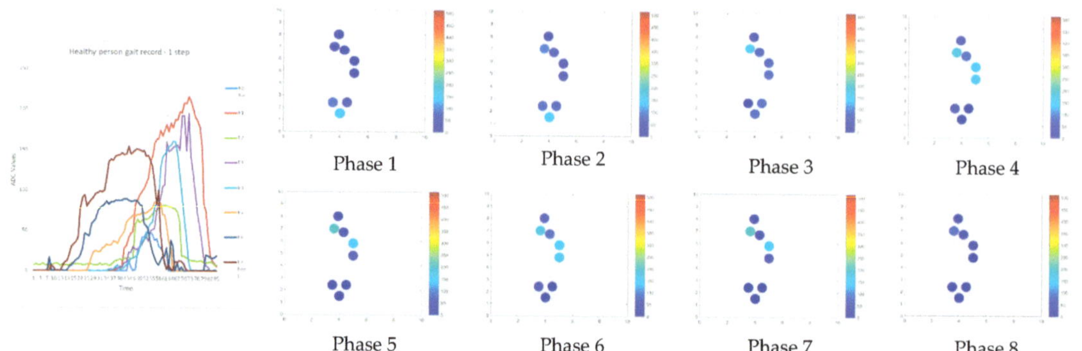

Phase 1 Phase 2 Phase 3 Phase 4

Phase 5 Phase 6 Phase 7 Phase 8

Figure 11. The eight-step phases are recorded data for one step, healthy person gait, right foot, and loading on the sensorial elements.

4. Discussion

The design and development of functional prostheses with limited capabilities (passive or quasi-passive protections) were considered. Then prostheses with more advanced features (semi-active or active) appeared, allowing the control of gait adaptable to different types of terrain with minimal difficulty.

There are several theoretical models proposed to be applied to the ankle-foot prosthesis. For example, a quasi-passive ankle prosthesis was designed by Collins and Kuo [42]. This prosthesis recycles energy when touched with the heel and returns it to the user in the terminal phase of the step. Another example of a prosthetic foot was developed by Grabowski et al. [43]. Although it had some durability limitations, this prosthesis could emulate the torques produced by the human ankle through cables, urethane springs, and fiberglass elements. However, it also had some durability limitations. Finally, Hansen and Nickel [44] proposed a passive ankle-foot prosthesis that operated in two states: standing and walking, by incorporating a locking mechanism to separate these phases.

Subsequently, the main challenge in developing ankle-foot prostheses was implementing active protections. In addition, these prostheses require considerable power and precise control of the impedance torque. Hitt et al. [45] presented a model of the robotic transtibial prosthesis called SPARKY to reduce the power demand. They used biomechanical energy regeneration technology to reduce energy consumption by combining spring and a mechanical tendon actuator. In 2012, Herr and Grabowski presented an ankle-foot prosthesis that closely mimicked the function of the human ankle and provided a net-positive power while walking. It is powered by a 200 W brushless DC motor and a series ball screw with a carbon composite lamellar spring [46]. Bergelin and Voglewede propose an active prosthesis using a four-bar mechanism and offering a range of motion similar to that of the human ankle, according to the results obtained from the simulations [47]. Finally, Cherelle et al. have a prosthesis that stores energy in the springs during the complete position phase, then released on push, and can provide all the power needed for forwarding propulsion with a low-power actuator [48].

There is currently a wide range of smart foot prostheses in which various sensors and electromechanical devices are installed to reproduce a more biological aspect of the human foot. In the case of prostheses, we should have a smart sole installed on the prosthetic foot so that it accurately captures real-time contact information through plantar pressure measurements. There are a variety of plantar pressure measuring systems, which can generally be classified into platform systems and shoe systems.

The advantages and limitations of foot pressure equipment were some of the health concerns of the subjects tested. The number of sensors for sufficient coverage of the foot regions is the fundamental condition to opt either for the plantar platform or the footwear system. In addition, a higher frequency sampling rate of the system can provide more accurate results. Variations in the morphology of the human foot depending on sex, age, and body mass must also be considered. Problems with foot pressure sensitivity, external load impact, and duration may occur during practical training. In most cases, qualitative data must be systematically collected through an appropriate experimental design when subjects perform a normal gait, considering those control variables, such as standard gait time (gait cycle) and cadence. Potential future directions of exploration would include different types of surface roughness to varying angles of inclination that affect the level of sensitivity to foot pressure.

Among the systems for measuring plantar pressure for gait analysis, the plantar insoles are widely used and cause many locomotion disorders. Although the soles of the feet provide information only in the positional phase, they are sufficient for most gait rehabilitation systems, posture control and balance [41]. For example, Razak et al. [49] specified in their study that the location of the sensors was determined, taking into account only the dominant pressure points of the plantar surface during walking. Most insoles obtain the necessary information from a discrete number of sensors (the minimum number varies from 2 to 16 or more, depending on the need).

5. Conclusions

This work presented a novel low-cost lower limb prosthesis based on sensors placed in a plantar insole. The proposed platform has the following advantages:

- It can be used for any type or size of foot.
- It is designed modularly, allowing most components to be reused when reconfiguring for another foot size.
- The soleplate allows you to reconfigure the number of sensors and their positioning.
- The sole allows you to adjust the position of the straps on the sandals.
- Clamping systems (sandals and gyroscopic sensors) allow an optimal adjustment.
- The system can be extended in terms of pressure and gyroscopic sensors.
- The purchasing system is light and similar to a regular pair of shoes, limiting errors as it does not disturb the gait.
- Data acquisition accuracy far exceeds the needs of such a system.
- The simplified and optimized system can be integrated into a prosthesis, constantly coordinating its configuration depending on the step phase and the type of movement.
- The cost of the data acquisition system is around 200 Euro, which can thus be considered low cost and high-efficiency.

6. Patents

Patent application—A/00213/2019: Spherical joint based on intelligent fluids [39].

Author Contributions: Conceptualization, C.F.P., L.F.M. and I.C.V.; methodology, L.F.M.; validation, L.F.M., I.C.V. and Ș.I.C.; formal analysis, C.F.P., L.F.M. and I.C.V.; investigation, C.F.P., L.F.M., I.C.V. and D.C.; resources, C.F.P., F.L.P. and Ș.I.C.; data curation, C.F.P., L.F.M., Ș.I.C. and F.L.P.; writing—original draft preparation, C.F.P., L.F.M. and I.C.V.; writing—review and editing, C.F.P., L.F.M. and I.C.V.; supervision, C.F.P., L.F.M., I.C.V., D.C. and N.B.; project administration, C.F.P., L.F.M. and I.C.V.; funding acquisition, N.B. All authors have read and agreed to the published version of the manuscript.

Funding: This research was funded by PNIII scheme, Program 2: Increasing the competitiveness of the Romanian economy through RDI, Subprogram 2.1: Competitiveness through research, development and innovation—Experimental project—demonstration, grant number 344PED/2020 and POC-Competitiveness Operational Program.

Institutional Review Board Statement: The study was conducted according to the guidelines of the Declaration of Helsinki. Ethical review and approval were waived for this study due to all the measurements being performed using non-invasive, contactless devices and no administration of medical drugs were required in the process. To assure that all the ethical principles were respected during our research work, we contracted an external consultant, RehabMed, which assisted us during all the research phases.

Informed Consent Statement: Informed consent was obtained from all subjects involved in the study.

Data Availability Statement: Not applicable.

Acknowledgments: This work was supported by a grant of the Romanian Ministry of Education and Research, CCCDI—UEFISCDI, project number PN-III-P2-2.1-PED-2019-0937, within PNCDI III;HUB-UCv-Support Center for International RD Projects for the Oltenia region-cod SMIS 107885.

Conflicts of Interest: The authors declare no conflict of interest.

References

1. McDonald, C.L.; Westcott-McCoy, S.; Weaver, M.R.; Haagsma, J.; Kartin, D. Global prevalence of traumatic non-fatal limb amputation. *Prosthet. Orthot. Int.* **2021**, *45*, 105–114. [CrossRef] [PubMed]
2. Hughes, W.; Goodall, R.; Salciccioli, J.D.; Marshall, D.C.; Davies, A.H.; Shalhoub, J. Trends in Lower Extremity Amputation Incidence in European Union 15+ Countries 1990–2017, Editor's Choice. *Eur. J. Vasc. Endovasc. Surg.* **2020**, *60*, 602–612. [CrossRef] [PubMed]
3. Walter, N.; Alt, V.; Rupp, M. Lower Limb Amputation Rates in Germany. *Medicina* **2022**, *58*, 101. [CrossRef] [PubMed]
4. Laferrier, J.Z.; Groff, A.; Hale, S.; Sprunger, N.A. A Review of Commonly Used Prosthetic Feet for Developing Countries: A Call for Research and Development. *J. Nov. Physiother.* **2018**, *8*, 380. [CrossRef]
5. Versluys, R.; Beyl, P.; van Damme, M.; Desomer, A.; van Ham, R.; Lefeber, D. Prosthetic feet: State-of-the-art review and the importance of mimicking human ankle-foot biomechanics. *Disabil. Rehabil. Assist. Technol.* **2009**, *4*, 65–75. [CrossRef]
6. Voloshina, A.S.; Collins, S. A Review of Design and Control Approaches in Lower-Limb Prosthetic Devices. *Biology* **2019**. Available online: https://www.semanticscholar.org/paper/A-Review-of-Design-and-Control-Approaches-in-Voloshina-Collins/90dc8 5af676c1c01301d941a2d0b8de4e1d9c9a1 (accessed on 8 January 2022).
7. Schmalz, T.; Blumentritt, S.; Jarasch, R. Energy expenditure and biomechanical characteristics of lower limb amputee gait: The influence of prosthetic alignment and different prosthetic components. *Gait Posture* **2002**, *16*, 255–263. [CrossRef]
8. Mattes, S.; Martin, P.; Royer, T. Walking symmetry and energy cost in persons with unilateral transtibial amputations: Matching prosthetic and intact limb inertial properties. *Arch. Phys. Med.* **2000**, *81*, 561–568. [CrossRef]
9. Tesio, L.; Lanzi, D.; Detrembleur, C. The 3-D motion of the centre of gravity of the human body during level walking. II. Lower limb amputees. *Clin. Biomech.* **1998**, *13*, 83–90. [CrossRef]
10. Klute, G.K.; Czernieki, J.; Hannaford, B. Development of powered prosthetic lower limb. In Proceedings of the 1st Natl Meeting, Veterans Affairs Rehabil. R&D Service, Washington, DC, USA, 1–3 October 1998.
11. Chen, L.; Wang, C.; Wang, J.; Song, X. Design of Hybrid Phase Sliding Mode Control Scheme for Lower Extremity Exoskeleton. *Appl. Sci.* **2019**, *9*, 3754. [CrossRef]
12. Vickers, D.R.; Palk, C.; McIntosh, A.S.; Beatty, K.T. Elderly uni-lateral transtibial amputee gait on an inclined walkway: A biomechanical analysis. *Gait Posture* **2008**, *27*, 518–529. [CrossRef] [PubMed]
13. Torburn, L.; Schweiger, G.P.; Perry, J.; Powers, C.M. Below-knee amputee gait in stair ambulation. *Clin. Orthop. Relat. Res.* **1994**, *303*, 185–192. [CrossRef]
14. Au, S.K.; Bonato, P.; Herr, H. An EMG-position controlled system for an active ankle-foot prosthesis: An initial experimental study. In Proceedings of the 9th International Conference on Rehabilitation Robotics, Chicago, IL, USA, 28 June–1 July 2005; pp. 375–379. [CrossRef]
15. Au, S.K.; Herr, H. Initial experimental study on dynamic interaction between an amputee and a powered ankle-foot prosthesis. In Proceedings of the Workshop on Dynamic Walking: Mechanics and Control of Human and Robot Locomotion, Ann Arbor, MI, USA, 6–8 May 2006.
16. Hitt, J.; Bellman, R.; Holgate, M.; Sugar, T.; Hollander, K. The sparky (spring ankle with regenerative kinetics) projects: Design and analysis of a robotic transtibial prosthesis with regenerative kinetics. In Proceedings of the ASME International Design Engineering Technical Conferences and Computers and Information in Engineering Conference, DETC2007, Las Vegas, NV, USA, 4–7 September 2007; pp. 1587–1598.
17. Au, S.K.; Herr, H.M. Powered ankle-foot prosthesis: The importance of series and parallel motor elasticity. *IEEE Robot. Autom. Mag.* **2008**, *15*, 52–59. [CrossRef]
18. Zhao, S.; Lei, Y.; Wang, Z.; Zhang, J.; Liu, J.; Zheng, P.; Gong, Z.; Sun, Y. Biomimetic Artificial Joints Based on Multi-Material Pneumatic Actuators Developed for Soft Robotic Finger Application. *Micromachines* **2021**, *12*, 1593. [CrossRef] [PubMed]
19. Tabucol, J.; Brugo, T.M.; Povolo, M.; Leopaldi, M.; Oddsson, M.; Carloni, R.; Zucchelli, A. Structural FEA-Based Design and Functionality Verification Methodology of Energy-Storing-and-Releasing Prosthetic Feet. *Appl. Sci.* **2022**, *12*, 97. [CrossRef]

20. Russo, M.; Chaparro-Rico, B.D.M.; Pavone, L.; Pasqua, G.; Cafolla, D. A Bioinspired Humanoid Foot Mechanism. *Appl. Sci.* **2021**, *11*, 1686. [CrossRef]
21. Fridman, A.; Ona, I.; Isakov, E. The influence of prosthetic foot alignment on trans-tibial amputee gait. *Prosthet. Orthot. Int.* **2003**, *27*, 17–22. [CrossRef]
22. Lemaire, E.D.; Samadi, R.; Goudreau, L.; Kofman, J. Mechanical and biomechanical analysis of a linear piston design for angular-velocity-based orthotic control. *J. Rehabil. Res. Dev.* **2013**, *50*, 43–52. [CrossRef]
23. Sewell, P.; Noroozi, S.; Vinney, J.; Andrews, S. Developments in the trans-tibial prosthetic socket fitting process: A review of past and present research. *Prosthet. Orthot. Int.* **2000**, *24*, 97–107. [CrossRef]
24. Hernández-Acosta, M.A.; Torres-San Miguel, C.R.; Piña-Díaz, A.J.; Paredes-Rojas, J.C.; Aguilar-Peréz, L.A.; Urriolagoitia-Sosa, G. Numerical Study of a Customized Transtibial Prosthesis Based on an Analytical Design under a Flex-Foot® Variflex® Architecture. *Appl. Sci.* **2020**, *10*, 4275. [CrossRef]
25. Faragó, P.; Grama, L.; Farago, M.-A.; Hintea, S. A Novel Wearable Foot and Ankle Monitoring System for the Assessment of Gait Biomechanics. *Appl. Sci.* **2021**, *11*, 268. [CrossRef]
26. Amaro, C.M.; Castro, M.A.; Roseiro, L.; Neto, M.A.; Amaro, A.M. Plantar Pressure Evaluation during the Season in Five Basketball Movements. *Appl. Sci.* **2020**, *10*, 8691. [CrossRef]
27. Deffenbaugh, B.W.; Herr, H.; Pratt, G.A.; Wittig, M.B. Electronically Controlled Prosthetic Knee. U.S. Patent WO2001054630A1, 2 August 2001. Available online: https://patents.google.com/patent/WO2001054630A1 (accessed on 12 January 2022).
28. Available online: https://www.ossur.com/en-gb (accessed on 12 January 2022).
29. Herr, H.; Wilkenfeld, A. User-adaptive control of a magnetorheological prosthetic knee. *Ind. Robot Int. J.* **2003**, *30*, 42–55. [CrossRef]
30. Thorarinsson, E.T.; Jonsdottir, F.; Palsson, H. Design of a Magnetorheological Prosthetic Knee. In Proceedings of the NordDesign 2006, Reykjavik, Iceland, 16–18 August 2006. Available online: https://www.semanticscholar.org/paper/Design-of-a-Magnetorheological-Prosthetic-Knee-Thorarinsson-Jonsdottir/2fb5d016910572203a45070b04e308c8033df2d1 (accessed on 12 January 2022).
31. Herr, H.M.; Weber, J.A.; Casler, R.J., Jr. Biomimetic Joint Actuators. U.S. Patent WO2012125562A1, 20 September 2012. Available online: https://patents.google.com/patent/WO2012125562A1/en (accessed on 12 January 2022).
32. de Andrade, R.M.; Martins, J.S.R.; Pinotti, M.; Filho, A.B.; Vimieiro, C.B.S. Novel active magnetorheological knee prosthesis presents low energy consumption during ground walking. *J. Intell. Mater. Syst. Struct.* **2021**, *32*, 1591–1603. [CrossRef]
33. Narang, Y.S.; Arelekatti, V.N.; Winter, A.G. The Effects of Prosthesis Inertial Properties on Prosthetic Knee Moment and Hip Energetics Required to Achieve Able-Bodied Kinematics *IEEE Trans. Neural Syst. Rehabil. Eng.* **2016**, *24*, 754–763. [CrossRef]
34. Murthy Arelekatti, V.N.; Winter, A.G. Design and Preliminary Field Validation of a Fully Passive Prosthetic Knee Mechanism for Users with Transfemoral Amputation in India. *ASME J. Mech. Robot.* **2018**, *10*, 031007. [CrossRef]
35. Pinto Maquilon, J.K.; Velasquez, M.C. Durability of lower limb prostheses with low-cost exoskeletal technology, in patients with lower limb amputatispeta to any cause. *Ann. Phys. Rehabil. Med.* **2018**, *61*, e470–e471. [CrossRef]
36. Vladu, I.C.; Pană, C.F.; Copiluşi, C.; Petrişor, I.; Vladu, I.; Bîzdoacă, N.G. *Controllable Passive Transtibial Prosthesis—Design, Book Mechanisms and Machine Science, Advances in Service and Industrial Robotics, Mechanisms and Machine Science*; Springer: Berlin/Heidelberg, Germany, 2021; Volume 102, pp. 176–183, ISSN1 2211-0984/ISSN2 2211-0992 (electronic). ISBN1 978-3-030-75258-3/ISBN2 978-3-030-75259-0 (eBook), 9-0 (eBook).
37. Powered Lower-Limb Prostheses. Available online: http://research.vuse.vanderbilt.edu/cim/research_leg.html (accessed on 10 January 2022).
38. Willick, S.E. The athlete with limb deficiency, Paralympic Athletes Targeted Topic, ASPETAR. Sports Medicine Journal. Available online: https://www.aspetar.com/journal/viewarticle.aspx?id=432#.Ynx8TuhBxPY (accessed on 10 January 2022).
39. Cristian, I.V.; Cristina, F.P.; Viorel, S.; Daniela, M.P.-P.; Ileana, V.; Dan Cristian, G.; Daniela, T.; Nicu, G.B. Spherical Joint Based on Intelligent Fluids. Patent Application A/00213/2019, 4 May 2019.
40. Lorkowski, J.; Gawronska, K.; Pokorski, M. Pedobarography: A Review on Methods and Practical Use in Foot Disorders. *Appl. Sci.* **2021**, *11*, 11020. [CrossRef]
41. Wafai, L.; Zayegh, A.; Woulfe, J.; Aziz, S.M.; Begg, R. Identification of Foot Pathologies Based on Plantar Pressure Asymmetry. *Sensors* **2015**, *15*, 20392–20408. [CrossRef]
42. Collins, S.H.; Ku, A.D. Recycling energy to restore impaired ankle function during human walking. *PLoS ONE* **2010**, *5*, e9307. [CrossRef]
43. Grabowski, A.M.; Rifkin, J.; Kra, R. K3 promoter prosthetic foot reduces the metabolic cost of walking for unilateral transtibial amputees. *J. Prosthet. Orthot.* **2010**, *22*, 113–120. [CrossRef]
44. Hansen, A.H.; Nicke, E.A. Development of a bimodal ankle-foot prosthesis for walking and standing/swaying *J. Med. Devices* **2013**, *7*, 035001. [CrossRef]
45. Hitt, J.K.; Sugar, T.G.; Holgate, M. An active foot-ankle prosthesis with biomechanical energy regeneration. *J. Med. Devices* **2010**, *1*, 011003. [CrossRef]
46. Herr, H.M.; Grabowski, A.M. Bionic ankle–foot prosthesis normalizes walking gait for persons with leg amputation. *Proc. R. Soc. Lond. B Biol. Sci* **2012**, *279*, 457–464. [CrossRef]

47. Bergelin, B.J.; Voglewede, P.A. Design of an active ankle-foot prosthesis utilizing a four-bar mechanism. *J. Mech. Des.* **2012**, *134*, 061004. [CrossRef]
48. Cherelle, P.; Grosu, V.; Matthys, A. Design and validation of the ankle mimicking prosthetic (AMP-) foot 2.0. *IEEE Trans. Neural Syst. Rehabilit. Eng.* **2014**, *22*, 138–148. [CrossRef]
49. Abdul Razak, A.H.; Zayegh, A.; Begg, R.K.; Wahab, Y. Foot plantar pressure measurement system: A review. *Sensors* **2012**, *12*, 9884–9912. [CrossRef]

Article

Analysis of Dynamic Behavior of ParReEx Robot Used in Upper Limb Rehabilitation

Daniela Tarnita [1], Ionut Daniel Geonea [1,*], Doina Pisla [2,*], Giuseppe Carbone [3], Bogdan Gherman [2], Nicoleta Tohanean [4], Paul Tucan [2], Cristian Abrudan [4] and Danut Nicolae Tarnita [5]

[1] Faculty of Mechanics, University of Craiova, 200512 Craiova, Romania
[2] Research Center for Industrial Robots Simulation and Testing CESTER, Technical University of Cluj-Napoca, 400641 Cluj-Napoca, Romania
[3] Department of Mechanical, Energy and Management Engineering DIMEG, University of Calabria, 87036 Cosenza, Italy
[4] Department of Neurology, University of Medicine and Pharmacy "Iuliu Hatieganu", 400012 Cluj-Napoca, Romania
[5] Emergency Hospital of Dolj County, 200528 Craiova, Romania
*   Correspondence: ionut.geonea@edu.ucv.ro (I.D.G.); doina.pisla@mep.utcluj.ro (D.P.)

Citation: Tarnita, D.; Geonea, I.D.; Pisla, D.; Carbone, G.; Gherman, B.; Tohanean, N.; Tucan, P.; Abrudan, C.; Tarnita, D.N. Analysis of Dynamic Behavior of ParReEx Robot Used in Upper Limb Rehabilitation. *Appl. Sci.* **2022**, *12*, 7907. https://doi.org/10.3390/app12157907

Academic Editors: Alessandro Gasparetto and Dimitris Mourtzis

Received: 22 June 2022
Accepted: 3 August 2022
Published: 7 August 2022

Publisher's Note: MDPI stays neutral with regard to jurisdictional claims in published maps and institutional affiliations.

Abstract: This paper presents a dynamic analysis of the ParReEx multibody mechanism, which has been designed for human wrist joint rehabilitation. The starting point of the research is a virtual prototype of the ParReEx multibody mechanism. This model is used to simulate the dynamics of the multibody mechanism using ADAMS in three simulation scenarios: (a) rigid kinematic elements without friction in joints, (b) rigid kinematic elements with friction in joints, and (c) kinematic elements as deformable solids with friction in joints. In all three cases, the robot is used by a virtual patient in the form of a mannequin. Results such as the connecting forces in the kinematic joints and the torques necessary to operate the ParReEx robot modules are obtained by dynamic simulation in MSC.ADAMS. The torques obtained by numerical simulation are compared with those obtained experimentally. Finite element structural optimization (FEA) of the flexion/extension multibody mechanism module is performed. The results demonstrate the operational safety of the ParReEx multibody mechanism, which is structurally capable of supporting the external loads to which it is subjected.

Keywords: dynamic simulation; spherical parallel robot; joint friction; flexible links; rehabilitation; wrist joint

1. Introduction

Although there has been an increasing emphasis on prevention in recent decades, strokes are one of the leading causes of permanent disability worldwide [1]. In recent decades, strokes have led to the disability and even death of a very large number of people; in recent years, the number of people affected annually is about 13 million [2].

Globally, the leading causes of death are ischemic heart disease, with an incidence of 14.8%, and stroke, with an incidence of 11.8% [3].

Recent studies show that the number of stroke cases is increasing. Taking into account the aging of the population worldwide, there are advanced fears regarding the increasingly limited capacity of European and global medical systems to meet patient care requirements [4]. It is estimated that in the near future, strokes will affect an upper limb in over 85% of cases [5], and the success rate for arm rehabilitation will be only 10% [6] due to the complexity of upper limb movement [7]. In less than 10 years, the annual medical cost per stroke will increase rapidly; for example, in the USA, it will be 2.5 times higher than it was in 2012 [8]. Currently, in developed countries, 3–4% of medical costs are allocated to strokes [9]. Physical rehabilitation is becoming less and less effective as each patient has to work with a physical therapist [10], but the use of robotic exoskeletons allows for more effective therapy through repetitive movements and adaptive training [11,12].

The use of rehabilitation technology and robotics optimizes the results, minimizes costs by conserving labour and maximizing progressive repetitions, and improves the outcome compared to the high costs of conventional therapy performed as one-on-one training with a professional therapist [13–15].

Numerous articles have been dedicated to the study of robotic structures used for upper limb rehabilitation. Robotic systems for upper limb rehabilitation are the subject of much research. Thus, the design and performance evaluation of the RUPERT exoskeleton for upper limbs is presented in [16]. An innovative robotic system with a spherical architecture is developed and presented in [17]. A parallel robotic system based on a spherical architecture, ASPIRE, was designed for three degrees-of-freedom (DOFs) shoulder rehabilitation, as described in this study. In [18], a study of torque monitoring is performed for the ASPIRE robotic system. Clinical trials based on the use of ASPIRE on post-stroke patients are presented. Further, design optimization is studied based on therapist feedback following clinical evaluation. A study of the ASPIRE robot for upper limb rehabilitation is presented in [19], where, with the help of ADAMS software, the dynamic analysis of the robotic system is performed. The study of hand joint movements is also presented in [20]. Classifications and analyses of these structures can be found in [21].

Over the past decade, many robotic devices have been designed to enhance the motor rehabilitation process. In [22–25], the authors noted an improvement in quality of life and concluded that robot-assisted therapy is beneficial for recovering the motor functions of the upper limbs in patients with early-stage stroke. Neurorehabilitation therapy including task-oriented training with the ARMin exoskeleton robot can enhance improvement of motor function in a chronically impaired paretic arm after stroke more effectively than conventional therapy [23].

A hybrid exoskeleton named EFW Exo II is developed for the motor function rehabilitation of elbow, forearm and wrist, and the ranges of motion meet the rehabilitation demands of daily living [24]. For clinical trials, an acquisition data system based on wearable sensors was used; due to their advantages, these systems are often used in evaluating biomechanical human data [25–31].

Several robotic devices have been developed to rehabilitate the wrist joint, which, after the shoulder, is the second-most complex joint of the human upper limb. The authors of [32] review the main wrist rehabilitation robots developed recently in terms of mechanical design and aspects of control, and highlight aspects of experimental tests of these robots on patients with neurological disorders.

Krebs et al. studied a wrist rehabilitation robot that offers three DOFs [33], which, mounted on the accompanying robot, MIT-MANUS [34,35], has become a very suitable tool for the rehabilitation of the shoulders and elbow. Another wrist robot [36] with three active DOFs was developed with a similar structure and an additional passive prismatic joint. RiceWrist [37], as an electrically operated forearm–wrist exoskeleton, uses a serial RRR manipulator and cable transmissions to rehabilitate the wrist and radio–ulnar joints. Robots presented in [38,39] allow the rehabilitation of two movements: forearm supination/pronation and wrist flexion/extension, while in [40], the human wrist has been simplified to a single DOF—namely flexion/extension. A parallel wrist rehabilitation robot driven by two pneumatic actuators is developed in [41] to rehabilitate two DOFs: flexion/extension and radial/ulnar deviation.

Patient safety is a critical milestone in the development of robotic systems for medical applications to avoid any unwanted damage. In this case, it can be achieved by implementing a dynamic model in the control system [42,43] using various principles such as: the Lagrange–Euler formalism, the principle of virtual work [44] or the Newton–Euler equations [45–47].

ADAMS View, software designed for the analysis of mobile systems dynamics, with an increased accuracy of kinematic and dynamic results obtained from numerical simulations has been used in several studies of robotic systems for rehabilitation of human movements, such as exoskeletons or flexible robotic structures for minimally invasive surgery [19,27,48,49].

For structural optimization of a kinematic element, one of the most powerful software programs is ANSYS. In [19], we presented finite element analysis of a kinematic element in the ASPIRE robot structure.

The aim of this study is to analyze from a kinematic, dynamic and experimental point of view a robotic system designed for the rehabilitation of the human wrist joint of post-stroke survivors. Numerical simulations are performed for the robot mounted on the wrist of a virtual patient. For validation of the dynamic model built in ADAMS, the results obtained by numerical simulation are compared with those obtained experimentally. Three distinct situations will be studied:

- The kinematic elements are considered as rigid bodies, and the friction in the kinematic joints is not considered;
- The kinematic elements are considered as rigid bodies, and the friction in the robot joints is considered;
- The kinematic elements are considered as flexible bodies, and the friction in the robot joints is considered.

The paper is structured as follows: after the introduction, a functional constructive description of the ParReEx robot for wrist rehabilitation is presented. Dynamic analysis is presented in ADAMS for the three considered operating hypotheses. Then, structural optimization is performed for the circular guidance of the ParReEx robot. At the end, experimental results, discussions, and conclusions are presented.

2. Constructive and Functional Description of the ParReEx-Wrist Robot

The ParReEx-wrist robotic system is designed for the rehabilitation of the wrist, namely flexion and extension of the wrist and ulnar and radial deviation of the wrist [49]. The robot has two degrees of freedom (DoF), one for each type of motion (Figure 1), namely: the revolute joint q_1 to perform flexion–extension and the revolute joint q_2 to perform radial–ulnar deviation. Each joint separately actuates a kinematic chain as follows: q_1 actuates an RR chain consisting of the R_1 and R_3 revolute joints connected through link d around the OZ axis; q_2 actuates link *e* around the OY axis, fixed between two bearings. Links d and e are connected via two revolute joints, namely R_2 and R_3, required to perform the circumduction motion (which is a combination of flexion–extension and ulnar–radial deviation motions). Notations used in Figure 1 represents: q_1—flexion–extension motion; q_2—radial–ulnar deviation motion; R_1, R_2, R_3, R_4—revolute joints; e, d—mechanism links.

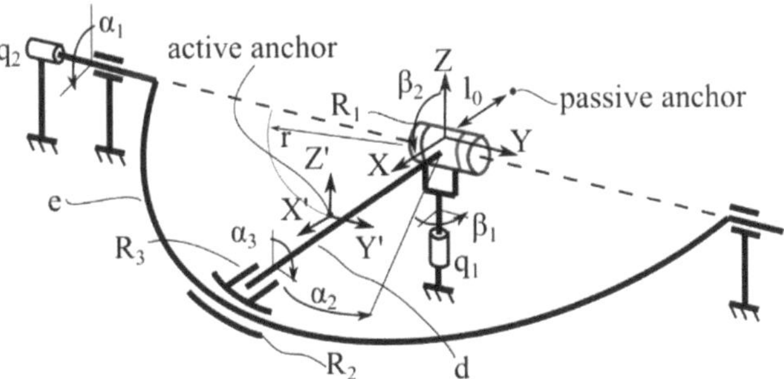

Figure 1. The ParReEx-wrist kinematic scheme [47].

The ParReEx-wrist CAD model is presented in Figure 2. Motor (1) achieves the flexion extension motion of the wrist through a set of bevel gears (3) and a custom-made circular rail and sledge guiding (4). Motor (2) performs radial–ulnar deviation using a timing belt and pulleys (5). The patient's hand is supported using the wrist support (6), which acts as active anchor, and the forearm support (7), which acts as passive anchor. Since the patient's wrist is placed above the fixed coordinate system (Figure 1), as the intersection between the flexion–extension and radial–ulnar deviation robot motion axis, an additional straight rail-and-sledge mechanism (8) is required, as well as a spherical joint (9) to perform the circumduction motion.

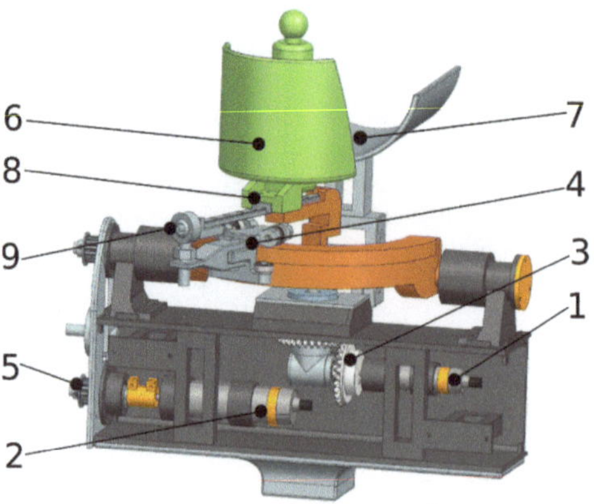

Figure 2. The ParReEx-wrist CAD model.

Notations used in Figure 2 represents: 1—motor for flexion–extension motion; 2—motor for radial–ulnar deviation motion; 3—bevel gears; 4—sledge pulleys; 5—belt transmission; 6—wrist support; 7—forearm support; 8—straight rail and sledge mechanism; 9—spherical joint.

The ParReEx-wrist experimental model is presented in Figure 3. Based on the virtual prototype of the ParReEx robot, the execution drawings for the component parts of the assembly were made. The housing of the robotic system and the component parts, such as those numbered 6, 7 and 8 in Figure 2, are manufactured by additive manufacturing. For this purpose, we used a 3D printer, which works on the fused deposit manufacturing principle. For the execution of the printed parts, we used 3D models, and, using Cura Ultimaker software, we made the G-code for 3D printing. The printer uses ABS (acrylonitrile butadiene styrene) wire, which is melted by the print head, and successive layers are deposited to achieve the shape of the parts. Thus, these parts are assembled to realize the prototype of the robotic system, as shown in Figure 3, where the robotic system is shown in the situation of its use by a patient.

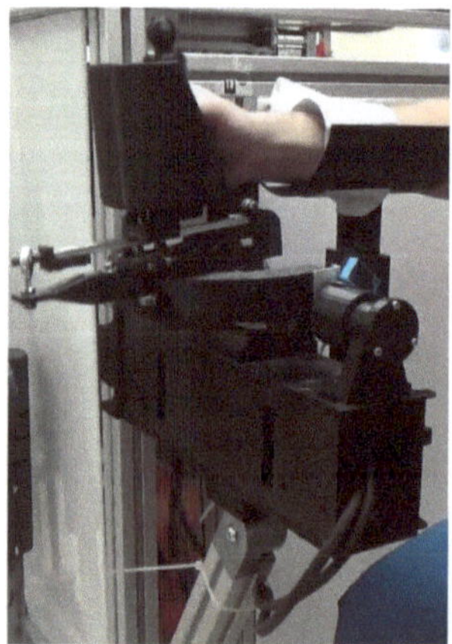

Figure 3. The ParReEx-wrist experimental model.

3. Dynamic Analysis in MSC.ADAMS of ParReEx-Wrist Robot

3.1. Dynamic Analysis in MSC.ADAMS of ParReEx-Wrist Robot, Considering the Kinematic Elements as Rigid Solids without Considering Joint Friction

Based on the structural scheme of the ParReEx robot, a virtual prototype is designed using Unigraphics NX software. The development of the virtual prototype is necessary from two aspects: it will be used to simulate the dynamic behavior of the robot, as well as for structural optimization with the finite element method. The development of the dynamic model in ADAMS involves the following steps:

✓ Definition of the materials of the robot subassemblies.

Robot element materials can be selected from the ADAMS database, or new materials can be defined by their properties. In this case, the materials of the component elements are defined in the database, and their mechanical properties are further explained. For each component element of the robot, we specify the material from which it is made. Thus, in the robot component, we have parts that are made of aluminum alloy with the following properties: density = 2470 kg/m^3, Young's Modulus = 7.17×10^{10} N/m^2 and Poisson's ratio = 0.33. The robot housing, the circular guide and the elements supporting the patient's hand are made of ABS. Acrylonitrile butadiene styrene, commonly known as ABS, is a polymer and opaque thermoplastic comprised of acrylonitrile, butadiene, and styrene. ABS is an incredibly versatile and durable plastic, making it an extremely popular material used in a variety of industries. The mechanical properties of this material are as follows: density = 1.15×10^3 kg/m^3, Young's Modulus=1.6×10^{10} N/m^2 and Poisson's ratio = 0.340.

✓ Kinematic joints related to the robot structure are defined.

In the structure of the robot, there are rotational joints and upper rolling joints. In Figure 4, the kinematic joints defined in the ADAMS dynamic model are detailed, namely:

- Joint A and Joint B are the rotational kinematic joints of the circular guide (4) with the robot housing. Joint A defines Motion 1, the law of motion for radial–ulnar deviation;
- Joint C is a rotational joint of the Hooke coupling structure;
- Joint D is a rotational joint of the bevel pinion with the system housing. This joint specifies Motion 2, the motion law for hand joint flexion–extension;
- Kinematic Joints R_1 and R_2, are superior joints that provide contact for a point on a curve. Special attention has been paid to the correct definition of the rolling joints between the circular guide and the slide. For this purpose, we have chosen to define a point-on-curve upper joint. In this case, for Joint R_1 we have specified the contact between the arc at the top of the guide and a point belonging to the slide. For Joint R_2, we have specified the contact between the arc of the circle at the bottom of the guide and a point belonging to the slide. These joints assure non-detaching contact between the circular guide (an arc of a circle) and the slide.

All other kinematic joints are defined according to the structural schematic shown in Figure 1, as detailed in Figure 4 on the ADAMS dynamic model.

✓ Another important step in the completion of the dynamic model is the definition of the toothed belt drive from the drive that performs ulnar–radial motion. For this purpose, we used the ADAMS machinery facility, which allows the definition of a toothed belt drive.

✓ Another step in finalizing the ParReEx robot model is to define the bevel gear.

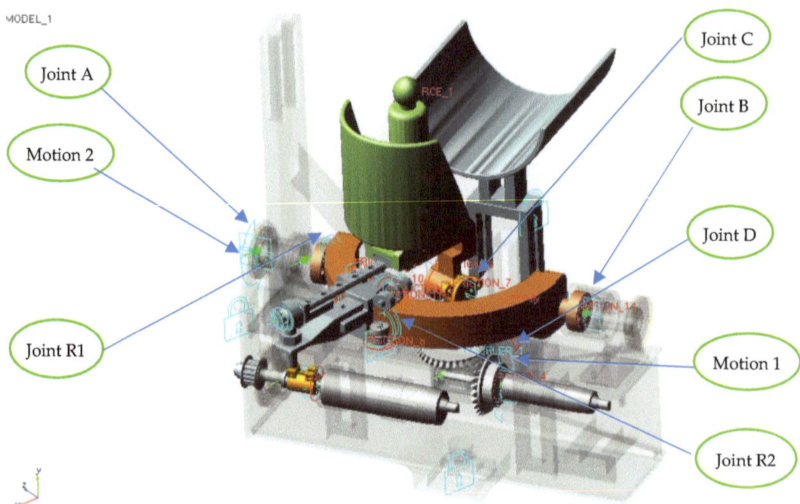

Figure 4. Kinematic joints defined in the ADAMS dynamic model.

For this purpose, we used the existing facility in Adams machinery to define gear drives. All the parameters required to define a bevel gear can be set in Table 1, namely: axis of rotation, number of teeth, face width, back cone distance, pitch angle, pitch apex, bore radius, outer module, addendum and dedendum factor.

Table 1. Bevel gear geometric parameters.

Parameter	Value	
	Gear 1	Gear 2
No. of teeth	39	39
Orientation of rotation axis	0, 90, 0	90, 90, 0
Face width (mm)	25.4	25.4
Back cone distance (mm)	10	15
Pitch angle (°)	19.747	70.253
Pitch apex	0	0
Bore radius (mm)	2	2
Mean spiral angle (°) and direction	35, LH	35, RH
Outer trans. module (mm)	4.535	4.535
Tooth depth and width	Data Type I-ISO23509	
Profile shift coefficient	0.505	−0.505
Dedendum factor	1.25	1.25
Thickness mod. coef.	3.7×10^{-2}	-5.5×10^{-2}

The material properties of bevel gears are shown in Table 2. Further, on the bevel gear, the contact parameters between the tooth flanks are set, as shown in Table 3, namely: contact stiffness, contact exponent and damping coefficient.

Table 2. Bevel gear material properties.

Parameter	Value
	Gear1, Gear 2
Material type	Steel
Density (kg/m^3)	7801
Young's Modulus (N/m^2)	2.07×10^{11}
Poisson ratio	0.33

Table 3. Bevel gear contact parameters.

Parameter	Value
Stiffness (N/mm^2)	5×10^5
Contact exponent	1.1
Damping coefficient	5000
Penetration (mm)	0.01
Static friction coefficient	0.11
Static friction transition velocity (mm/s)	0.2
Dynamic friction coefficient	0.10
Dynamic friction transition velocity (mm/s)	0.5

The law-of-motion time-history diagram is shown in Figure 5. The two parameters, the amplitude of motion and the time period, were chosen taking into account the similarity between the laws of motion in numerical simulation and those recommended by therapists for proper physical rehabilitation. The two laws of motion that will be assisted by the robot are named in MSC.ADAMS as MOTION_1 and MOTION_2. These movements are defined using the STEP function, as shown below.

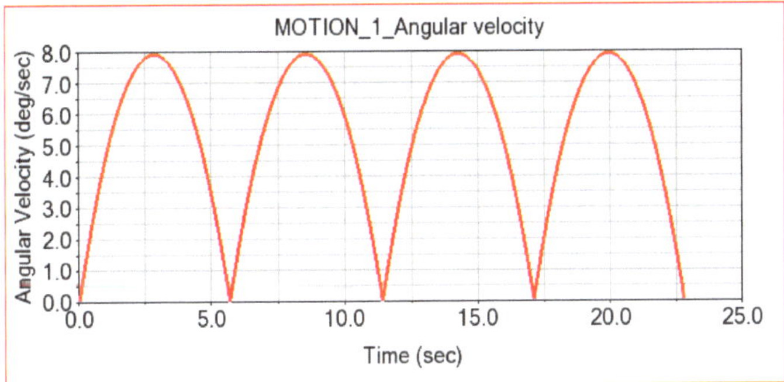

Figure 5. Law of motion imposed in ADAMS on ParReEx-wrist robot motor Joint E.

The law of MOTION 1, which corresponds to the flexion–extension movement, has a variation of $\pm 50°$, with one step of the flexion movement performed in 5.7 s.

The law of MOTION 2 corresponds to ulnar–radial motion; it is specified on the same 5.7 s step, but the amplitude is 30 degrees.

MOTION 1: STEP (time, 0.0, 0.0d, 5.7, −50.0d) + STEP (time, 5.7, 0d, 11.4, 50.0d) + STEP (time, 11.4, 0.0d, 17.1, −50.0d) + STEP (time, 17.1, 0d, 22.8, 50.0d).

MOTION 2: STEP (time, 0.0, 0.0d, 5.7, 30.0d) + STEP (time, 5.7, 0d, 11.4, −30.0d) + STEP (time, 11.4, 0.0d, 17.1, 30.0d) + STEP (time, 17.1, 0d, 22.8, −30.0d).

The mannequin-rehabilitation robot assembly is dynamically analyzed in MSC.ADAMS with the two modules of the ParReEx-wrist robot (flexion/extension and radial–ulnar motion) active simultaneously for a period of 22.8 s. Figure 6 presents the two extreme operating positions of the ParReEx robot corresponding to the combined flexion–extension and ulnar–radial movements.

In the first phase, a dynamic simulation with the specified laws of motion and without considering the friction in the kinematic joints of the robot is performed. The results obtained for this simulation—meaning the variations of the motor torque calculated by numerical simulation in ADAMS when simultaneously performing both types of movement: flexion–extension and radial–ulnar motion of the wrist—are presented in Figure 7.

Figure 6. The two extreme positions during the operation of the virtual model of the ParReEx-wrist robot, corresponding to combined flexion–extension and radial–ulnar motion.

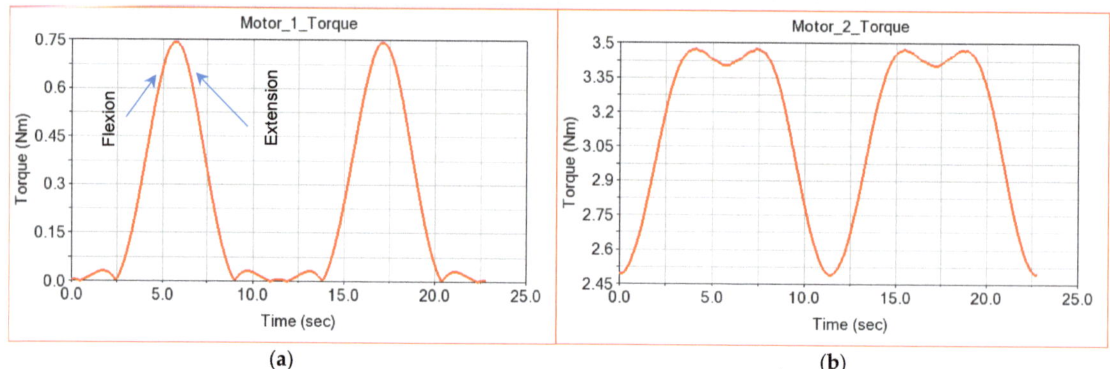

Figure 7. ADAMS simulation computed torque for: (**a**) the flexion–extension module and (**b**) the radial–ulnar module.

Analyzing the graph in Figure 7a computed for the flexion–extension motion, it is found that the maximum value of the actuation moment of this flexion/extension module is 0.7436 Nm, while the minimum value is near 0 Nm.

The motor torque obtained by numerical simulation for the radial–ulnar motion (Figure 7b) has a maximum value around 3.4748 Nm and a minimum of 2.47 Nm. In conclusion, the radial–ulnar movement requires an actuation moment greater than about four times the moment required for the flexion–extension movement, so the flexion–extension module will be the most structurally loaded. For the two modules, the graphs of the connecting forces that develop in Joints A, B, R1, R2 and C are presented.

Figure 8 shows the orientation of the local reference axis systems attached to the kinematic torques of the ParReEx robot for clear identification of the components of the connecting forces.

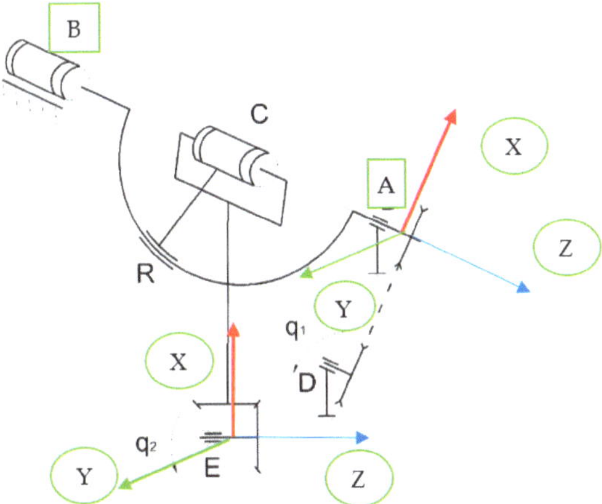

Figure 8. Positioning of the axes of the kinematic joints of the ParReEx robot.

Graphs of time variation for the components of the connecting force in Joint A in relation to the local reference system attached to this joint are shown in Figure 9a–c. Analysis of these graphs shows that the maximum value recorded along the Y-axis of the local reference system in Joint A is 111.7138 N, and the minimum is 60.8693 N, with an RMS (round mean square) equal to 95.1698 N (Figure 9b). The component along the X-axis reaches an absolute maximum value of 20.8049 N, and the minimum value is equal to 0.6315 N, with an RMS of 13.9479 N (Figure 9a). The component along the Z-axis reaches 88.6393 N with an RMS of 49.6273 N (Figure 9c). The value of the resulting force in this joint has a maximum of 142.1042 N and a minimum of 64.3436 N according to Figure 9d.

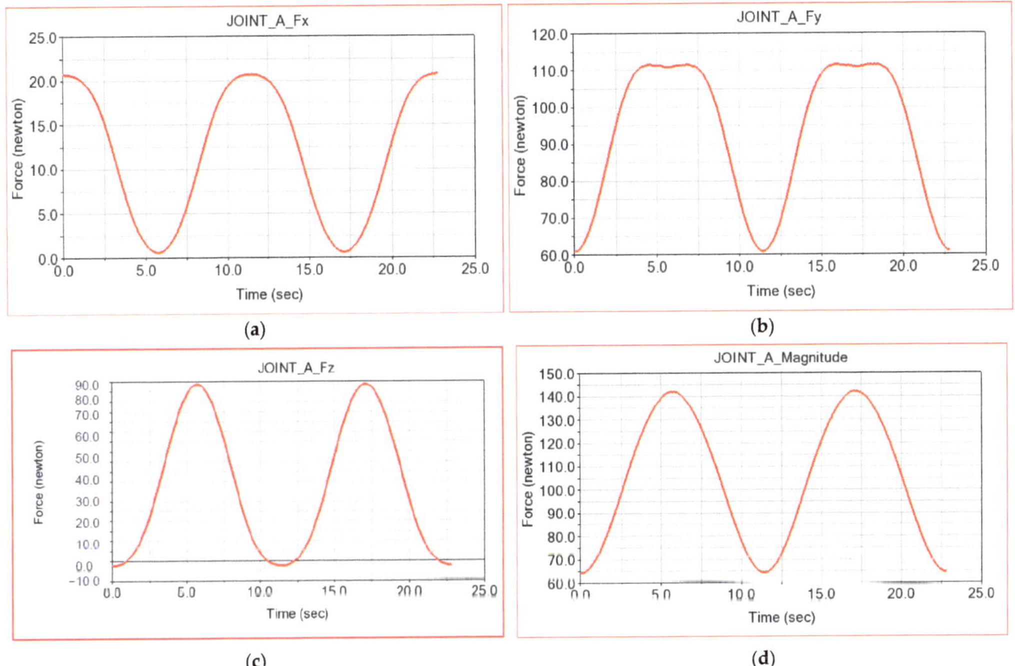

Figure 9. Connecting forces in rotation Joint A: (**a**) in the x-direction; (**b**) in the y-direction; (**c**) in the z-direction; and (**d**) the resultant.

These results are also presented in Table 4.

Table 4. Joint A connection force components.

Component Parameter	Joint A Fx	Joint A Fy	Joint A Fz	Joint A MAG
Min	0.6315	60.8693	−2.4821	64.3436
Max	20.8049	111.7138	88.6393	142.1042
Avg	11.8015	93.2991	37.1483	104.7257
RMS	13.9479	95.1698	49.6273	108.2345

Graphs of time variation for the components of the connecting force in Joint B are shown in Figure 10a–c. The maximum value reported for the X-axis component is 25.8542 N, and the minimum is 11.002 N, with an RMS equal to 20.1699 N (Figure 10a). The Y-axis component has a maximum value equal to 0.3771 N, and the minimum value is equal to 0.2556 N, with an RMS of 0.3118 N (Figure 10b) The component along the Z-axis reaches 29.4384 N with an RMS of 24.3394 N (Figure 10c). The value of the resulting force in this joint reaches a maximum of about 32.6581 N, a minimum of 30.7333 N and an RMS equal to 31.6122 N, as we can see in Figure 10d.

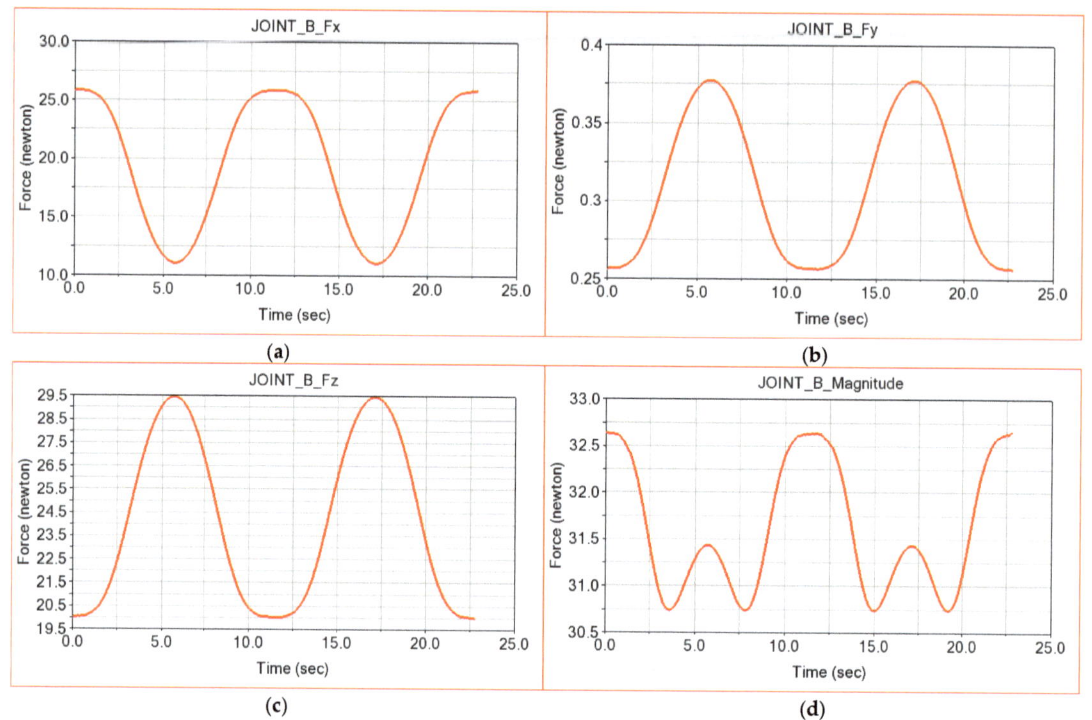

Figure 10. Connecting forces in rotational Joint B: (**a**) in the x-direction; (**b**) in the y-direction; (**c**) in the z-direction; and (**d**) the resultant.

These results are also shown in Table 5.

Table 5. Joint B connection force components.

Component Parameter	Joint B Fx	Joint B Fy	Joint B Fz	Joint B MAG
Min	11.002	0.2556	19.9511	30.7333
Max	25.8542	0.3771	29.4384	32.6581
Avg	19.391	0.3084	24.0797	31.6122
RMS	20.1699	0.3118	24.3394	31.6122

There is a higher load distribution on Coupling A, i.e., the RMS value is 3.48 times higher for the force resulting from Joint A.

To model the contact between the radial bearings and the rolling path of the circular guide, we defined in ADAMS primitives such as the upper rolling torque between the outer ring of the bearing on the rolling path of the circular guide. The ParReEx robot has four bearings in the structure that make contact between the circular guide and the adjacent element.

Similar graphs are obtained for Joints R1 and R2 (Figures 11 and 12). The values of these results are presented in tabular form in Tables 6 and 7.

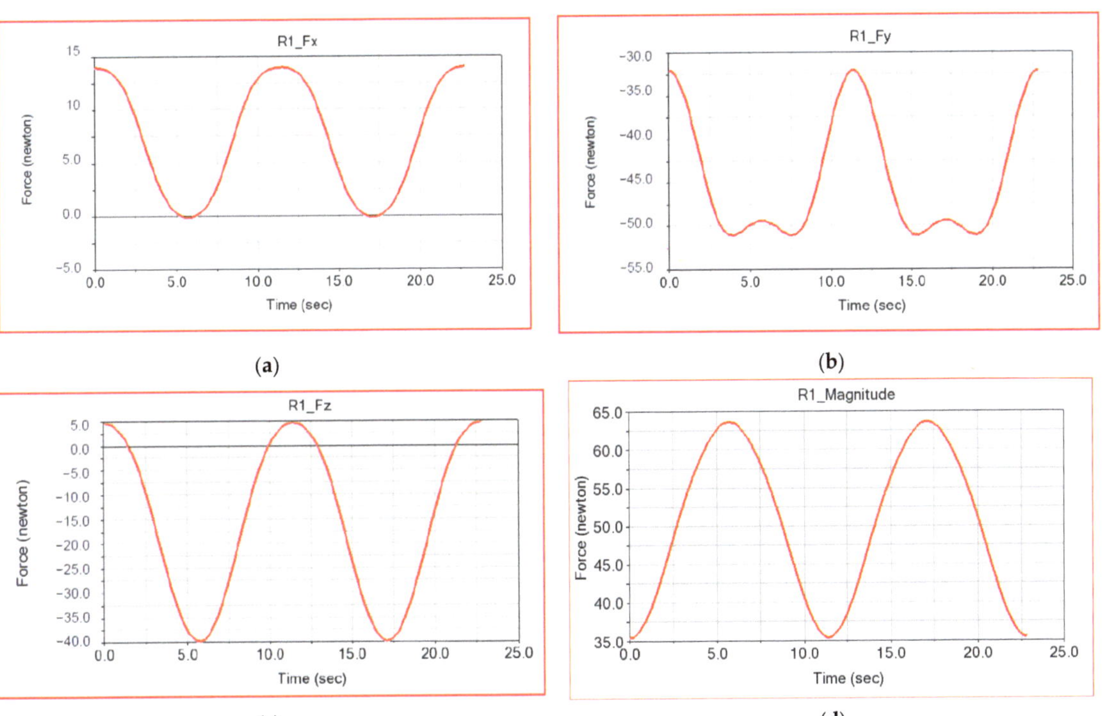

Figure 11. Connecting forces in rotational Joint R1: (**a**) in the x-direction; (**b**) in the y-direction; (**c**) in the z-direction; and (**d**) the resultant.

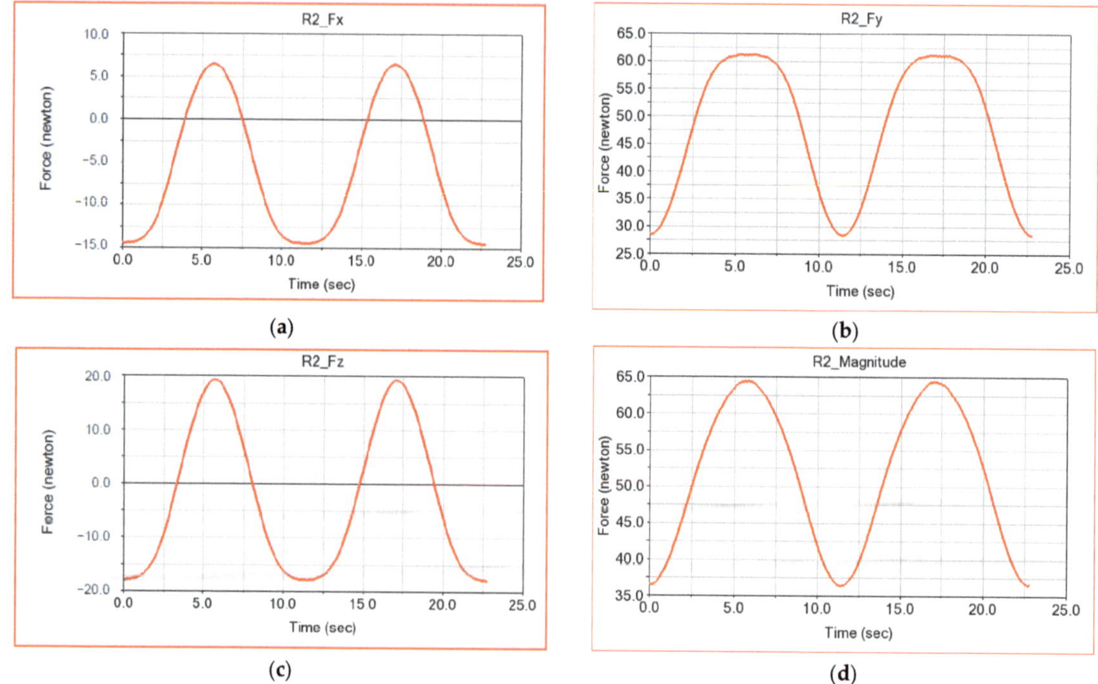

Figure 12. Connecting forces in rotational Joint R2 of the flexion/extension module: (**a**) in the x-direction; (**b**) in the y-direction; (**c**) in the z-direction; and (**d**) the resultant.

Table 6. Joint R1 connection force components.

Parameter \ Component	Joint R1 Fx	Joint R1 Fy	Joint R1 Fz	Joint R1 MAG
Min	−0.1407	−51.1415	−39.9417	35.325
Max	13.994	−32.1514	4.533	63.5811
Avg	7.2739	−44.6805	−15.7091	50.263
RMS	8.9958	45.1932	22.3775	51.226

Table 7. Joint R2 connection force components.

Parameter \ Component	Joint R2 Fx	Joint R2 Fy	Joint R2 Fz	Joint R2 MAG
Min	−14.4231	28.4593	−17.9002	36.502
Max	6.4731	61.2824	−19.281	64.5064
Avg	−5.6765	48.3102	−2.6405	21.6015
RMS	9.5789	49.7893	13.7125	52.39

For the situation in which the ParReEx robot is used by a patient and the frictional forces in the kinematic torques are not taken into account, the following results are obtained for the connecting forces in the Joint C, as presented in Figure 13.

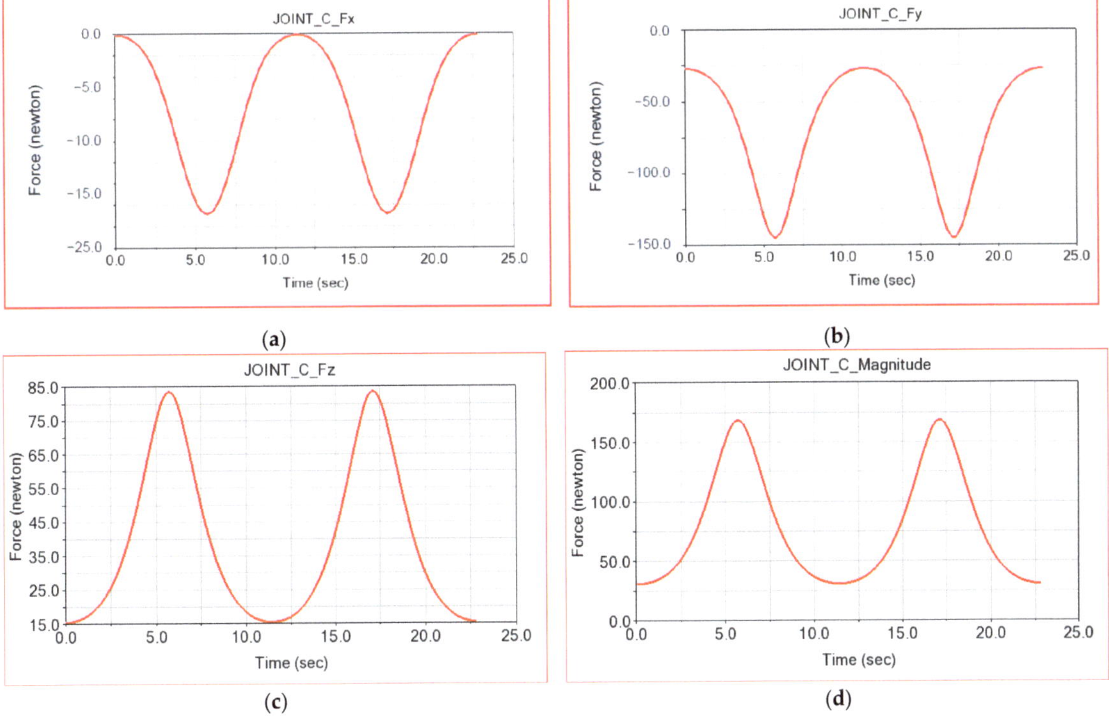

Figure 13. Connecting forces in rotational Joint C: (**a**) in the x-direction; (**b**) in the y-direction; (**c**) in the z-direction; and (**d**) the resultant.

Table 8 shows the values of the components and the resultant reaction in coupling C for the first case of study, i.e., without friction in the coupling and with the kinematic elements considered as rigid solids.

Table 8. Joint C connection force components.

Parameter \ Component	Joint C Fx	Joint C Fy	Joint C Fz	Joint C MAG
Min	−20.9865	−144.671	15.4783	30.957
Max	−0.0759	−26.8092	85.5258	168.3648
Avg	−8.4707	−67.4532	38.9441	78.4077
RMS	11.2895	78.3566	45.2392	91.18

3.2. Dynamic Analysis in MSC.ADAMS of the ParReEx-Wrist Robot Considering Joint Friction

This subchapter is dedicated to a virtual dynamic study in ADAMS of the actuation moments and the connection forces in the kinematic Joints A, B, C, R1 and R2 considering the friction in the kinematic joints—a case that simulates the real situation and is different from the previous simulation in which simplistic hypotheses were used.

In principle, we are interested in simulating the dynamics of the robot in the context of its use by a patient. Therefore, in all dynamics studies, we consider this context.

To define the parameters required for the study of dynamics considering friction, existing data in the literature were used [50]. The following values were adopted for this study: static friction coefficient of 0.2 and dynamic friction coefficient of 0.15 (Table 9).

Table 9. Joint friction parameters.

Parameter	Value
Mu static	0.2
Mu dynamic	0.15
Friction arm (mm)	10
Bending friction arm (mm)	10
Pin radius (mm)	5
Stiction Transition velocity	0.1
Max. stiction deformation	0.01
Friction torque preload	0
Effect	Stiction only

Using numerical simulation in ADAMS considering the friction in joints, results were obtained in the form of laws of variation of the moments necessary to operate the flexion–extension module and the radial–ulnar deviation module (Figure 14).

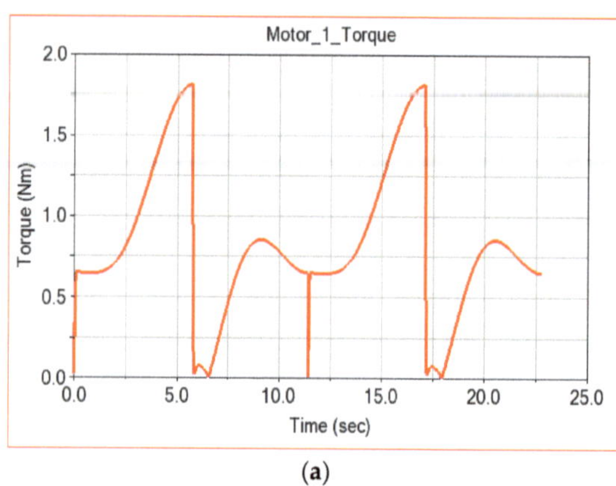
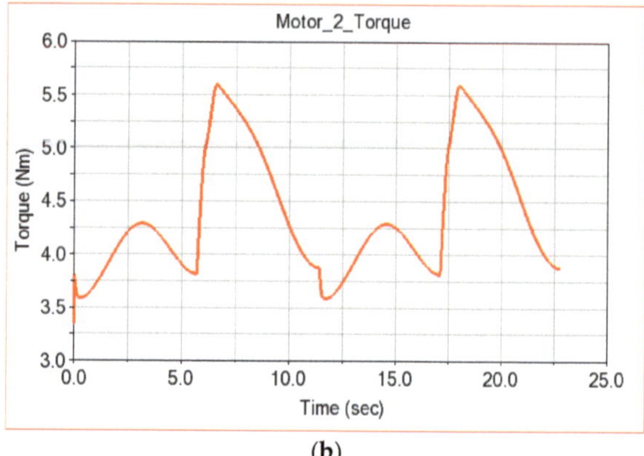

(a)   (b)

Figure 14. Second case—ADAMS-simulation-computed torque for: (**a**) the flexion–extension module and (**b**) the radial–ulnar deviation module.

Table 10 shows the characteristic values for the actuation moments of the two robot modules for the second case study, namely when we considered the friction in the kinematic couplings.

Table 10. Robot motion parameters.

Parameter / Component	Motion 1 Torque (Nm)	Motion 2 Torque (Nm)
Min	0	0
Max	1.8086	5.5927
Avg	0.8143	4.3763
RMS	0.9317	4.4156

There is an increase in the motor torque required to operate the modules in this case, as well as a steeper variation, which is similar to that determined experimentally [18].

The graphs of the laws of variation of the connection forces considering friction as calculated in ADAMS are presented in Figure 15 for Joint A and in Figure 16 for Joint B.

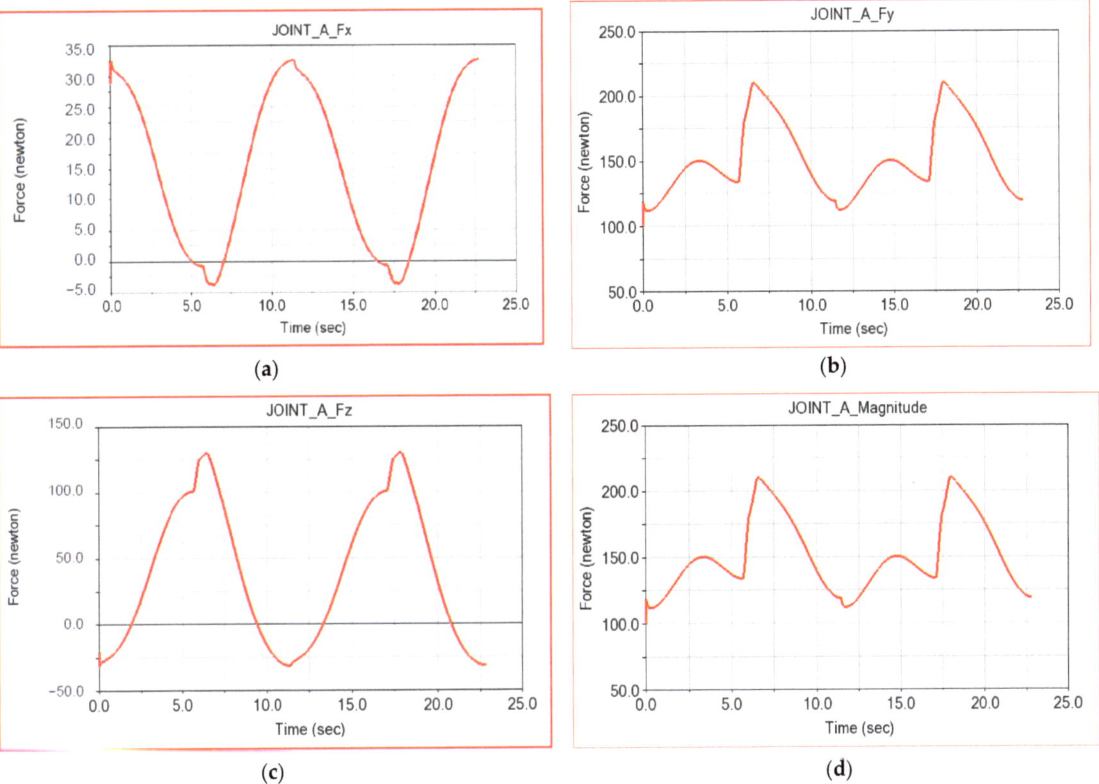

Figure 15. Connecting forces in rotation Joint A computed with friction consideration: (**a**) in the x-direction; (**b**) in the y-direction; (**c**) in the z-direction; and (**d**) the resultant.

The relevant results for the connecting force in Coupling A and its components are also given in Table 11.

Table 11. Joint A connection force components.

Component Parameter	Joint A Fx	Joint A Fy	Joint A Fz	Joint A MAG
Min	−3.7855	99.3604	−32.0278	99.3604
Max	32.514	210.0206	129.7281	210.0206
Avg	15.0504	150.641	38.4032	150.641
RMS	19.5913	153.1624	65.5017	153.1624

We compare the results of the second simulation in the Discussion. The results obtained for Joint B in Case 2 are presented in Table 12.

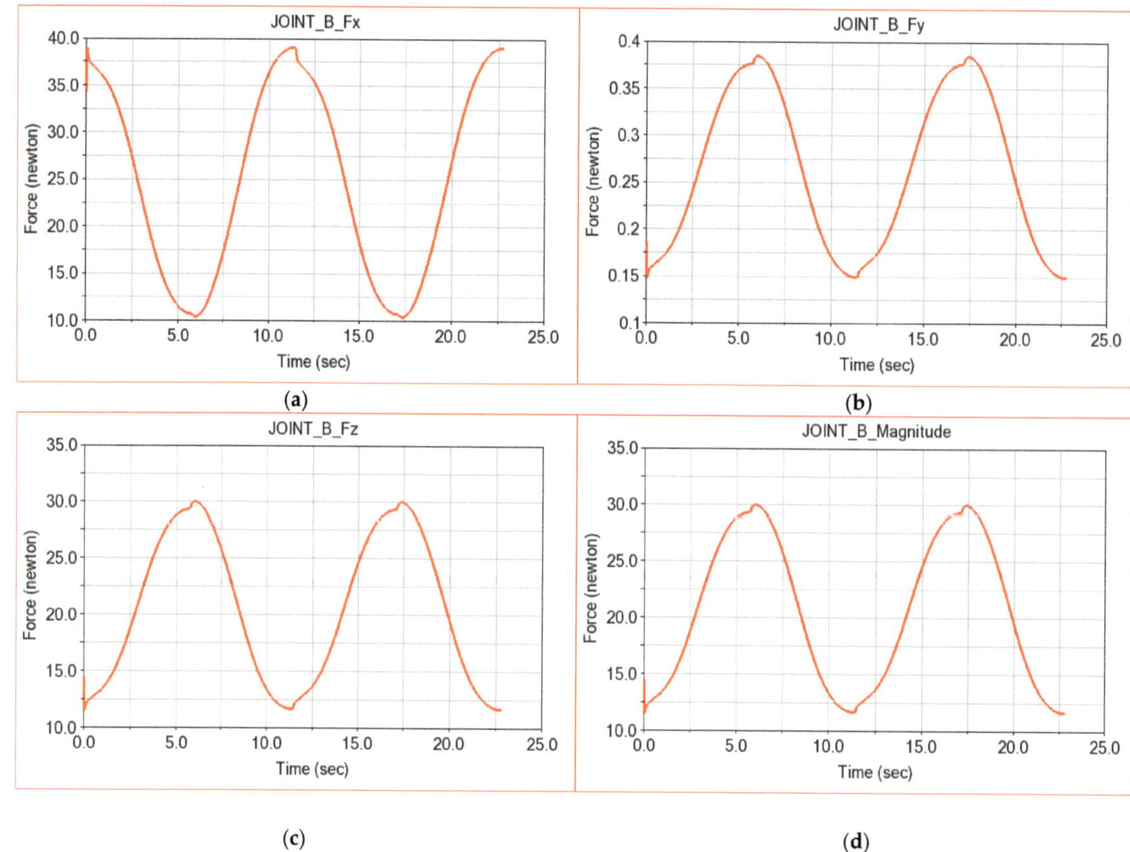

Figure 16. Connecting forces in Joint B calculated assuming friction: (**a**) in the x-direction; (**b**) in the y-direction; (**c**) in the z-direction; and (**d**) the resultant.

Table 12. Joint B connection force components.

Parameter \ Component	Joint B Fx	Joint B Fy	Joint B Fz	Joint B MAG
Min	10.4354	0.1475	11.5174	11.5174
Max	39.1768	0.3846	30.0259	30.0259
Avg	24.8299	0.2639	20.6054	20.6054
RMS	26.8299	0.2767	21.6025	21.6025

To model the contact between the radial bearings and the circular guideway, we defined in ADAMS superior rolling couplings of the outer ring of the bearing on the circular guideway. The ParReEx robot has four bearings that ensure contact between the circular guide and the conjugate element.

Figure 17 shows the numerical results obtained by numerical simulation in ADAMS. View for the connection forces in the R1 coupling.

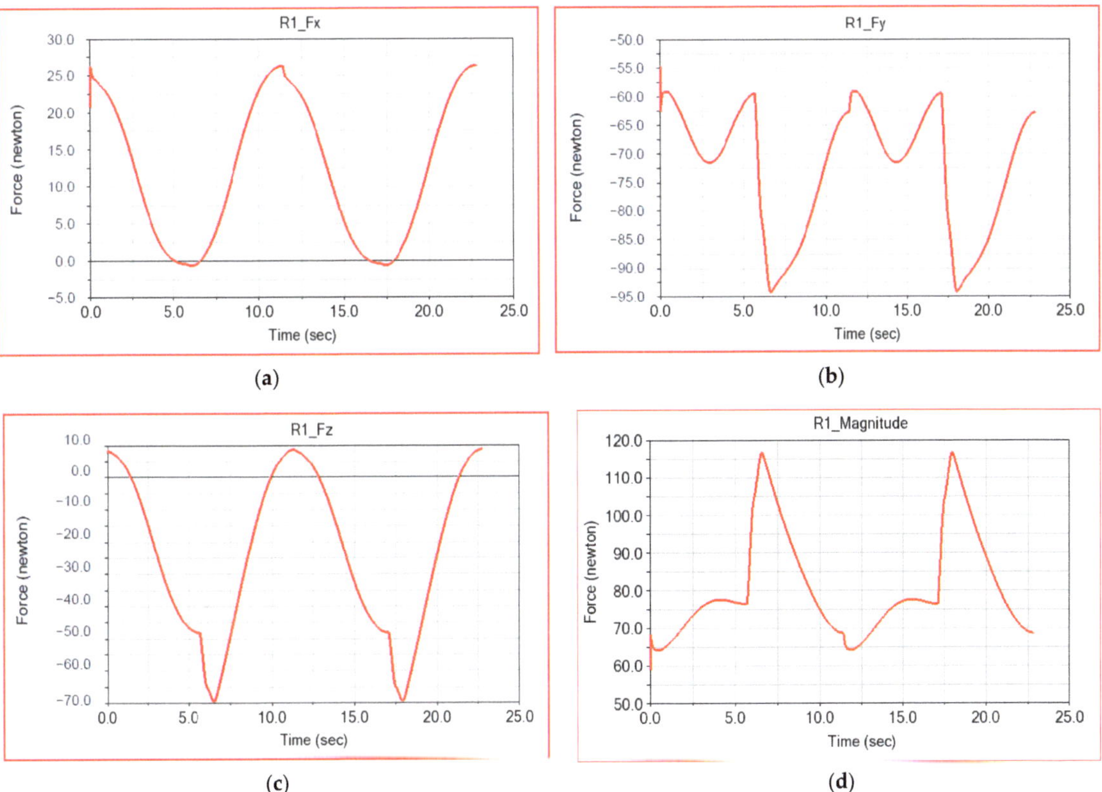

(a)

(b)

(c)

(d)

Figure 17. Connecting forces in rotational Joint R1: (**a**) in the x-direction; (**b**) in the y-direction; (**c**) in the z-direction; and (**d**) the resultant.

Table 13 shows the values of the components of the connection forces as well as the resulting reaction in the kinematic joint R1 for the second case study.

Table 13. Joint R1 connection force components.

Component / Parameter	Joint R1 Fx	Joint R1 Fy	Joint R1 Fz	Joint R1 MAG
Min	−0.6321	−94.2472	−69.4304	58.9881
Max	26.2981	−54.6674	8.8217	116.7565
Avg	11.9904	−72.4773	−23.7857	81.0712
RMS	15.3428	73.2824	34.1712	82.3006

Figure 18 shows the numerical results obtained by numerical simulation in ADAMS. View for the connection forces in the R2 coupling.

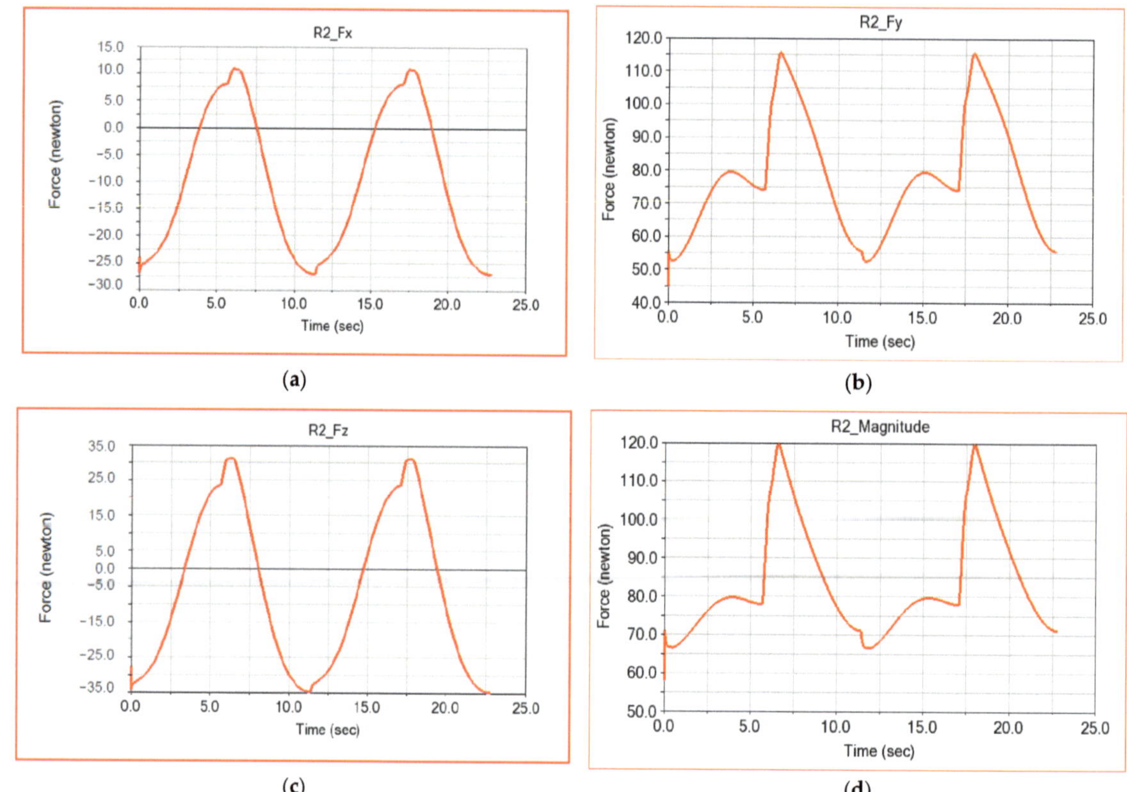

Figure 18. Connecting forces in rotational Joint R2: (**a**) in the x-direction; (**b**) in the y-direction; (**c**) in the z-direction; and (**d**) the resultant.

Table 14 shows the values of the components of the connection forces as well as the resulting reaction in the kinematic joint R2 for the second case study.

Table 14. Joint R2 connection force components.

Parameter	Joint R2 Fx	Joint R2 Fy	Joint R2 Fz	Joint R2 MAG
Min	−27.1507	44.5072	−34.8471	57.757
Max	10.8219	115.4258	31.153	119.6574
Avg	−9.6479	77.8998	−5.9878	83.4607
RMS	16.3354	79.8368	23.0639	84.6918

For the situation when the ParReEx robot is used by a patient and we consider the frictional forces in the kinematic joints, the following results are obtained for connecting forces in Joint C, as presented in Figure 19.

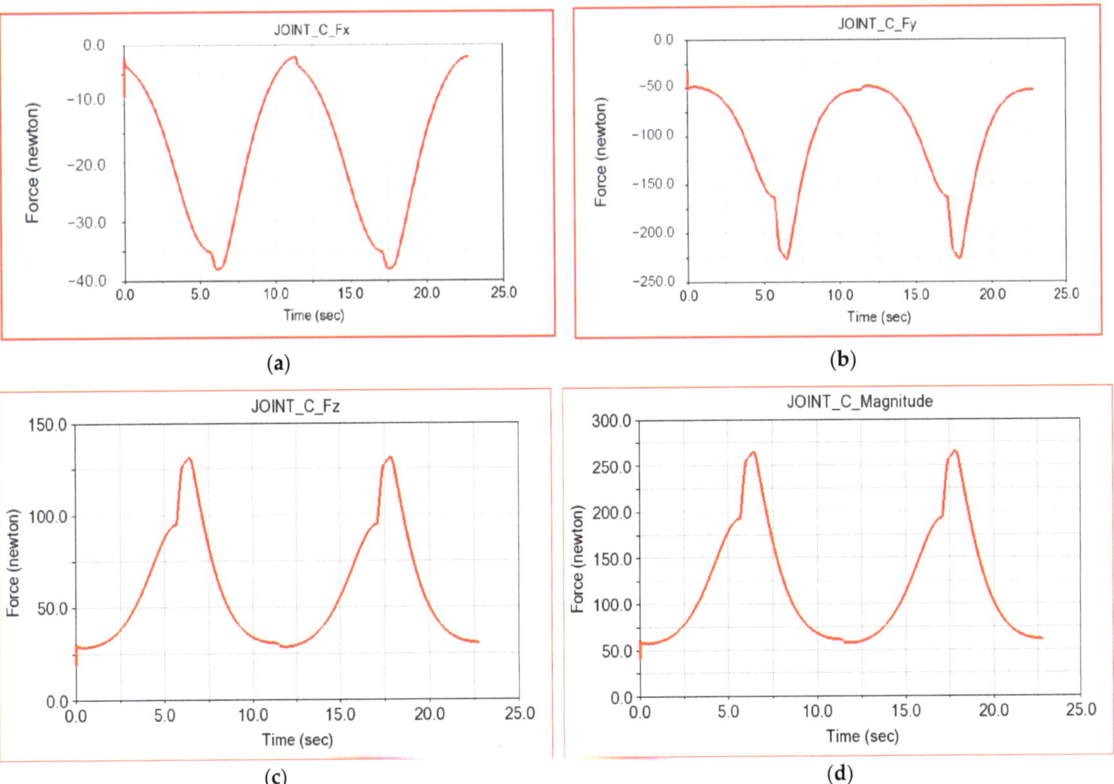

Figure 19. Connecting forces in rotational Joint C: (**a**) in the x-direction; (**b**) in the y-direction; (**c**) in the z-direction; and (**d**) the resultant.

Table 15 shows the values of the components of the connection forces as well as the resulting reaction in the kinematic joint C for the second case study.

Table 15. Joint C connection force components.

Component Parameter	Joint C Fx	Joint C Fy	Joint C Fz	Joint C MAG
Min	−38.152	−226.9065	18.9705	39.0085
Max	−2.303	−32.8579	131.0045	264.6976
Avg	−18.3145	−101.217	58.4376	118.4064
RMS	21.8939	114.4402	66.0721	133.9455

For Joint A, the maximum RMS values increase from 108.2345 N in Case 1 (no joint friction) to 153.1624 N in Case 2 (with joint friction).

Numerical simulation in ADAMS is performed for the manikin-robot rehabilitation assembly considering friction in the rotating joints. The connection forces in the joints have obvious increases compared to the previously studied cases because friction plays an important role in the real operation of the robot.

3.3. Dynamic Analysis in MSC.ADAMS of the ParReEx-Wrist Robot Considering Flexible Elements and Joint Friction

This subchapter is dedicated to the virtual dynamic study in ADAMS of the actuation moments and the connection forces in kinematic joints A, B, C, R1 and R2 considering friction in the kinematic joints and element flexibility, a case that simulates the real situation and is different from the first one, in which simplifying hypotheses have been used.

This simulation assumption is necessary because, in operation, the robot elements deform, and this influences the dynamic operating parameters. The way to transform a rigid body into a deformable one in ADAMS involves:

- Discretizing the body into finite elements and performing modal dynamics analysis, in which vibration modes are calculated based on modal superposition;
- Dynamic analysis considering the solid deformable body;
- Determination of torsional and flexural deformations of the flexible body.

Table 16 shows the vibration modes of the flexible element, namely the circular guide. Since in operation this element has the highest elasticity, it is very important to study the dynamics of the robot considering this circular guide as a deformable solid. Figure 20 shows that Vibration Mode 8 is flexural and occurs at 1538 Hz. Vibration modes consider the external connections of the element and can be flexural, torsional or a combination of the two.

Table 16. Circular guide modal parameters.

Mode Number	Frequency (Hz)	Mode Number	Frequency (Hz)
7	597.685	16	1.299×10^4
8	1538.863	17	1.546×10^4
9	1828.696	18	2.315×10^4
10	2804.835	19	2.827×10^4
11	3835.677	20	2.987×10^4
12	4032.088	21	6.923×10^4
13	5993.852	22	8.053×10^4
14	6743.001	23	1.343×10^5
15	9246.1609	24	1.808×10^5

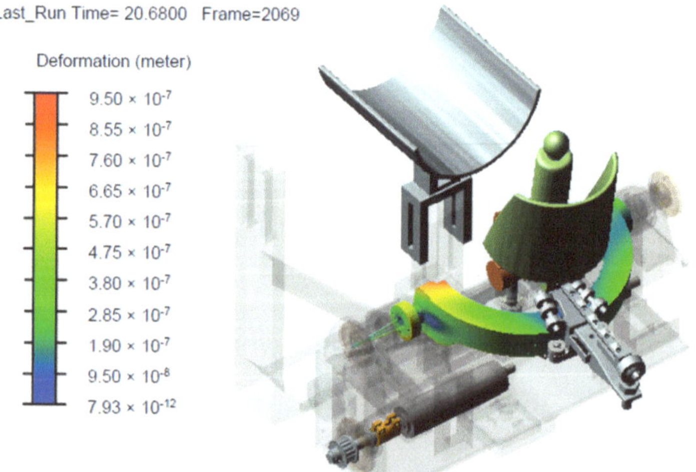

Figure 20. The deformable body in ADAMS and view of body deformation map.

Numerical simulation in ADAMS considering the friction in joints and the flexibility of the circular guide element resulted in laws of variation of the torque necessary to operate the flexion–extension module as well as the radial–ulnar module (Figure 21).

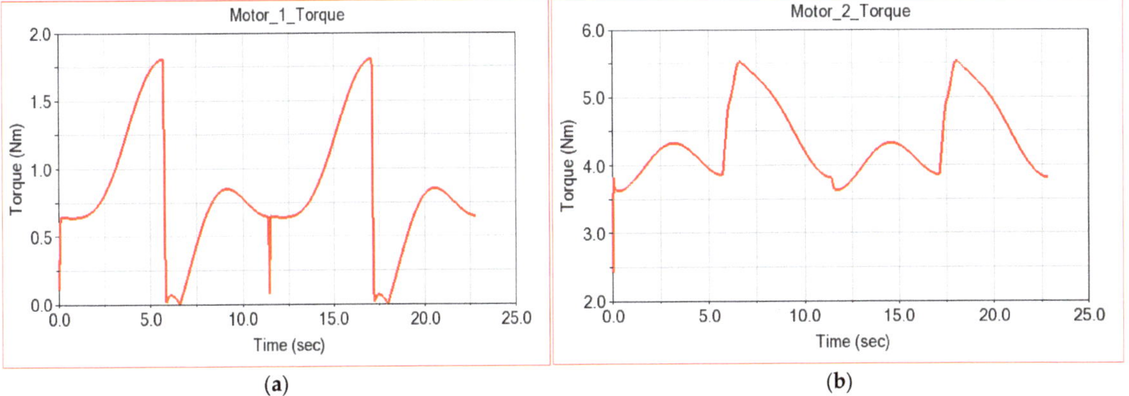

Figure 21. The laws of variation of computed torque for: (**a**) the flexion–extension module and (**b**) the radial–ulnar module.

There is an increase in the motor torque required to operate the modules in this case as well as a steeper variation, similar to values determined experimentally [18].

In Table 17, the characteristic values for the actuation moments of the two modules of the robot are presented for case 3 of the study, i.e., considering the flexibility of the elements and the friction in the couplings.

Table 17. Robot motion parameters in Case 3.

Parameter	Component	Motion 1 Torque	Motion 2 Torque
	Min	0.0013	2.431
	Max	1.8086	5.5339
	Avg	0.8143	4.3712
	RMS	0.9316	4.4062

Figure 22 shows the laws of variation for the components of the connection forces in coupling A as well as for the resultant reaction. These results are obtained in case 3 of the simulation, i.e., the one in which we considered simultaneously the friction in the kinematic couplings and the flexibility of the elements.

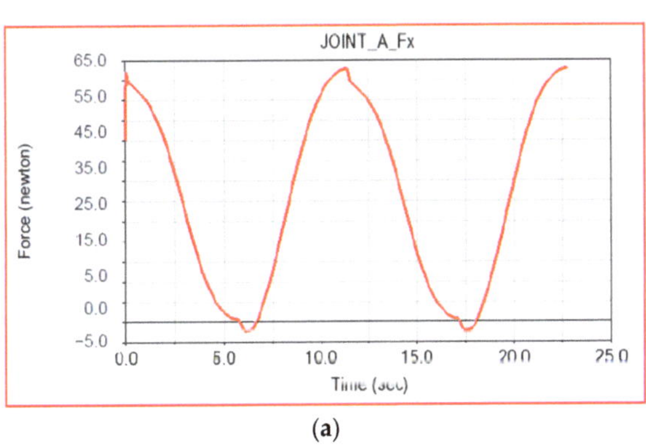

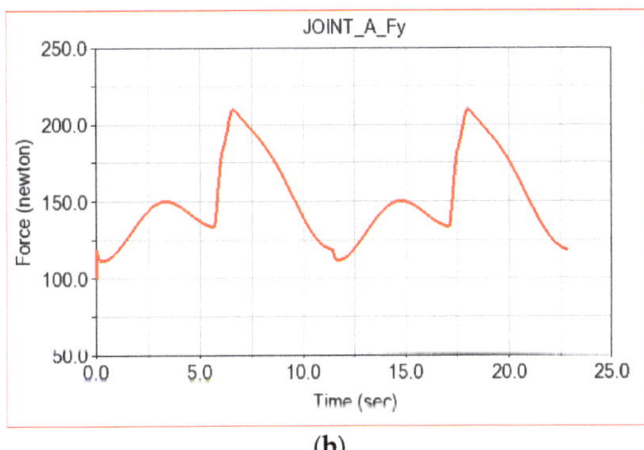

Figure 22. *Cont.*

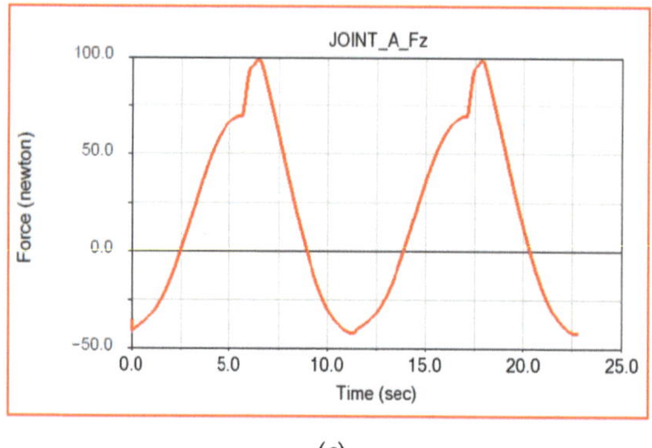
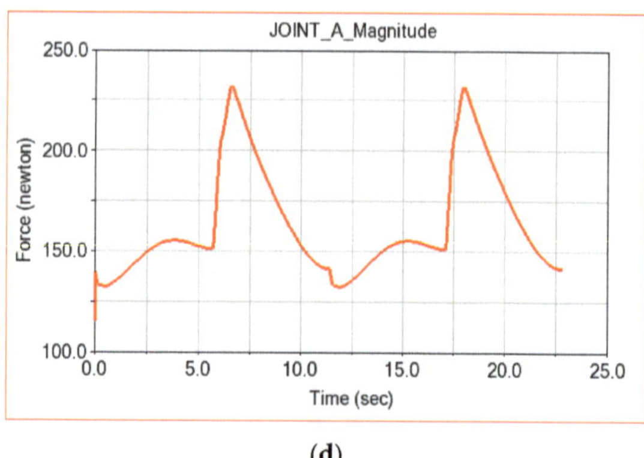

(c)   (d)

Figure 22. Connecting forces in rotation Joint A computed with friction and flexible links: (**a**) in the x-direction; (**b**) in the y-direction; (**c**) in the z-direction; and (**d**) the resultant.

Table 18 shows the numerical results obtained for the components of the forces and the resulting reaction in joint A for simulation case 3.

Table 18. Joint A connection force components.

Parameter	Component Joint A Fx	Joint A Fy	Joint A Fz	Joint A MAG
Min	−2.2236	99.6565	−42.8897	115.1857
Max	62.8521	210.1286	98.7351	231.5797
Avg	30.9324	150.7614	17.3479	163.231
RMS	38.4147	153.2828	48.9964	165.4441

Figure 23 shows the laws of variation for the components of the connection forces in coupling B as well as for the resultant reaction, for the third simulation case.

Table 19 shows the numerical results obtained for the components of the forces and the resulting reaction in joint B for simulation case 3.

Table 19. Joint B connection force components.

Parameter	Component Joint B Fx	Joint B Fy	Joint B Fz	Joint B MAG
Min	0.4806	−0.8625	−1.6387	0.4815
Max	9.1875	0.4868	1.0018	9.2307
Avg	8.8505	0.4033	−0.4287	8.91
RMS	8.853	0.4095	0.9483	8.9131

Figure 24 shows the laws of variation for the components of the connection forces in coupling R1 as well as for the resultant reaction, for the third simulation case.

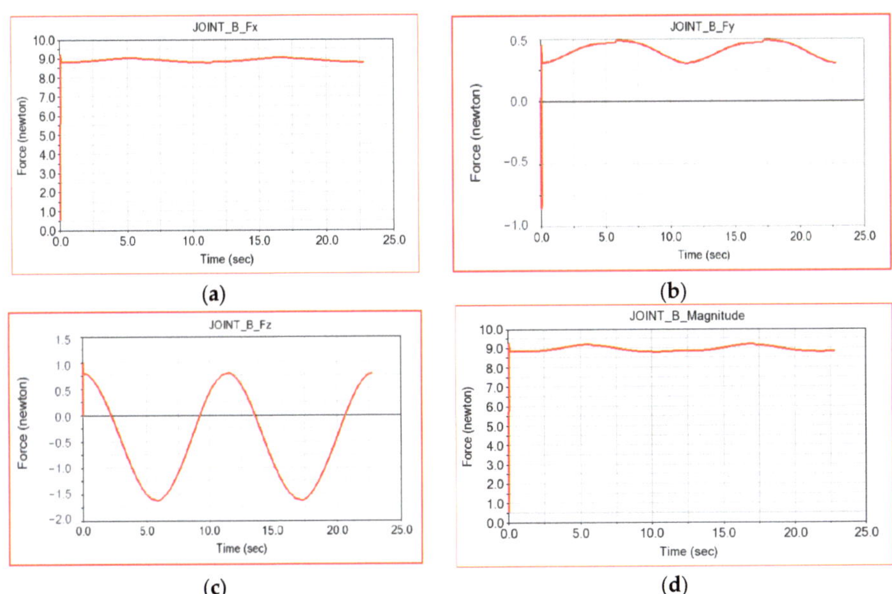

Figure 23. Connecting forces in Joint B calculated assuming friction and flexible links: (**a**) in the x-direction; (**b**) in the y-direction; (**c**) in the z-direction; and (**d**) the resultant.

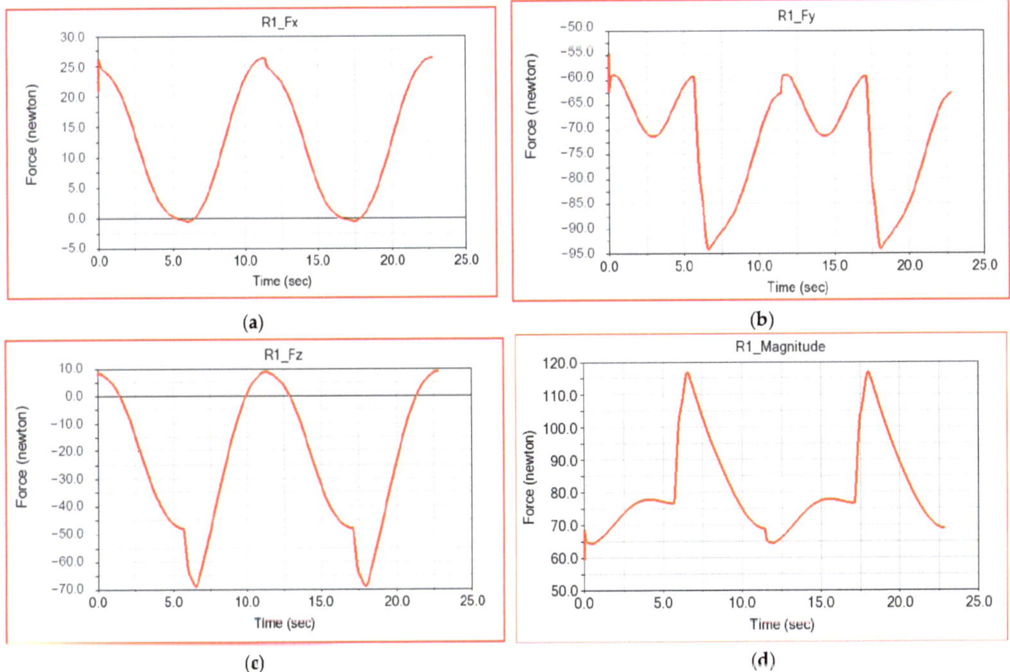

Figure 24. Connecting forces in rotational Joint R1 in Case 3: (**a**) in the x-direction; (**b**) in the y-direction; (**c**) in the z-direction; and (**d**) the resultant.

Table 20 shows the numerical results obtained for the components of the forces and the resulting reaction in joint R1 for simulation case 3.

Table 20. Joint R1 connection force components.

Component Parameter	Joint R1 Fx	Joint R1 Fy	Joint R1 Fz	Joint R1 MAG
Min	−0.6319	−94.247	−69.4296	59.116
Max	26.2983	−54.7853	8.8217	116.7565
Avg	11.2983	−72.4702	−23.7732	81.0625
RMS	15.3456	73.4702	34.1647	82.2926

Figure 25 shows the laws of variation for the components of the connection forces in coupling R2 as well as for the resultant reaction, for the third simulation case.

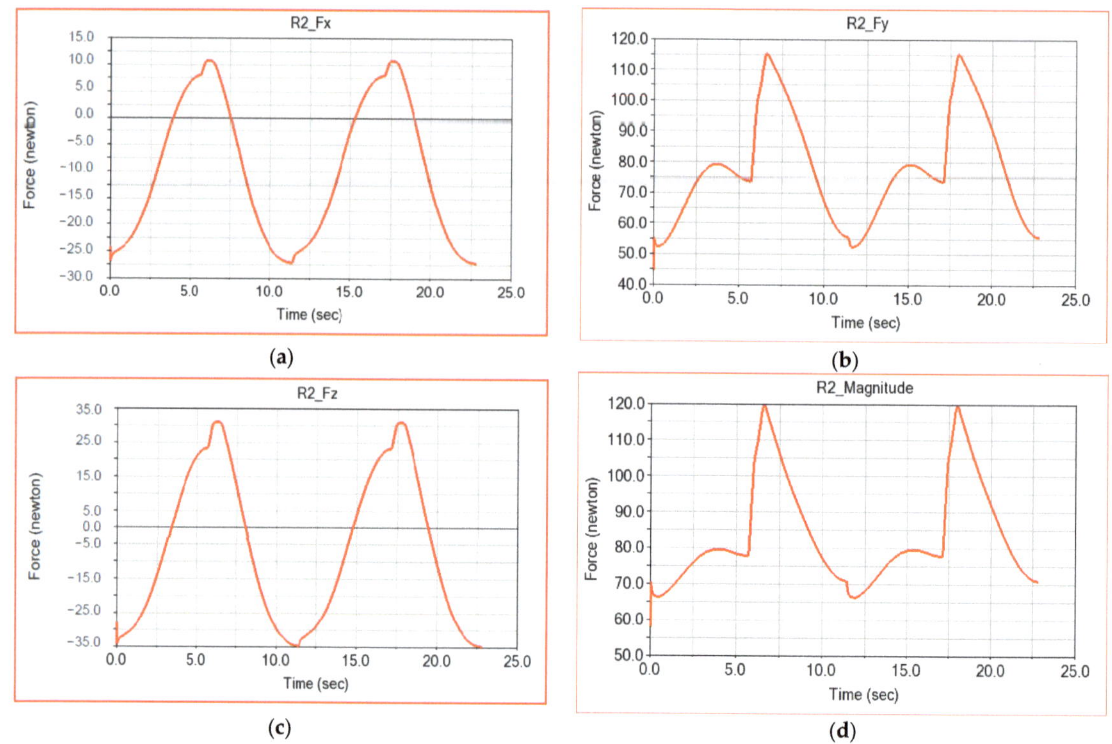

Figure 25. Connecting forces in rotational Joint R2 in Case 3: (**a**) in the x-direction; (**b**) in the y-direction; (**c**) in the z-direction; and (**d**) the resultant.

Table 21 shows the numerical results obtained for the components of the forces and the resulting reaction in joint R2 for simulation case 3.

Table 21. Joint R2 connection force components.

Component Parameter	Joint R2 Fx	Joint R2 Fy	Joint R2 Fz	Joint R2 MAG
Min	−27.1509	44.6085	−34.8469	57.8858
Max	10.8215	115.425	31.1528	119.6571
Avg	−9.6539	77.8865	−5.9965	83.4504
RMS	16.3394	79.8258	23.066	84.6827

For the situation when the ParReEx robot is used by a patient considering the frictional forces in the kinematic joints and the flexibility of the elements, the following results are obtained for connecting forces in Joint C, as shown in Figure 26.

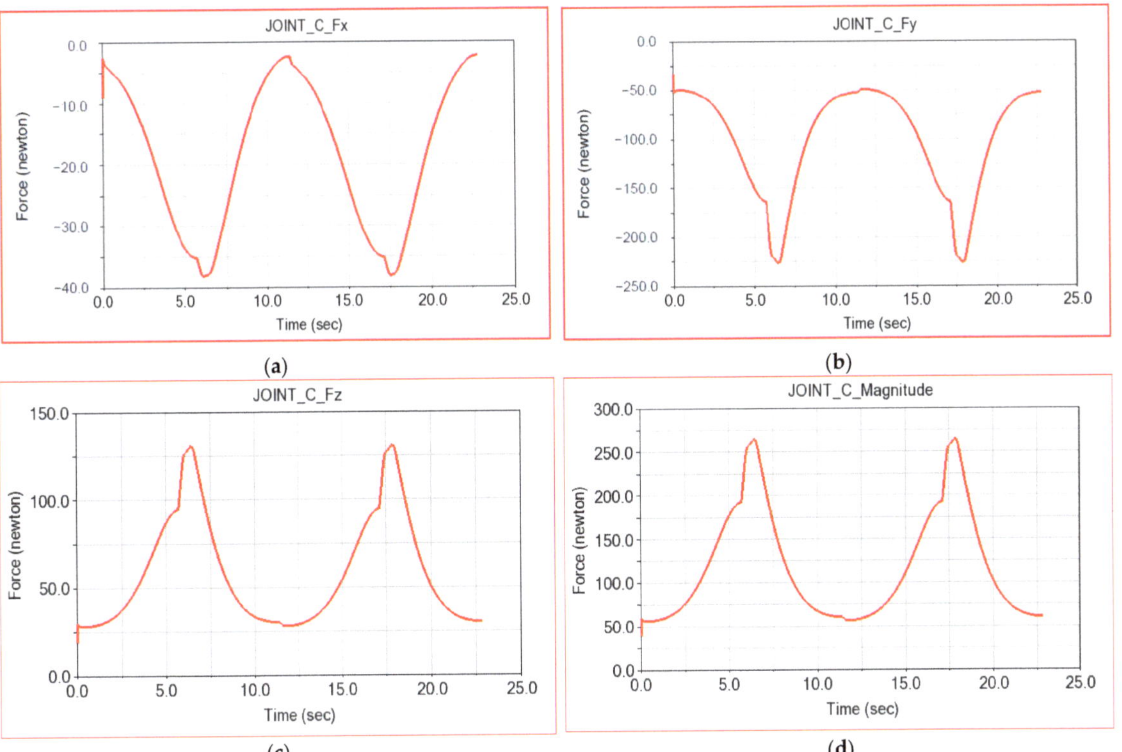

Figure 26. Connecting forces in rotational Joint C in Case 3: (**a**) in the x-direction; (**b**) in the y-direction; (**c**) in the z-direction; and (**d**) the resultant.

Table 22 shows the numerical results obtained for the components of the forces and the resulting reaction in joint C for simulation case 3.

Table 22. Joint C connection force components.

Parameter	Component	Joint C Fx	Joint C Fy	Joint C Fz	Joint C MAG
Min		−38.152	−226.9063	19.031	39.1179
Max		−2.3031	−32.9627	131.0044	264.6962
Avg		−18.3109	−101.1899	58.422	118.375
RMS		21.8903	114.4194	66.0601	133.9213

As previously specified, in this simulation case, we are interested in the translational displacements associated with the marker attached to the center of mass of the circular guide relative to its undeformed location. In addition to the displacements, we are also interested in the specific deformations of this marker. These results are shown in Figures 27 and 28.

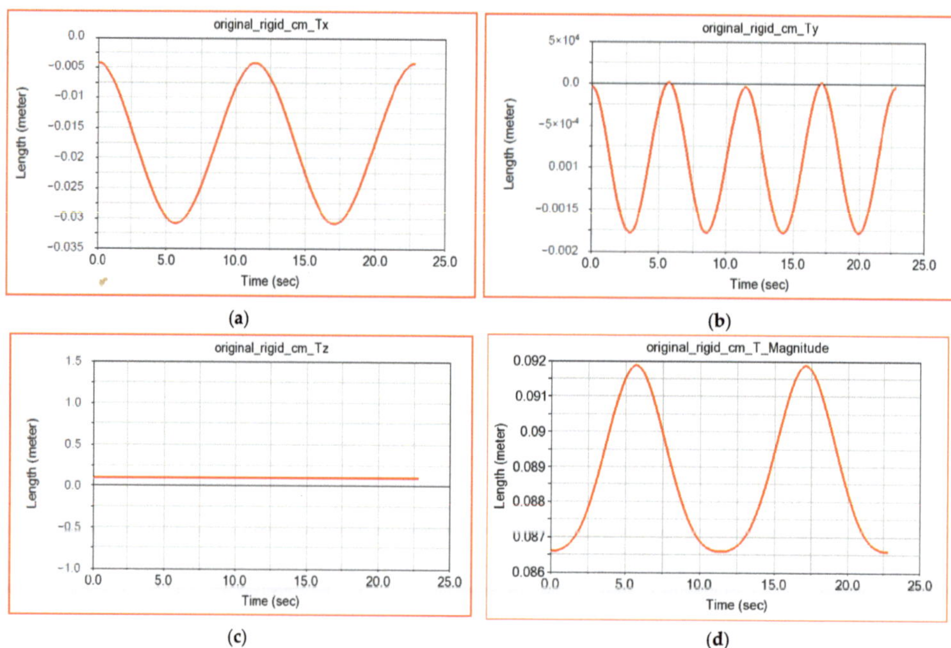

Figure 27. Displacement of the center of mass of the circular guide: (**a**) in the x-direction; (**b**) in the y-direction; (**c**) in the z-direction; and (**d**) the resultant.

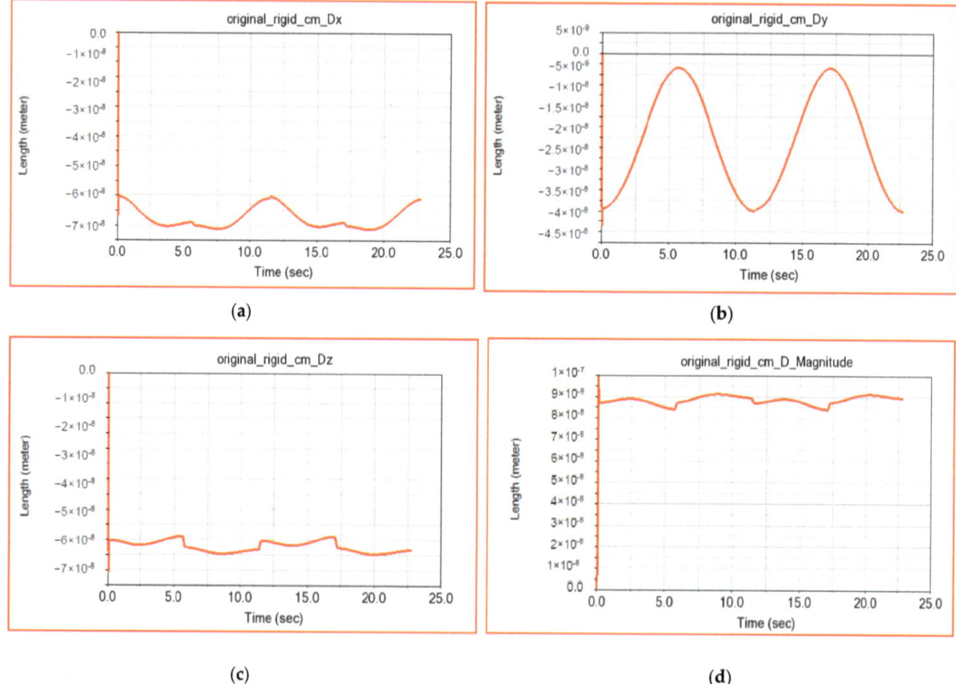

Figure 28. Deformation of the center of mass of the circular guide: (**a**) in the x-direction; (**b**) in the y-direction; (**c**) in the z-direction; and (**d**) the resultant.

Table 23 shows the numerical results obtained for the components and the resulting displacement of the marker attached to the center of mass of the circular guide.

Table 23. Displacement of the center of mass of the circular guide.

Component Parameter	Depl Tx (m)	Depl Ty(m)	Depl Tz(m)	Depl T Mag(m)
Min	−0.031	−0.0018	0.0865	0.0866
Max	−0.0042	7.620×10^{-6}	0.0865	0.0919
Avg	−0.0176	-9.243×10^{-4}	0.0865	0.0887
RMS	0.0199	0.0011	0.0865	0.0887

Table 24 shows the numerical results obtained for the components and the resulting deformations of the marker attached to the center of mass of the circular guide.

Table 24. Deformation of the center of mass of the circular guide.

Component Parameter	Def Dx(m)	Def Dy(m)	Def Dz(m)	Def D Mag(m)
Min	-6.638×10^{-8}	-4.132×10^{-8}	-6.603×10^{-8}	2.46×10^{-11}
Max	-1.948×10^{-11}	1.405×10^{-11}	-3.737×10^{-12}	9.960×10^{-8}
Avg	-6.256×10^{-8}	-2.094×10^{-8}	-5.735×10^{-8}	8.83×10^{-8}
RMS	6.268×10^{-8}	2.412×10^{-8}	5.739×10^{-8}	8.834×10^{-8}

A map of the specific strain distribution of the robot's circular guide is shown in Figure 29. These deformations have very small values, so this guide is designed with sufficient stiffness. Its mass can be reduced by structural stiffening, which will be discussed in the next paragraph.

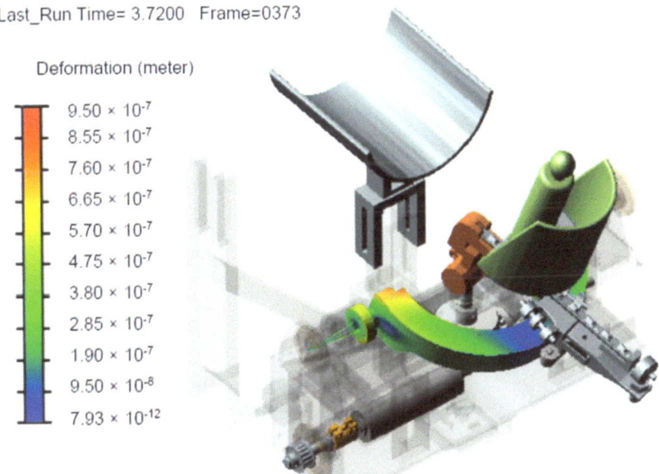

Figure 29. Deformation map of the robot circular guide.

4. Structural Optimization of the Circular Guide of the ParRcEx Robot

The circular guide in the structure of the ParReEx robot is an important part that must maintain a certain torsional and bending rigidity and a corresponding resistance to fatigue. To achieve structural optimization of the circular guide, we start from its existing geometry, which we model parametrically with the help of the ANSYS Workbench-Design Modeler preprocessor, as shown in Figure 30.

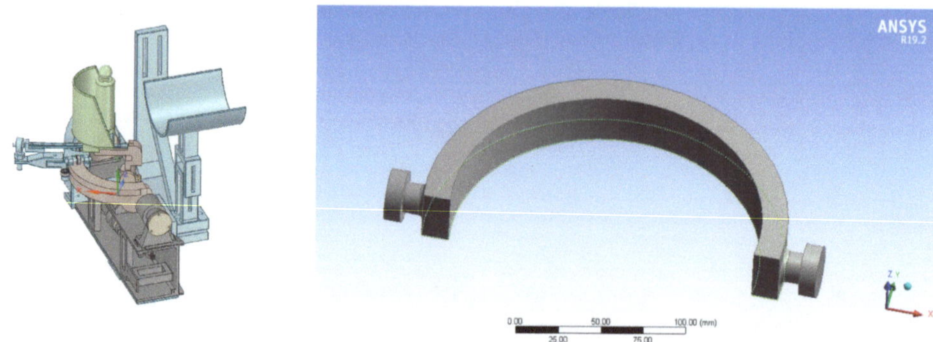

Figure 30. Parameterized model of the circular guide in ANSYS Design Modeler.

Structural optimization involves two objectives:

- Reducing the mass of the circular guide, which is made of aluminum alloy, by respecting the constraint of a maximum equivalent stress;
- Increased fatigue resistance by reducing peak tensions in areas with stress concentrators. These areas are the two spindles through which the guide is supported in the bearing housings.

The optimization block diagram is shown in Figure 31.

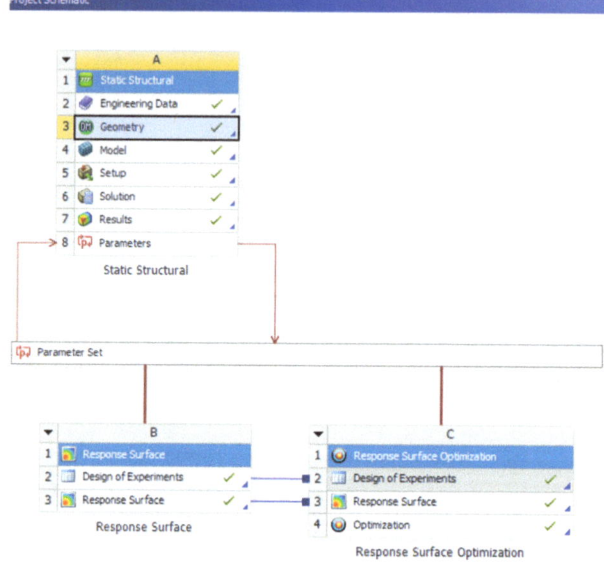

Figure 31. Block diagram of the structural optimization process of the guide.

The steps of the structural optimization process of the circular guide in accordance with the block diagram in Figure 31 are described in the following.

4.1. Parameterized Geometric Modelling of Circular Guidance Using the Design Modeler Preprocessor

Three geometrical parameters of the guide are selected as input data, namely: the radii of the two spindles connecting the circular guide, and the inner radius of the guide, named R1, as shown in Figure 32. At this stage, the physically possible ranges of variation of the geometrical parameters included in the optimization process are checked and determined. Thus, the range of variation of the two radii is between 2.7 and 3.3 mm. For the inner radius of the circular guide, the variation limits are from 82.8 to 101.2 mm.

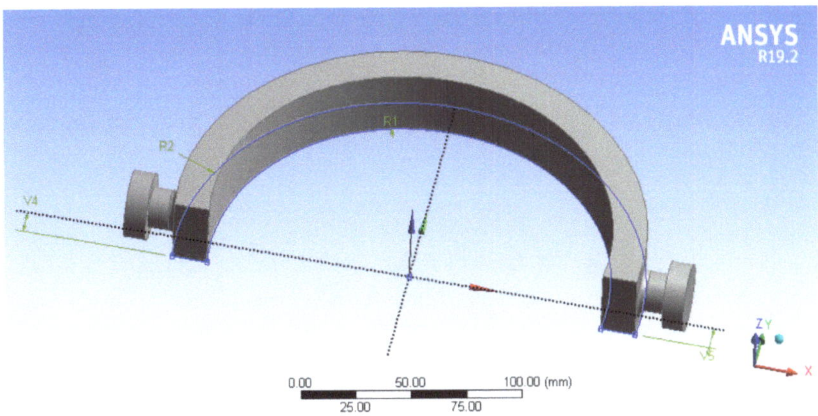

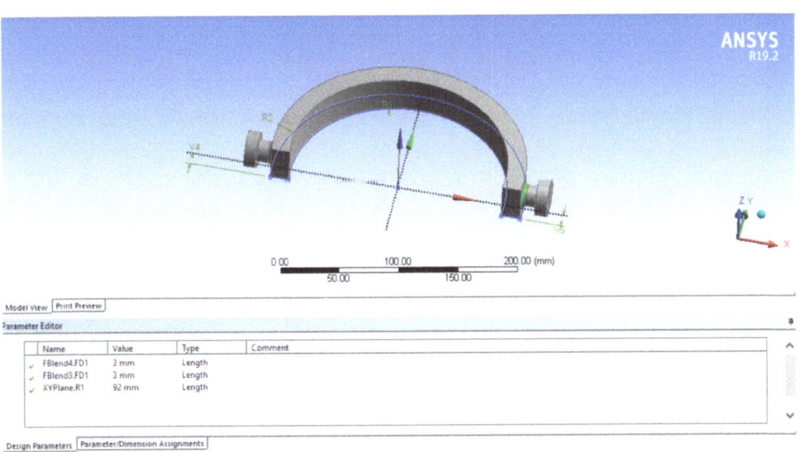

Figure 32. Parameters selected as input in the optimization process.

The output parameters are the guide mass as well as the von Mises equivalent stress. The initial guide mass is calculated based on the definition of the guide material and has a value of 0.54397 kg. The von Mises equivalent stress is calculated after running structural static analysis.

The worst-case loading condition of the circular guide is when the element that rests on the guide is in the extreme position, in which case the guide is subject to a compound stress of both twisting and bending. For finite element analysis, we specified the boundary conditions and loads according to the real model. Thus, we defined rotation Joints A and B between the circular guide and the casing. A torque of 5000 N mm acts on the drive end of the guide, while a fixed support is defined on the opposite end. For discretization of the guide into finite elements, we used hexahedral elements and specified a finite element size of 2 mm. The discretized structure comprises 130,489 elements and 31,668 nodes.

Figure 33 shows two important aspects for the construction of the finite element model, namely the definition of the torsional moment and the finite element discretization of the circular guide.

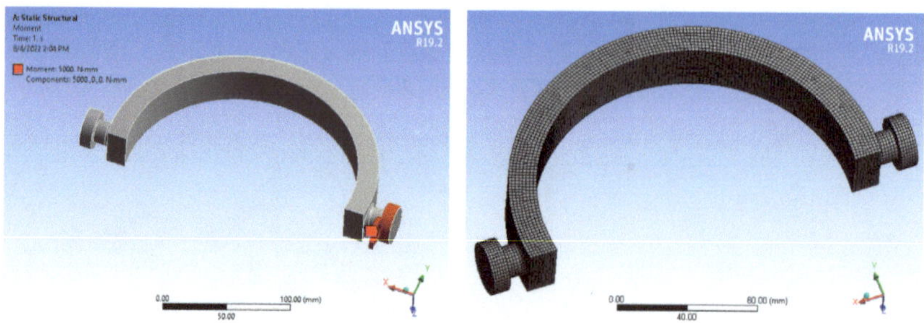

Figure 33. Definition of boundary conditions and discretization into finite elements.

4.2. Running Transient Structural Analysis

Thus, Figure 34 shows the distribution of the equivalent von Mises stresses, the maximum principal stresses, the distribution of the equivalent specific strains, and the total displacements.

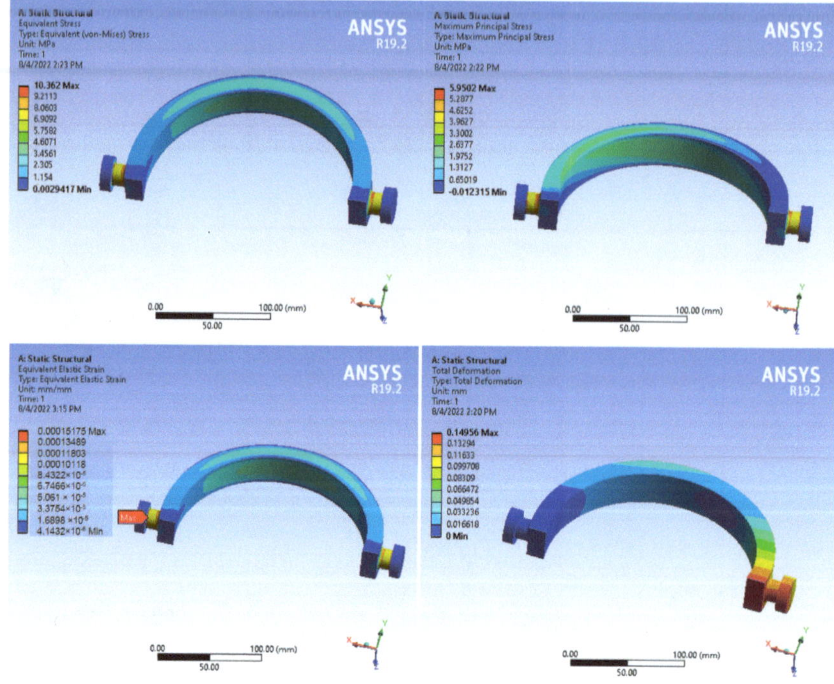

Figure 34. Results of mechanical stresses, displacements, and strains for non-optimized structure.

At this stage, the maximum von Mises equivalent stress of 10.362 MPa is defined as the output parameter.

4.3. Conduct a Design-of-Experiments Study

Based on a design-of-experiments (DE) study, central composite design, several possible design variants are calculated, as shown in Table 25. There are 15 possible design solutions of the considered structure. For each design point of the structure, we present on the line related to the respective DP the values of the input parameters (P1, P2-radii of the connection (Blend), and P3-radius R1) and the output parameters (P4-Solid Mass and P5-Equivalent Stress Maximum). From these results, the following is found:

-    For the output parameter P4-Solid Mass, the minimum calculated value is 0.27269 kg, and the maximum is 0.79132 kg;
-    For the output parameter P5-Equivalent Stress, the minimum calculated value is 10.125 MPa, and the maximum is 19.837 MPa.

Table 25. Design points of design-of-experiments.

Param. No.	Design Point	P1—FBlend4.FD1 (mm)	P2—FBlend3.FD1 (mm)	P3—XYPlane.R1 (mm)	P4—Solid Mass (kg)	P5—Equivalent Stress Maximum (MPa)
1	1 DP	3	3	92	0.54397	10.362
2	2	2.7	3	92	0.54391	10.159
3	3	3.3	3	92	0.54403	10.3
4	4	3	2.7	92	0.54391	10.125
5	5	3	3.3	92	0.54403	10.304
6	6	3	3	82.8	0.79132	10.205
7	7	3	3	101.2	0.27269	19.837
8	8	2.7561	2.7561	84.52	0.7468	10.415
9	9	3.2439	2.7561	84.52	0.7469	10.297
10	10	2.7561	3.2439	84.52	0.7469	10.302
11	11	3.2439	3.2439	84.52	0.74699	10.364
12	12	2.7551	2.7561	99.48	0.32514	13.127
13	13	3.2439	2.7561	99.48	0.32523	13.126
14	14	2.7561	3.2439	99.48	0.32523	13.131
15	15	3.2439	3.2439	99.48	0.32533	13.127

5. Obtaining a Response Surface

A response surface is a 2D or 3D graph showing the variation of output parameters relative to an input parameter. Table 26 shows the results of a Min–Max Search analysis. The table shows the minimum (lines 3 and 4) and maximum (lines 6 and 7) values of the output parameters P4 and P5. Obviously, increasing the input parameter P1 causes a decrease in output parameter P4 (Mass) and an increase in parameter P5 (Equivalent Stress).

Table 26. Min–Max Search.

Name	P1—FBlend4.FD1 (mm)	P2—FBlend3.FD1 (mm)	P3—XYPlane.R1 (mm)	P4—Solid Mass (kg)	P5—Equivalent Stress Maximum (MPa)
Output Parameter Minimums					
P4-Solid Mass	2.7	2.7	101.2	**0.27258**	19.837
P5—Equivalent Stress Maximum	3.033	3.2561	89.656	0.60933	**10.211**
Output Parameter Maximums					
P4—Solid Mass	3.3	3.3	82.8	**0.79144**	10.351
P5—Equivalent Stress Maximum	3.183	2.7264	101.2	0.27268	**19.837**

The response surfaces obtained are shown in Figures 35 and 36. Calculation of response surfaces is necessary because the next step is to optimize these surfaces in order to obtain an optimal solution.

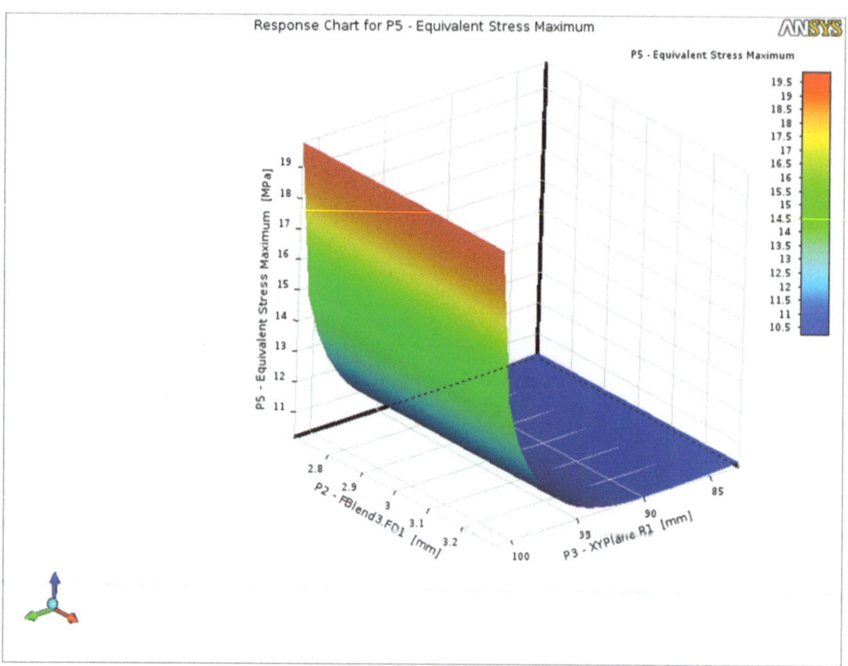

Figure 35. Response chart for P6—Equivalent Stress Maximum vs. P1 and P3.

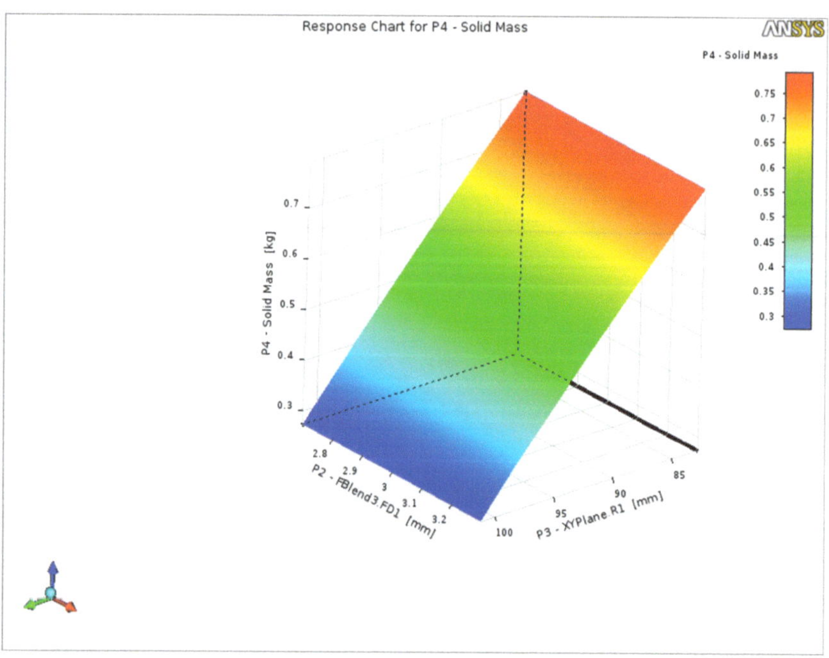

Figure 36. Response chart for P4—Solid Mass vs. P1 and P3.

The computed response surface design parameters (Solid Mass and Equivalent Stress Maximum) characteristics are shown in the Table 27.

Table 27. Response surface parameters.

Parameter / Characteristic	P4—Solid Mass	P5—Equivalent Stress Maximum
Coefficient of Determination (Best Value = 1)		
Learning Points	⋆⋆⋆ 1	⋆⋆⋆ 0.99915
Cross-Validation on Learning Points	⋆⋆⋆ 1	⋆⋆⋆ 0.99857
Root Mean Square Error (Best Value = 0)		
Learning Points	3.32288×10^{-7}	0.073305
Cross-Validation on Learning Points	7.1713×10^{-7}	0.094999
Relative Maximum Absolute Error (Best Value = 0%)		
Learning Points	⋆⋆⋆ 0	5.6275
Cross-Validation on Learning Points	⋆⋆⋆ 0.0010358	7.434
Relative Average Absolute Error (Best Value = 0%)		
Learning Points	⋆⋆⋆ 0	⋆⋆ 1.954
Cross-Validation on Learning Points	⋆⋆⋆ 0.00029053	⋆ 2.7089

⋆ —not acceptable; ⋆⋆ —acceptable; ⋆⋆⋆ —excellent.

Response Surface Optimization

Optimization involves minimizing the output parameter P5, maximum equivalent von Mises stress, and minimizing parameter P4, mass. The defined optimization constraint assumes that the equivalent von Mises stress does not exceed 14 MPa.

Table 28 shows the purpose of the optimization, i.e., the minimization of the input parameters P4 and P5, specifies the optimization method used, and details the 3 points of the optimization surface that are candidates for the optimal solution.

Table 28. Optimization study.

Optimization Study	
Minimize P5; P5 ≤ 14 MPa	Goal, Minimize P5 (Default importance); Strict constraint, P5 values less than or equal cu 14 MPa (Default importance)
Minimize P4	Goal, Minimize P4 (Default importance)
Optimization Method	
MOGA	The MOGA method (Multi-Objective Genetic Algorithm) is a variant of the popular NSGA-II (Non-dominated Sorted genetic Algorithm-II) based on controlled elitism concept. It supports multiple objectives and constraints and aims at finding the global optimum.
Configuration	Generate 300 samples initially, 600 samples per iteration and find 3 candidates in a maximum of 20 iterations.
Status	Converged after 3509 evaluations.

Candidate Points	Candidate Point 1	Candidate Point 2	Candidate Point 3
P1—FBlend4. FD1 (mm)	2.8361	3.0327	3.0587
P2—FBlend3. FD2 (mm)	2.7499	3.2963	3.0008
P3—XYPlane.R1 (mm)	97.543	98.151	95.869
P4—Solid Mass (kg)	⋆⋆ 0.3833	⋆⋆ 0.36531	⋆ 0.43281
P5—Equivalent Stress Maximum (MPa)	⋆⋆ 11.539	⋆⋆ 11.919	⋆⋆ 10.834

⋆ —not acceptable; ⋆⋆ —acceptable.

Table 29 shows the results obtained for the three design solutions (candidate points). Candidate Point 2 is identified as the optimal solution, for which the values of the input parameters are: P1 = 3.0327 mm, P2 = 3.2963 mm, and P3 = 98.151 mm, and the values obtained for the output parameters are: P4 = 0.36531 mm and P5 = 11.919 MPa.

Table 29. Optimization candidate points.

Name	P1—FBlend4.FD1 (mm)	P2—FBlend3.FD1 (mm)	P3—XYPlane.R1 (mm)	P4—Solid Mass (kg)		P5—Equivalent Stress Maximum (MPa)	
				Parameter Value	Variation from Reference	Parameter Value	Variation from Reference
Candidate Point 1	2.8361	2.7499	97.543	★★ 0.3833	4.92%	★★ 11.539	−3.18%
Candidate Point 2	3.0327	3.2963	98.151	★★ 0.36531	0.00%	★★ 11.919	0.00%
Candidate Point 3	3.0587	3.0008	95.869	★ 0.43281	18.48%	★★ 10.834	−9.10%

★ —not acceptable; ★★ —acceptable.

The goal of structural optimization is to achieve more-uniform distribution of stresses and strains to eliminate stress concentrators, to increase fatigue strength, and, most importantly, to reduce the mass of the guide.

As shown in Figure 37 this objective has been achieved. Symmetrical stress distribution is observed, and the optimization constraint of keeping the maximum stresses below the imposed limit is fulfilled.

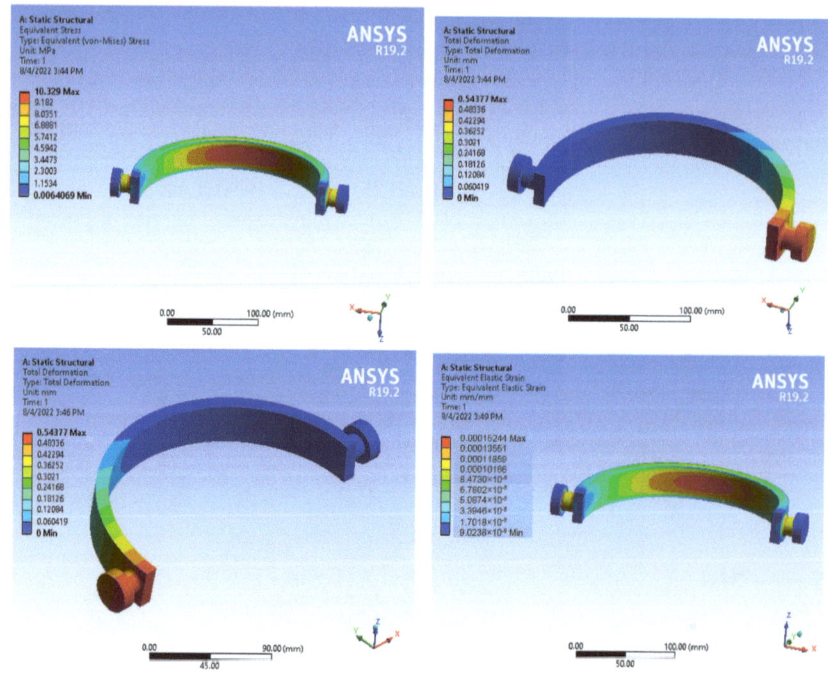

Figure 37. Results of mechanical stresses, displacements, and deformations for the optimized structure.

6. Experimental Determination of the Robot Module Actuation Torques

The robotic system is tested in laboratory conditions using a healthy subject (with informed consent). The tested subject is 1.85 m height and weighs 88 kg, which leads to hand weight (according to [51]) similar to the simulation loads presented in Figure 21. The results are shown in Figure 38, which presents the time history diagram of the torque required for the flexion–extension motion for seven cycles with the following motion parameters: motion amplitude (range of motion) of $60°$ (both for flexion and extension) and speed of $30°/s$. The torque is estimated using the MC_ReadActualTorque (FB) function in Automation Studio, B & R Automation [47], which provides an approximate value of the required motor torque. The maximum required value of the torque is 4.5 Nm, while the average is 0.0821 Nm. Figure 39 presents the time history diagram of the radial–ulnar deviation motion for seven cycles using the following motion parameters: radial motion amplitude of $20°$, ulnar motion amplitude of $30°$ and speed of $30°/s$. The maximum required torque value is -26 Nm, while the average is -10.887 Nm.

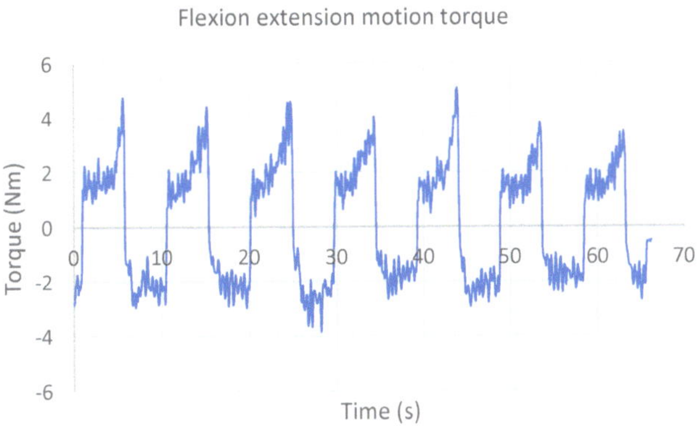

Figure 38. The time history diagram for the flexion–extension motion of the ParReEx-wrist robotic system.

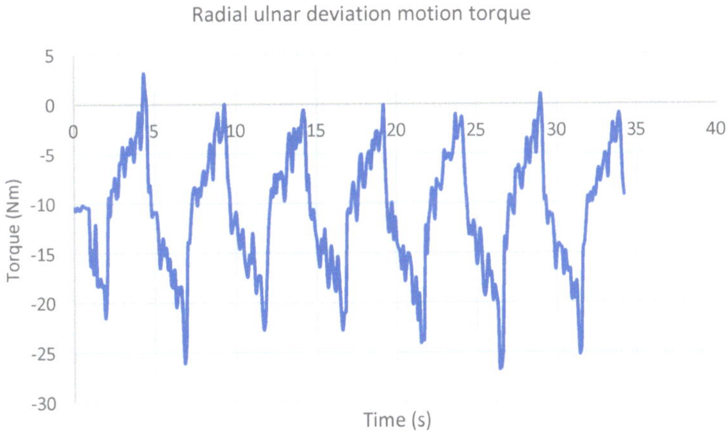

Figure 39. The time history diagram for the radial–ulnar deviation motion of the ParReEx-wrist robotic system.

Very good agreement is observed between the smoothness of the actuation moments determined by numerical simulation and experimentally, as shown in Figures 21, 38 and 39. This validates the results obtained by numerical simulation in ADAMS.

7. Discussion

Simulations of ParReEx robot dynamics were performed in three simulation cases: (1) without friction in the couplings, (2) with friction, and (3) with friction and flexible elements.

The evolution of the RMS value of the resulting connecting forces in Couplings A and B for the three simulation scenarios is shown in Figure 40. The RMS value of a set of values (or a continuous-time waveform) is the square root of the arithmetic mean of the squares of the values, or the square of the function that defines the continuous waveform.

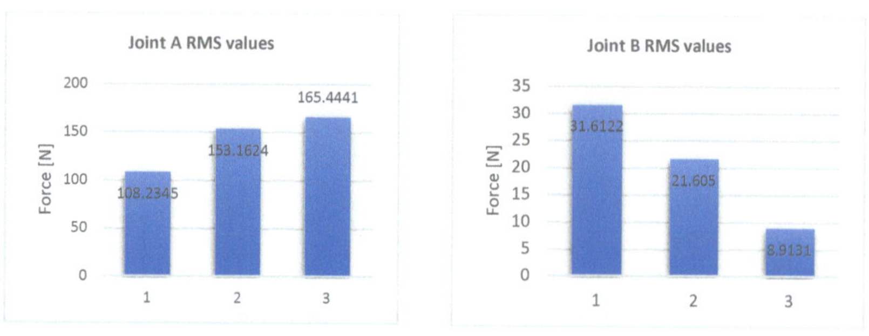

Figure 40. Evolution of the RMS values for the connection forces in Couplings A and B.

The presence of friction in the coupling increases the reaction in Coupling A and decreases the reaction in Coupling B. This can be explained by the fact that charging Coupling A unloads Coupling B. The presence of a toothed belt drive in Coupling A causes additional loading of this driving joint. In this case, the bearing in Coupling B will be a driving one, i.e., it will also take up the axial forces that occur in the circular guide element.

For the other couplings, friction increases the RMS values of the resulting connecting forces. Thus, Figure 41 shows the evolution of the RMS values for the resultant coupling forces in R1 and R2.

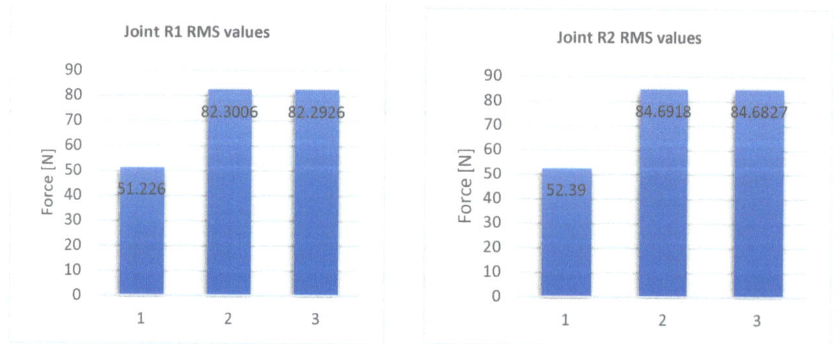

Figure 41. Evolution of the RMS values for the connection forces in Joints R1 and R2.

Further, friction increased the actuation moment for flexion–extension (Torque 1) and radial–ulnar deviation (Torque 2), as Figure 42 illustrates.

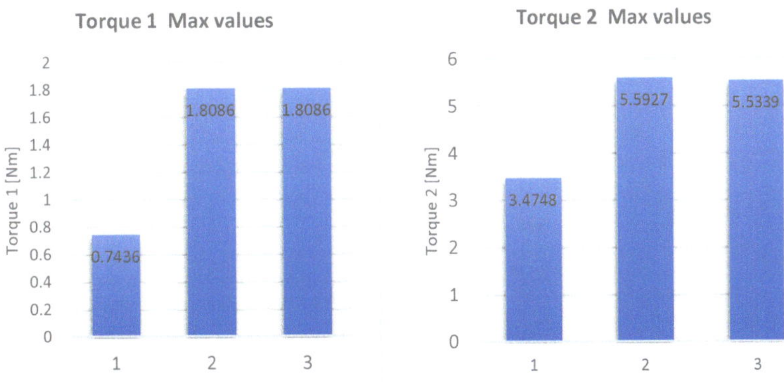

Figure 42. Evolution of the maximum values for the moments of actuation for flexion–extension and radial–ulnar deviation motions.

Numerical simulation in ADAMS (Figure 21) showed that flexion–extension motion of the wrist joint required a maximum moment of 1.8086 Nm, and radial–ulnar deviation required a maximum moment of 5.5339 Nm. Experimentally, higher values were obtained for these parameters: 4.5 Nm for flexion–extension and 20 Nm for radial–ulnar deviation. As a remark on the experimentally determined values, it is found that these quantities do not start from zero, so in the initial starting position, the system is preloaded by the human subject's arm. The same happened in the numerical simulation in ADAMS, only in the simulation we considered the load produced by the patient's arm of 30 N as a constant load.

We also found that the experimentally measured values for the actuation moments have similar variation. The differences between the experimentally measured values and those obtained by numerical simulation can be explained as follows:

✓ In the case of numerical simulation, we defined the motion law by the STEP function, and at the end of the stroke the angular velocity is reduced, which is not the case in the experimental case;

✓ The virtual prototype does not consider the presence of clearance in the kinematic joints. In the case of the experimental prototype, these clearances are inherent, so the components could not be assembled. The presence of these clearances produces jumps of moments at the end of the stroke, which is not the case with the virtual simulation.

8. Conclusions

In this paper, we conducted a study on the dynamics of the ParReEx robot used for the wrist joint. Dynamic analysis was performed using MSC.ADAMS in three simulation scenarios:

(a) Kinematic elements were considered rigid bodies and friction in the kinematic couplings was not considered;

(b) Kinematic elements were considered rigid bodies and friction in the robot couplings was considered;

(c) Both flexibility of kinematic elements and friction in the couplings were considered.

Structural optimization of the circular guide of the ParReEx robot using ANSYS software was performed.

The differences between the experimentally measured values and those obtained by numerical simulation are acceptable for the following reasons:

- Parts made by 3D printing have tolerances within 0.2 mm. For this reason, the experimental model is in real assembly conditions;

- Similar variations of torsional moments were determined experimentally and by numerical simulation in ADAMS, which validates the two models.

Dynamic analysis allows us to draw conclusions about the operational safety of a robotic system. A next step is dynamic optimization of the robotic structure. Structural optimization can eliminate stress concentrators in the robot structure, which contributes to increased operational safety.

The results obtained by numerical simulation were validated by experimental measurements. Thus, the actuation moments for the two movements of the wrist joint—flexion–extension and radial–ulnar deviation—were determined experimentally. Concordance between the variations determined by numerical simulation and those determined experimentally was observed.

Author Contributions: Conceptualization, D.T., I.D.G., D.P. and G.C.; methodology, D.T., I.D.G. and D.P.; software, I.D.G. and D.T.; validation, P.T., B.G., N.T., C.A. and D.N.T.; formal analysis, I.D.G., P.T. and B.G.; investigation, I.D.G., D.T. and D.P.; resources, D.T.; writing—original draft preparation, I.D.G. and D.T.; writing—review and editing, D.P., B.G. and G.C.; visualization, D.N.T., N.T. and C.A.; supervision, D.T., D.P. and G.C.; project administration, D.P. All authors have read and agreed to the published version of the manuscript.

Funding: This work was supported by a grant from the Romanian Ministry of Research and Innovation, CCCDI—UEFISCDI, project number PN-III-P2-2.1-PED-2019-3022/546PED/2020 (NeuroAssist) within PNCDI III, and by the project POCU/380/6/13/123927—ANTREDOC, "Entrepreneurial competencies and excellence research in doctoral and postdoctoral studies programs", co-funded from the European Social Fund through the Human Capital Operational Program 2014–2020.

Institutional Review Board Statement: Not applicable.

Informed Consent Statement: All subjects gave their informed consent for inclusion before they participated in the study. The study was conducted in accordance with the Declaration of Helsinki, and the protocol was approved by the Ethics Committee of NeuroAssist project.

Data Availability Statement: The data presented in this study are openly available in the reference number.

Conflicts of Interest: The authors declare no conflict of interest.

References

1. Klamroth-Marganska, V.; Blanco, J.; Campen, K.; Curt, A.; Dietz, V.; Ettlin, T.; Felder, M.; Fellinghauer, B.; Guidali, M.; Kollmar, A.; et al. Three-dimensional, task-specific robot therapy of the arm after stroke: A multicentre, parallel-group randomised trial. *Lancet Neurol.* **2014**, *13*, 159–166. [CrossRef]
2. Lindsay, M.P.; Norrving, B.; Sacco, R.L.; Brainin, M.; Hacke, W.; Martins, S.; Pandian, J.; Feigin, V. World stroke organization (WSO): Global stroke fact sheet 2019. *Int. J. Stroke* **2019**, *14*, 806–817. [CrossRef] [PubMed]
3. Feigin, V.L.; Norrving, B.; Mensah, G.A. Global Burden of Stroke. *Circ. Res.* **2017**, *120*, 439–448. [CrossRef] [PubMed]
4. Virani, S.S.; Alonso, A.; Benjamin, E.J.; Bittencourt, M.S.; Callway, C.W.; Carson, A.P.; Chamberlain, A.M.; Chang, A.R.; Cheng, S.; Delling, F.N.; et al. Heart disease and stroke statistics-2020 update: A report from the Americanheart association. *Circulation* **2020**, *141*, e1–e458. [CrossRef] [PubMed]
5. Skilbeck, C.E.; Wade, D.T.; Hewer, R.L.; Wood, V.A. Recovery after stroke. *J. Neurol. Neurosurg. Psychiatry* **1982**, *46*, 5–8. [CrossRef]
6. Kwakkel, G.; Kollen, B.J.; Van der Grond, J.; Prevo, A.J. Probability of regaining dexterity in the flaccid upper limb: Impact of severity of paresis and the time since onset in acute stroke. *Stroke* **2003**, *34*, 2181–2186. [CrossRef]
7. Sivan, M.; Gallagher, J.; Makower, S.; Keeling, D.; Bhakta, B.; O'Connor, R.J.; Levesley, M. Home-based computer assisted arm rehabilitation robotic device for upper limb exercise after stroke: Result of feasibility study in home setting. *J. Neuroeng. Rehabil.* **2014**, *11*, 163. [CrossRef]
8. Gorelick, P.B. The global burden of stroke: Persistent and disabling. *Lancet Neurol.* **2019**, *18*, 417–418. [CrossRef]
9. Murie-Fernandez, M.; Irimia, P.; Martinez-Vila, E.; Meyer, M.J.; Teasell, R. Neuro-rehabilitation after stroke. *Neurol.* **2010**, *25*, 189–196. [CrossRef]
10. Van Peppen, R.P.; Kwakkel, G.; Wood-Dauphinee, S.; Hendriks, H.J.; Van der Wees, P.J.; Dekker, J. The impact of physical therapy on functional outcomes after stroke: What's the evidence? *Clin. Rehabil.* **2004**, *18*, 833. [CrossRef]
11. Jarrassé, N.; Proietti, T.; Crocher, V.; Robertson, J.; Sahbani, A.; Morel, G.; Roby-Brami, A. Robotic exoskeletons: A perspective for the rehabilitation of arm coordination in stroke patients. *Front. Hum. Neurosci.* **2014**, *8*, 1845–1846. [CrossRef] [PubMed]
12. Yamamoto, I.; Inagawa, N.; Matsui, M.; Hachisuka, K.; Wada, F.; Hachisuka, A. Research and development of compact wrist rehabilitation robot system. *Bio-Med. Mater. Eng.* **2014**, *24*, 123–128. [CrossRef] [PubMed]
13. Loureiro, R.C.V.; Harwin, W.; Nagai, K.; Johnson, M. Advances in upper limb stroke rehabilitation: A technology push. *Med. Biol. Eng. Comput.* **2011**, *49*, 1103–1118. [CrossRef] [PubMed]

14. Volpe, B.T.; Huerta, P.T.; Zipse, J.L.; Rykman, A.; Edwards, D.; DiPietro, L.; Hogan, N.; Krebs, H.I. Robotic Devices as Therapeutic and Diagnostic Tools for Stroke Recovery. *Arch. Neurol.* **2009**, *66*, 1086–1090. [CrossRef]
15. Reinkensmeyer, D.J. Robotic Assistance for Upper Extremity Training after Stroke. In *Advanced Technologies in Rehabilitation*; Gaggioli, A., Keshner, E.A., Weiss, P.L., Riva, E., Eds.; IOS Press: Amsterdam, The Netherlands, 2009; pp. 25–39.
16. Huang, J.; Tu, X.; He, J. Design and evaluation of RUPERT wearable upper extremity exoskeleton robot for clinical and in home therapies. *IEEE Trans. Syst. Man Cybern. Syst.* **2016**, *46*, 926–935. [CrossRef]
17. Vaida, C.; Plitea, N.; Carbone, G.; Birlescu, I.; Ulinici, I.; Pisla, A.; Pisla, D. Innovative development of a spherical parallel robot for upper limb rehabilitation. *Int. J. Mech. Robot. Syst.* **2018**, *4*, 256–276. [CrossRef]
18. Pisla, D.; Tarnita, D.; Tucan, P.; Tohanean, N.; Vaida, C.; Geonea, I.D.; Bogdan, G.; Abrudan, C.; Carbone, G.; Plitea, N. A Parallel Robot with Torque Monitoring for Brachial Monoparesis Rehabilitation Tasks. *Appl. Sci.* **2021**, *11*, 9932. [CrossRef]
19. Geonea, I.D.; Tarnita, D.; Pisla, D.; Carbone, G.; Bolcu, A.; Tucan, P.; Georgescu, M.; Tarniță, D.N. Dynamic Analysis of a Spherical Parallel Robot Used for Brachial Monoparesis Rehabilitation. *Appl. Sci.* **2021**, *11*, 11849. [CrossRef]
20. Tarnita, D.; Marghitu, D. Analysis of a hand arm system. *Robot. Comput. Integr. Manuf.* **2013**, *29*, 493–501. [CrossRef]
21. Gull, M.A.; Bai, S.; Bak, T. A Review on Design of Upper Limb Exoskeletons. *Robotics* **2020**, *9*, 16. [CrossRef]
22. Veerbeek, J.M.; Langbroek-Amersfoort, A.C.; Van Wegen, E.E.; Meskers, C.G.; Kwakkel, G. Effects of robot-assisted therapy for the upper limb after stroke. *Neurorehabilit. Neural Repair* **2017**, *31*, 107–121. [CrossRef] [PubMed]
23. Zengin-Metli, D.; Özbudak-Demir, S.; Eraktaş, İ.; Binay-Safer, V.; Ekiz, T. Effects of robot assistive upper extremity rehabilitation on motor and cognitive recovery, the quality of life, and activities of daily living in stroke patients. *J. Back Musculoskelet. Rehabil.* **2018**, *31*, 1059–1064. [CrossRef] [PubMed]
24. Kim, G.; Lim, S.; Kim, H.; Lee, B.; Seo, S.; Cho, K.; Lee, W. Is robot-assisted therapy effective in upper extremity recovery in early stage stroke?—A systematic literature review. *J. Phys. Therapy Sci.* **2017**, *29*, 1108–1112. [CrossRef] [PubMed]
25. Shahar, N.; Schwartz, I.; Portnoy, S. Differences in muscle activity and fatigue of the upper limb between Task-Specific training and robot assisted training among individuals post stroke. *J. Biomech.* **2019**, *89*, 28–33. [CrossRef] [PubMed]
26. Bian, H.; Chen, Z.; Wang, H.; Zhao, T. Mechanical design of EFW Exo II: A hybrid exoskeleton for elbow-forearm-wrist rehabilitation. In Proceedings of the 2017 International Conference on Rehabilitation Robotics (ICORR), London, UK, 17–20 July 2017; pp. 689–694. [CrossRef]
27. Geonea, I.D.; Tarnita, D. Design and evaluation of a new exoskeleton for gait rehabilitation. *Mech. Sci.* **2017**, *8*, 307–321. [CrossRef]
28. Tarnita, D.; Geonea, I.; Petcu, A.; Tarnita, D.N. Numerical simulations and experimental human gait analysis using wearable sensors. In *International Workshop on Medical and Service Robots*; Springer: Cham, Switzerland, 2016; pp. 289–304.
29. Berceanu, C.; Tarnita, D.; Filip, D. About an experimental approach used to determine the kinematics of the human movement. *J. Solid State Phenom. Robot. Autom. Syst.* **2010**, *166–167*, 45–50. [CrossRef]
30. Tarnita, D.; Pisla, D.; Geonea, I.; Vaida, C.; Catana, M.; Tarnita, D.N. Static and Dynamic Analysis of Osteoarthritic and Orthotic Human Knee. *J. Bionic Eng.* **2019**, *16*, 514–525. [CrossRef]
31. Tarnita, D.; Catana, M.; Dumitru, N.; Tarnita, D.N. Design and simulation of an orthotic device for patients with osteoarthritis. In *New Trends in Medical and Service Robots*; Springer: Cham, Switzerland, 2016; pp. 61–77.
32. Hussain, S.; Jamwal, P.K.; Van Vliet, P.; Ghayesh, M.H. State-of-the-Art Robotic Devices for Wrist Rehabilitation: Design and Control Aspects. *IEEE Trans. Hum. Mach. Syst.* **2020**, *50*, 361–372. [CrossRef]
33. Krebs, H.I.; Volpe, B.T.; Williams, D.; Celestino, J.; Charles, S.K.; Lynch, D.; Hogan, N. Robot-Aided Neurorehabilitation: A robot for wrist rehabilitation. *IEEE Trans. Neural Syst. Rehabil. Eng.* **2007**, *15*, 327. [CrossRef]
34. Hogan, N.; Krebs, H.I.; Charnnarong, J.; Srikrishna, P.; Sharon, A. MIT-MANUS: A workstation for manual therapy and training II. In *Proceedings IEEE International Workshop on Robot and Human Communication*; IEEE: New York, NY, USA, 1992; Volume 1833, pp. 161–165.
35. Krebs, H.I.; Celestino, J.; Williams, D.; Ferraro, M.; Volpe, B.; Hogan, N. A Wrist extension for MIT-MANUS. *Adv. Rehabil. Robot.* **2004**, *306*, 377–390.
36. Cappello, L.; Elangovan, N.; Contu, S.; Khosravani, S.; Konczak, J.; Masia, L.; Konczak, J. Robot-aided assessment of wrist proprioception. *Front. Hum. Neurosci.* **2015**, *9*, 198. [CrossRef] [PubMed]
37. Pehlivan, A.U.; Sergi, F.; Erwin, A.; Yozbatiran, N.; Francisco, G.E.; O'Malley, M.K. Design and validation of the RiceWrist-S exoskeleton for robotic rehabilitation after incomplete spinal cord injury. *Robotica* **2014**, *32*, 1415–1431. [CrossRef]
38. Amirabdollahian, F.; Ates, S.; Basteris, A.; Cesario, A.; Buurke, J.; Hermens, H.; Hofs, D.; Johansson, E.; Mountain, G.; Nasr, N.; et al. Design, development and deployment of a hand/wrist exoskeleton for home-based rehabilitation after stroke—SCRIPT project. *Robotica* **2014**, *32*, 1331–1346. [CrossRef]
39. Yin, L.; Huang, L.; Huang, J.; Tian, L.; Li, F. Solution-region-based synthesis approach for selecting optimal four-bar linkages with the Ball–Burmester point. *Mech. Sci.* **2019**, *10*, 25–33. [CrossRef]
40. Saadatzi, M.; Long, D.C.; Celik, O. Comparison of human–robot interaction torque estimation methods in a wrist rehabilitation exoskeleton. *J. Intell. Robot. Syst.* **2018**, *94*, 565–581. [CrossRef]
41. Bae, S.J.; Jang, S.H.; Seo, J.P.; Chang, P.H. The optimal speed for cortical activation of passive wrist movements performed by a rehabilitation robot: A functional NIRS study. *Front. Hum. Neurosci.* **2017**, *11*, 194. [CrossRef]
42. Zhang, L.; Li, J.; Cui, Y.; Dong, M.; Fang, B.; Zhang, P. Design and performance analysis of a parallel wrist rehabilitation robot (PWRR). *Robot. Auton. Syst.* **2020**, *125*, 103390. [CrossRef]

43. Carbone, G.; Gherman, B.; Ulinici, I.; Vaida, C.; Pisla, D. Design issues for an inherently safe robotic rehabilitation device. *Mech. Mach. Sci.* **2018**, *49*, 1025–1032.
44. Bhattacharya, S.; Nenchev, D.N.; Uchiyama, M. A recursive formula for the inverse of the inertia matrix of a parallel manipulator. *J. Mech. Mach. Theory* **1998**, *33*, 957–964. [CrossRef]
45. Codourey, A.; Burdet, E. A body oriented method for finding a linear form of the dynamic equatiojns of fully parallel robot. In Proceedings of the International Conference on Robotics and Automation, Albuquerque, NM, USA, 25 April 1997; pp. 1612–1619.
46. Dasgupta, B.; Mruthyunjaya, T.S. Closed-form dynamic equations of the general Stewart platform through the Newton–Euler approach. *J. Mech. Mach. Theory* **1998**, *33*, 993–1012. [CrossRef]
47. Tucan, P.; Gherman, B.; Major, K.; Vaida, C.; Major, Z.; Plitea, N.; Carbone, G.; Pisla, D. Fuzzy Logic-Based Risk Assessment of a Parallel Robot for Elbow and Wrist Rehabilitation. *Int. J. Environ. Res. Public Health* **2020**, *17*, 654. [CrossRef]
48. Tarnita, D.; Geonea, I.; Petcu, A.; Tarnita, D.-N. Experimental Characterization of Human Walking on Stairs Applied to Humanoid Dynamics. *Adv. Robot. Des. Intell. Control.* **2017**, *540*, 293–301.
49. Gherman, B.; Pisla, D.; Plitea, N.; Vaida, C.; Carbone, G.; Pisla, A.; Banica, A. Parallel Robotic System for Upper Limb Medical Recovery. Patent Number OSIM: 132,234, 14 June 2017.
50. Pennestri, E.; Rossi, V.; Salvini, P.; Valentini, P.P. Review and Comparison of Dry Friction Force Models. *Nonlinear Dyn.* **2016**, *83*, 1785–1801. [CrossRef]
51. Contini, R.; Drillis, R.J.; Bluestein, M. Determination of Body Segment Parameters. *Hum. Factors* **1963**, *5*, 493–504. [CrossRef] [PubMed]

Article

Dynamic Analysis of a Spherical Parallel Robot Used for Brachial Monoparesis Rehabilitation

Ionut Daniel Geonea [1], Daniela Tarnita [1,*], Doina Pisla [2,*], Giuseppe Carbone [3], Alexandru Bolcu [1], Paul Tucan [2], Marius Georgescu [1] and Danut Nicolae Tarniță [4]

1 Faculty of Mechanics, University of Craiova, 200512 Craiova, Romania; ionut.geonea@edu.ucv.ro (I.D.G.); alexandru.bolcu@edu.ucv.ro (A.B.); georgescumarius@live.com (M.G.)
2 Research Center for Industrial Robots Simulation and Testing CESTER, Technical University of Cluj-Napoca, 400641 Cluj-Napoca, Romania; paul.tucan@mep.utcluj.ro
3 Department of Mechanical, Energy and Management Engineering DIMEG, University of Calabria, 87036 Cosenza, Italy; giuseppe.carbone@unical.it
4 Emergency Hospital of Dolj County, 200528 Craiova, Romania; dan_tarnita@yahoo.com
* Correspondence: daniela.tarnita@edu.ucv.ro (D.T.); doina.pisla@mep.utcluj.ro (D.P.)

Abstract: This paper presents studies on the dynamic analysis of the ASPIRE robot, which was designed for the medical recovery of brachial monoparesis. It starts from the virtual model of the existing version of the ASPIRE robot, which is analysed kinematically and dynamically by numerical simulations using the MSC.ADAMS software. For this purpose, this paper presents theoretical aspects regarding the kinematics and dynamics of the markers attached to the flexible bodies built in a specifically developed MSC.ADAMS model. Three simulation hypotheses are considered: (a) rigid kinematic elements without friction in couplings, (b) rigid kinematic elements with friction in couplings, and (c) kinematic elements as deformable solids with friction in couplings. Experimental results obtained by using the physical prototype of ASPIRE are presented. Results such as the connecting forces in the kinematic joints and the torques necessary to operate the ASPIRE robot modules have been obtained by dynamic simulation in MSC.ADAMS and compared with those determined experimentally. The comparison shows that the allure of the variation curve of the moment obtained by simulation is similar to that obtained experimentally. The difference between the maximum experimental value and that obtained by simulation is less than 1%. A finite element analysis (FEA) of the structurally optimized flexion/extension robot module is performed. The results demonstrate the operational safety of the ASPIRE robot, which is structurally capable of supporting the stresses to which it is subjected.

Keywords: dynamic simulation; spherical parallel robot; joint friction; flexible links; rehabilitation; brachial monoparesis

Citation: Geonea, I.D.; Tarnita, D.; Pisla, D.; Carbone, G.; Bolcu, A.; Tucan, P.; Georgescu, M.; Tarniță, D.N. Dynamic Analysis of a Spherical Parallel Robot Used for Brachial Monoparesis Rehabilitation. *Appl. Sci.* **2021**, *11*, 11849. https://doi.org/10.3390/app112411849

Academic Editor: Manuel Armada

Received: 12 November 2021
Accepted: 8 December 2021
Published: 13 December 2021

Publisher's Note: MDPI stays neutral with regard to jurisdictional claims in published maps and institutional affiliations.

1. Introduction

One of the most important factors of global interest is maintaining quality of life in the coming years by increasing life expectancy [1]. Globally, the leading cause of long-term disability in adults is stroke or spinal cord injury [2,3]. Stroke is the second leading cause of death, with an incidence of 11.8%, after ischemic heart disease, with an incidence of 14.8% [4]. The most common consequences that can occur after a stroke are cognitive impairment (VCI), post-stroke fatigue (PSF) and post-stroke depression (PSD) [5]. In more than 85% of cases, stroke causes upper limb injuries, [6] and only in 10% of cases, the subjects regain arm mobility [7]. Strokes can cause pure brachial monoparesis, which supposes partial paralysis of the upper limb, an unusual cerebrovascular syndrome that can be misdiagnosed due to the lack of any other neurological deficits. Few reports in the literature adequately describe the syndrome or provide a substantial clinical or anatomical correlation [8,9]. The main problem is the difficult recovery due to the complexity of the movement of the upper limbs [10]. The patient must follow a repetitive and task-oriented functional training [11], which translates into a high consumption of time and energy for therapists [12].

Issues of annual medical cost and stroke trends were discussed in [13]. In the US, the annual medical cost for stroke will increase up to 2.5 times in 2030 compared to 2012. In the UK, stroke is the fourth leading cause of death [14]. The annual cost of health and social care for patients with stroke is over 5 billion GBP [15,16]. Currently, in developed countries, 3–4% of health expenditures are dedicated to stroke cases [17].

The number of strokes in the elderly is increasing. In addition, the age of people suffering from stroke is decreasing, reaching 50 years or even 40 years, the medical cost for the treatment of post-stroke trauma is high, it requires expensive medical care and, in addition, the number of therapists able to provide physical training is low compared to the number of patients [18].

In recent decades, it has been shown that exoskeleton-based rehabilitation can be used as an alternative to regular manual therapy to improve motor function [19]. Studies on the development of robotic structures for medical rehabilitation of the lower limb and upper limb have been intensively developed in recent years [20–30]. A complex classification, taking into account defining criteria, of robotic devices for upper limb rehabilitation can be found in [31,32], while in [33], a review of the literature investigating upper body kinematics in people without disabilities through optical motion capture can be found.

Li J. et al. [34] propose a new 4-DOF upper limb exoskeleton that has an ergonomic design and is kinematically compatible with the upper limb. Culmer P.R. et al. [35] developed a novel system for quantitatively measuring arm movement. In order to obtain upper limb joint angles and torques, a series of data was analysed with respect to Cartesian and upper limb coordinate systems to better understand arm movement, in particular to objectively evaluate physical therapy treatments and support the development of robotic devices to facilitate upper limb rehabilitation. In [36], the authors studied the activity and fatigue of upper extremity muscles, pain levels, subject satisfaction levels, and number of repetitions in task-specific training compared with robot-assisted training in individuals post-stroke. The research [37] concluded that virtual or actual task-specific robot training (TSRT) performed with a robotic orthosis or a physical therapist significantly reduced arm impairments around the shoulder and elbow without significant gains in fine motor hand control, activities of daily living or independence.

The exoskeleton technology is evolving quickly but still needs research to solve technical challenges and kinematic compatibility and promote the development of effective human–robot interaction. In [38], researchers have addressed a design problem of an exoskeleton by applying cable transmission, which allows the exoskeleton to place the actuators at a fixed base.

Hamida et al. [39] proposed and applied a combined methodology to optimize a cable-driven parallel robot for upper limb rehabilitation exercises by minimizing the tensions in cable and achieving the smallest footprint. In [40], the authors presented a new mechanism for providing variable stiffness for a 4 DOFs robot used for upper limb rehabilitation.

ASPIRE [22] is a device designed as an innovative structure, a parallel robotic system with 3 DOFs based on a spherical architecture designed for upper limb rehabilitation. The main purpose of its design was to solve problems related to shoulder flexion/extension as well as adduction/abduction and pronation/supination of the forearm. Its kinematic structure and motion were presented in [23]. Studies related to how it worked in clinical trials can be found in [30,41]. In [41], the ASPIRE interchangeability was investigated with classical physiotherapy using quantitative electroencephalogram (EEG), motor driving times and turn/amplitude analysis. A significant effect of the therapy was found for various pathologies. Design optimization following clinical evaluation and clinician feedback was presented in [31]. After careful monitoring and analysis of the robotic system, significant differences were revealed compared to the initial laboratory tests performed on healthy subjects. An improved design was elaborated based on the critical characteristics of clinical trials.

The aim of this study is to analyze from a kinematic, dynamic and experimental point of view the ASPIRE robot for the medical recovery of brachial monoparesis. For this purpose, a dynamic simulation of the robot in the ADAMS virtual environment when a virtual patient for rehabilitation uses the robot will be performed. Three distinct situations will be studied:

- The kinematic elements will be considered as rigid bodies, and the friction in the kinematic couplings will not be considered;
- The kinematic elements will be considered as rigid bodies, and the friction in the robot couplings will be considered;
- Both the flexibility of the kinematic elements and the friction in the couplings will be considered.

Results obtained by numerical simulation will be compared with those obtained from MSC.ADAMS. Finally, a verification of the structural integrity of the flexion/extension module of the robot, which is subject to the highest mechanical stresses, will be performed.

2. Constructive and Functional Description of the ASPIRE Robot

ASPIRE is a parallel robotic system based on a spherical architecture dedicated to be used in rehabilitation and it is the subject of a patent [42]. The movements that the robot can perform are flexion-extension and adduction-abduction of the shoulder joint, as well as pronation-supination of the forearm joint with the help of a third motor. The main design feature of the robotic system consists of two circular guides that lead to a spherical motion of the characteristic point of the mechanism.

The robotic structure's frame was built out of aluminum profiles, while the robot's housing was made of Plexiglas. The sliding carriages were 3D printed and appropriately lubricated to maintain a low friction coefficient, and the circular guides were made of aluminum. The shoulder rest was also 3D printed to guarantee that the shoulder was supported throughout the flexion and extension movements. The forearm support was 3D printed and integrated into an adjustable mechanism to accommodate varying forearm lengths. The metal palm rest was integrated into the pronation/supination mechanism. The horizontal circular guide was connected to the vertical circular guide through a passive revolute joint between the circular guides. An actuated height adjustment mechanism was added to allow easy setup of the robot between patients of different heights. To make it easier to set up the robot between patients of different heights, an actuated height adjustment system was incorporated.

The kinematic scheme based on which the virtual and the experimental model is built is shown in Figure 1 [43]. As shown in the kinematic diagram, the ASPIRE robot is composed of two modules. The first module, framed by the green line border, is intended for the rehabilitation movement of the shoulder. The XOYZ Cartesian reference system (frame) is placed in the centre of the shoulder joint. By rotating the circular segment, G_1, around the Z axis of the reference system, the adduction-abduction movement q_1 is performed.

The flexion-extension movement q_2 is performed by rotating the circular portion G_2 around the Y axis of the same frame. The two circular portions G_1 and G_2 are connected by means of the passive kinematic coupling (r_f), so that a spherical trajectory of the reference system $O'X'Y'Z'$ is obtained with respect to the point O, as shown in Figure 1b. The pronation-supination movement is defined by the q_3 coordinate. The second module allows the vertical adjustment of the rehabilitation mechanism so that, with the help of a screw-nut system, the translation movement q_4 is performed.

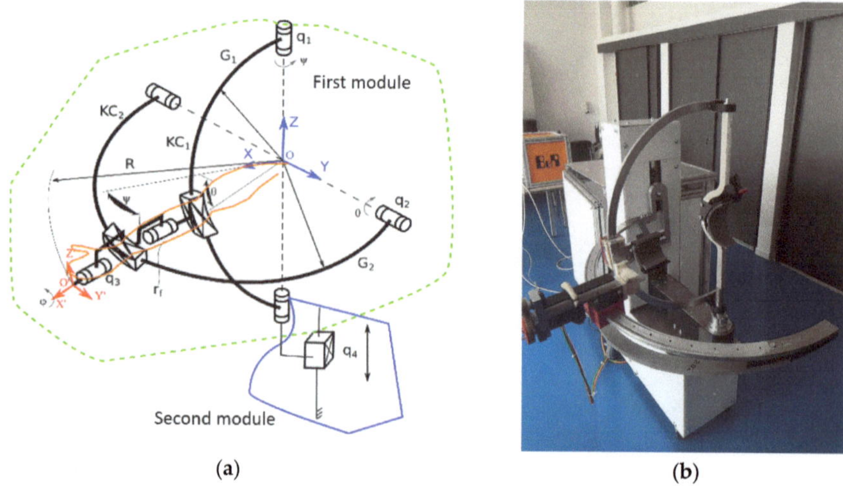

Figure 1. (**a**) Kinematic scheme; (**b**) physical prototype of ASPIRE [43].

3. Theoretical Aspects Regarding the Kinematic and Dynamic Analysis of the ASPIRE Robot in MSC.ADAMS

For the dynamic study of the robotic systems, the most used software is MSC.ADAMS [44], due to the accuracy of results obtained. Thus, the use of this software in various research fields in the industry is noticeable, among which can be mentioned exoskeleton robotic systems or flexible robotic units for minimally invasive surgery [45–52]. MSC.ADAMS uses the kinematics of markers in three essential areas: kinematics, dynamics and optimization of mobile mechanical systems. Marker kinematics deals with the study of the position, orientation, velocity and acceleration of markers [53].

The instantaneous position vector of a marker that is attached to a node, noted P, on a flexible body is the sum of three vectors (Figure 2).

$$\vec{r}_p = \vec{x} + \vec{s}_p + \vec{u}_p, \tag{1}$$

where $\vec{x}$ is the position vector of the origin of the local reference system with the origin in B, attached to the flexible body, relative to the origin of fixed reference system, G; $\vec{s}_p$ represents the position vector of the point P in the initial position on the undeformed body in relation to the local reference system attached to the flexible body; and $\vec{u}_p$ is the position vector of the point P', the new position of P on the deformed body, in relation to the initial position of the point P.

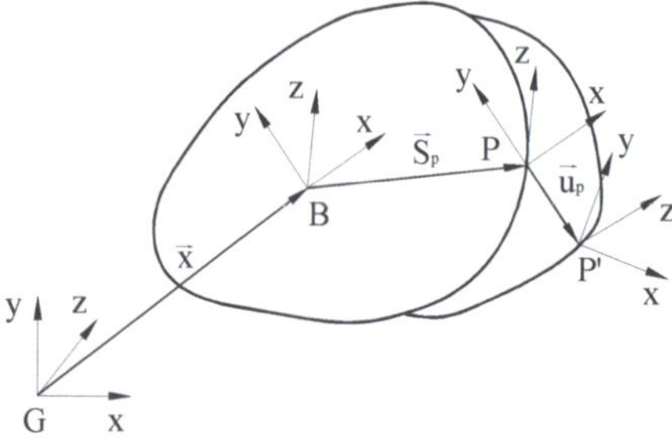

Figure 2. The position vector of the point P' of a deformed flexible body relative to a reference system attached to the flexible body and the fixed base G.

We write Equation (1) in a matrix form expressed in relation to the basic coordinate system:

$$[r_p] = [x] + [^G A^B]([s_p] + [u_p]),\tag{2}$$

where $[x]$ is matrix associated with the position vector $\vec{x}$, having as elements the generalized coordinates x, y and z of the origin B of the system associated with the flexible body.

$[s_p]$ is a matrix associated with the position vector $\vec{s}_p$ having as elements the constant coordinates of the point P in relation to the local system of the body.

$[^G A^B]$ is the transformation matrix from the local reference system in B to the basic system, G, being composed of cosine directors of the origin of the local reference system in relation to the fixed one. The orientation is calculated using Euler's angles, ψ, θ and φ, which are used to define the generalized coordinates of the flexible body.

$[u_p]$ is a matrix associated with the position vector of point P' with respect to P, expressed by coordinates related to the local coordinates system attached to the body. Matrix $[u_p]$ can be written:

$$[u_p] = [\Phi_p] \cdot [q],\tag{3}$$

where $[\Phi_p]$ is the portion of the form function matrix that corresponds to the translation degrees of freedom of the node P, being a $3 \times M$ matrix type, where M is the number of form functions. Coordinates of form functions q_i, $(i = 1...M)$ also represent the generalized coordinates of the flexible body.

Therefore, the complete matrix form of the generalized coordinates of the flexible body is:

$$\xi = \left\{ \begin{array}{c} x \\ y \\ z \\ \psi \\ \theta \\ \varphi \\ q_{i,\ (i=1...M)} \end{array} \right\} = \left\{ \begin{array}{c} X \\ \Psi \\ q \end{array} \right\},\tag{4}$$

3.1. Velocities

To calculate the kinetic energy, the instantaneous translation velocity of the point in relation to the global reference system, which is obtained by differentiating the relationship Equation (2) with respect to time, has been computed, resulting in:

$$[v_p] = [\dot{x}] + [^G \dot{A}^B]([s_p] + [u_p]) + [^G A^B][\dot{u}_p],\tag{5}$$

Considering the property of the vector product of two matrices, of which the matrix $[a]$ is antisymmetric:

$$[a] \times [b] = \begin{bmatrix} 0 & -a_z & a_y \\ a_z & 0 & -a_x \\ -a_y & a_x & 0 \end{bmatrix} [b] = [\tilde{a}][b] = -[b][\tilde{a}],\tag{6}$$

it can be written:

$$[^G \dot{A}^B][s_p] = [^G A^D]([^G \omega_B^B] \times [s_p]) = [^G A^R][^G \tilde{\omega}_B^R][s_p] - \\ = -[^G A^B][s_p][^G \tilde{\omega}_B^B]\tag{7}$$

where $[^G \omega_B^B]$ is the matrix of the angular velocity of the body relative to the fixed system (expressed in the coordinates of the body).

Replacing Equation (7) in Equation (5) results in:

$$[v_p] = [\dot{x}] - [{}^G A^B]\left([s_p] + [u_p]\right)[B][\dot{\psi}] + [{}^G A^B][\Phi_p][\dot{q}],\tag{8}$$

where:

$$[{}^G \widetilde{\omega}_B^B] = [B][\dot{\psi}],\tag{9}$$

representing the relationship that makes the connection between the angular velocity and the derivative in relation to the time of the orientation state.

3.2. Angular Velocities

To meet the angular constraints, the markers orientation of a flexible body during deformation is instantly evaluated. As the body deforms, the marker rotates at relatively small angles to the reference frame. Like translational deformations, these angles are obtained using a modal overlap, similar to Equation (3),

$$[\theta_p] = [\Phi_p^*][q],\tag{10}$$

where $[\Phi_p^*]$ is the part of form functions matrix that corresponds to the degrees of freedom of rotation of the node P. The matrix size of $[\Phi_p^*]$ is $3 \times M$, where M is the number of form functions.

The J marker orientation in relation to the base is represented by the Euler transformation matrix, $[{}^G A^J]$. This matrix is the product of the three transformation matrices:

$$\left[{}^G A^J\right] = \left[{}^G A^B\right] \cdot \left[{}^B A^P\right] \cdot \left[{}^P A^J\right],\tag{11}$$

where $[{}^G A^B]$ is the matrix for transforming the coordinates from the local reference system attached in B to the body to basic or global coordinates; $[{}^B A^P]$ is the transformation matrix during the change of orientation by deforming the flexible body of the pole P coordinates in relation to the local reference system attached to the body; and $[{}^P A^J]$ is the constant transformation matrix, defined by the user when placing the marker J on the flexible body.

Matrix $[{}^B A^P]$ is defined by the cosine directors θ_{pz}, θ_{py} of a vector with small angles, θ_p:

$$\left[{}^B A^P\right] = \begin{bmatrix} 1 & -\theta_{pz} & \theta_{py} \\ \theta_{pz} & 1 & -\theta_{pz} \\ -\theta_{py} & \theta_{pz} & 1 \end{bmatrix} = I + \widetilde{\theta}_p,\tag{12}$$

where the sign ~ represents the antisymmetric operator.

3.3. Angular Velocities

The angular velocity of a marker, J, of a flexible body is the sum of the angular velocities of the body and the angular velocities during deformation:

$$[{}^G \omega_B^J] = [{}^G \omega_B^P] = [{}^G \omega_B^B] + [{}^B \omega_B^P] = [{}^G \omega_B^B] + [\Phi_P^*][\dot{q}],\tag{13}$$

The equations governing a flexible body are derived from Lagrange's equations:

$$\begin{cases} \dfrac{d}{dt}\left(\dfrac{\partial L}{\partial \dot{\xi}}\right) - \dfrac{\partial L}{\partial \xi} + \dfrac{\partial F}{\partial \dot{\xi}} + \left[\dfrac{\partial \Gamma}{\partial \xi}\right]^T \lambda - Q = 0 \\ \Gamma = 0 \end{cases},\tag{14}$$

where L is the Lagrangian, defined by the relationship $L = T - V$, where T and V represents the kinetic and potential energy, respectively; F represents the energy dissipation function, defined by the relationship:

$$[F] = \frac{1}{2}[\dot{q}]^T[D][\dot{q}]. \tag{15}$$

Γ represents the constraint equations; λ represents Lagrange's multipliers for constraints; ζ represents the generalized coordinates defined in the relationships Equation (4); Q represents the generalized applied forces (applied forces projected on ζ); and $[D]$ represents the damping matrix.

The velocity in relation to Equation (5) can be expressed in terms of the derivative with respect to time of the state vector, ζ.

$$[v_p] = \left[[I] - [^G A^B]([s_p] + [u_p])[B] + [^G A^B][\Phi_p]\right][\dot{\zeta}]. \tag{16}$$

The kinetic energy for a flexible body is given by:

$$T = \frac{1}{2}\int_V \rho v^T v \, dV \approx \frac{1}{2}\sum_p [[m_p][v_p^T][v_p] + [^G\omega_P^B]^T[I_p][^G\omega_P^B]], \tag{17}$$

where $[m_p]$ and $[I_P]$ are the nodal mass matrix and the nodal inertia tensor of the node P.

$[I_P]$ is sometimes a negligible amount, as, for example, bars, beams, housings are used in the model with flexible components.

By substituting v_p and ω_p and simplifying, we obtain the kinetic energy in the generalized coordinates and the matrix of the generalized masses.

$$T = \frac{1}{2}[\dot{\zeta}]^T[M(\zeta)][\dot{\zeta}], \tag{18}$$

For clarity of presentation, the matrix of generalized masses is partitioned into $M(\zeta)$, in block matrices 3×3 in size.

$$M(\zeta) = \begin{bmatrix} M_{tt} & M_{tr} & M_{tm} \\ M_{tr}^T & M_{rr} & M_{rm} \\ M_{tm}^T & M_{rm}^T & M_{mm} \end{bmatrix}, \tag{19}$$

where indices t, r and m represent the translation, rotation and modal degrees of freedom, respectively.

The expression for the mass matrix $[M(\zeta)]$ simplifies to an expression in 9 inertia invariants:

$$\begin{gathered} M_{tt} = L^{-1}I; \quad M_{tr} = -A\left[L^2 + L_j^3 q_j\right]B \\ M_{tm} = AL^3; \quad M_{rr} = B^T\left[L^7 - \left[L_j^8 + L_j^{8T}\right]q_j - T_{ij}^9 q_i q_j\right]B \\ M_{rm} = B^T\left[L^4 + L_j^5 q_j\right]; \quad M_{mm} = L_6 \end{gathered} \tag{20}$$

The explicit dependence of the mass matrix on the modal coordinates is obvious. The dependence on the orientation coordinates of the system is due to the transformation matrices A and B. Inertia invariants are calculated from the N nodes of the finite element model based on information about each node m_p, its undeformed location s_p, and its participation in the components of form functions Φ_p. The discrete shape of the inertia components is provided in Table 1.

Position and orientation constraints for flexible body markers are satisfied using the kinematic properties of the previously presented markers.

The final form of the differential equations of motion is:

$$[M][\ddot{\zeta}] + [\dot{M}][\dot{\zeta}] - \frac{1}{2}\left[\frac{\partial M}{\partial \zeta}\dot{\zeta}\right]^T[\dot{\zeta}] + [K][\zeta] + [D][\dot{\zeta}] + \left[\frac{\partial \psi}{\partial \zeta}\right]^T \lambda = [Q], \tag{21}$$

where $[\xi]$, $\left[\dot{\xi}\right]$, and $\left[\ddot{\xi}\right]$ are the matrices associated with the generalized coordinates of the flexible body and their derivatives with respect to the time; $[M]$ represents the flexible body mass matrix; and $\left[\dot{M}\right]$ represents the derivative with respect to time of the flexible body mass matrix.

Table 1. The discrete shape of the inertia components.

Invariant	Observations	Dimension
$L^2 = \sum\limits_{p=1}^{N} m_p s_p$		3×1
$L^3_j = \sum\limits_{p=1}^{N} m_p \Phi_p$	$j = 1, \ldots, M$	$3 \times M$
$L^4 = \sum\limits_{p=1}^{N} m_p \tilde{s}_p \Phi_p + I_p \Phi'_p$		$3 \times M$
$L^5_j = \sum\limits_{p=1}^{N} m_p \tilde{\Phi}_{pj} \Phi_p$	$j = 1, \ldots, M$	$3 \times M$
$L^6 = \sum\limits_{p=1}^{N} m_p \Phi^T_p \Phi_p + \Phi'^T_p I_p \Phi'_p$		$M \times M$
$L^7 = \sum\limits_{p=1}^{N} m_p \tilde{s}^T_p \tilde{s}_p + I_p$		3×3
$L^8_j = \sum\limits_{p=1}^{N} m_p \tilde{s}_p \tilde{\Phi}_{pj}$	$j = 1, \ldots, M$	3×3
$L^9_{jk} = \sum\limits_{p=1}^{N} m_p \tilde{\Phi}_{pj} \tilde{\Phi}_{pk}$	$j, k = 1, \ldots, M$	3×3

4. Dynamic Analysis of the ASPIRE Robot Using MSC.ADAMS Software

In this paragraph, by using MSC.ADAMS [44] a software frequently used for numerical simulations in scientific researches [45–52], the virtual dynamic analysis of the ASPIRE robot when used by a virtual patient who needs to rehabilitate the movements of the upper limb is performed. The aim of this analysis is to determine the external loads of the kinematic elements of the ASPIRE robot. These loads will be the input data to perform structural analysis of the robot elements with the ANSYS finite element analysis program. A virtual model of the ASPIRE robot assembly—a human virtual mannequin that respects the average anthropometric dimensions of the sample of human subjects considered in the experimental tests, namely a mass of 72 kg and a height of 1.68 m—has been made in Solid Works. The patient is incorporated into the design of the robotic structure, and the center of the sphere, which is defined by two circular guides, is put in the center of the patient's shoulder, allowing for full arm motion while undergoing shoulder rehabilitation. The targeted motions of the ASPIRE robot are:

- Shoulder flexion/extension (± 45 degrees);
- Shoulder adduction/abduction (± 45 degrees);
- Forearm pronation/supination (± 45 degrees).

Figure 3 shows the assembly of the dummy together with the ASPIRE robot and the performed motions relevant for rehabilitation.

The geometric model of the robot assembly was imported into MSC.ADAMS, specifying the mechanical characteristics of the material of each element of the robot so that, based on these properties, the MSC.ADAMS software could calculate the mass, inertial properties and position of the centre of mass of each component. The kinematic joints of the robot were defined, as shown in Figure 4.

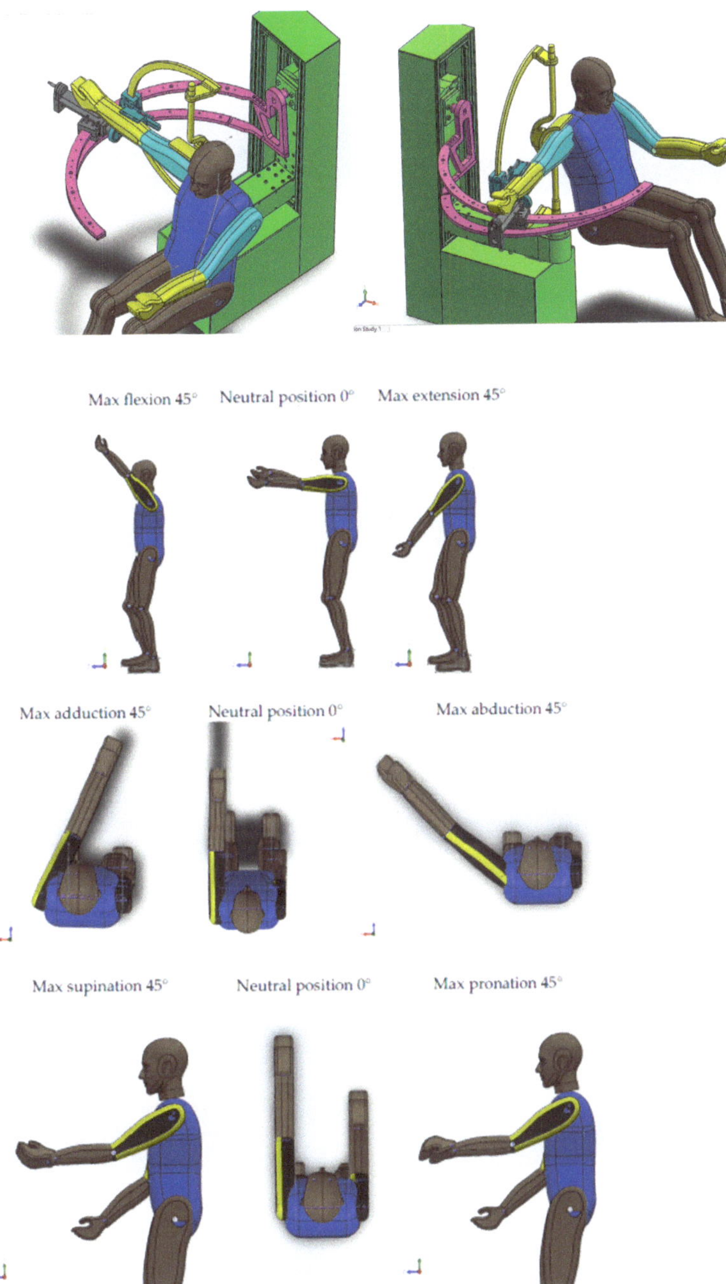

Figure 3. ASPIRE robot assembly—virtual mannequin in neutral position in Solid Works.

Elements 1 and 2 are connected to the base by joints A and B, respectively. For these couplings, we will present the connection forces and moments as calculated using the MSC.ADAMS program. A rotational joint is defined between elements 3 and 4, as shown in Figure 4. Element 3 (yellow) slides on the circular portion of element 1 by means of a guidance system.

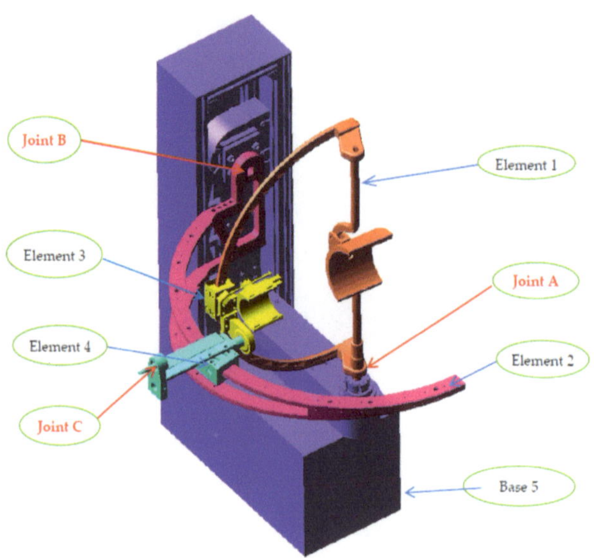

Figure 4. Model of the ASPIRE robot transferred to ADAMS.

For a correct simulation of the contact between elements 3 and 4 and the circular portions of elements 1 and 2, the contact of 2 points on the load arm of element 3 on the circular guidance system of element 1 was defined, as shown in Figure 5. The dynamic model is finalized taking into account this aspect, namely the contact between the elements running on the circular portions of the multi-body model, as well as the external forces given by the weight of the upper limb. For the dynamic analysis, we joined the dummy's arm with the mounting brackets of the forearm and the human arm. The WSTIFF solver and the SI2 integration algorithm were used to solve the dynamic model.

All three modules, the extension flexion module, the adduction/abduction module as well as the pronation/supination module, are geometrically shaped and have defined material properties, which allows us to calculate the inertial properties and their masses.

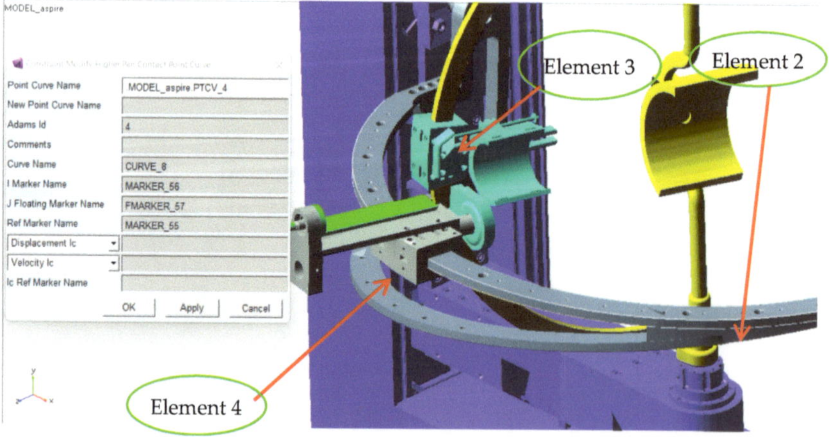

Figure 5. Defining the contact (by point-on-curve coupling) between slide 4 and the circular guide 2.

Regarding the dummy used in the virtual simulation, the mass of the upper limb segments are as follows: 0.52 kg for hand weight and 160 mm length, 1.52 kg for forearm weight with 240 mm length and 2.63 kg for upper arm weight with 260 mm length. The patient position in the MSC.ADAMS virtual simulation is shown in Figure 6.

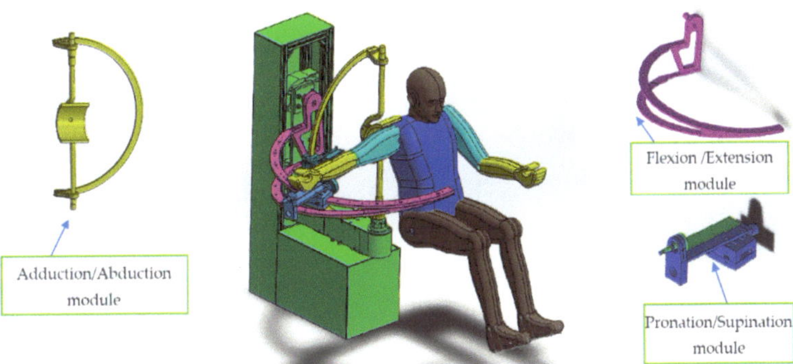

Figure 6. ASPIRE robot modules.

4.1. Dynamic Analysis in MSC.ADAMS of the Robot, Considering the Kinematic Elements as Rigid Solids

In order to analyze the dynamic behavior of the robot/mannequin assembly, harmonic laws of motion of the form: 0.4 × cos (1 × time), with an amplitude of 45 degrees and the period of 6.3 s were defined for the drive motors of each module. The time variation graph for the law of motion of the motors performing the rehabilitation movements is presented in Figure 7. In MSC.ADAMS, the implemented law of motion is defined as MOTION_1 or MOTION_2 for the second module, and in Figure 7, a measurement of the angle from the motor joint has been made. These parameters that define the amplitude and period of the movement were chosen because the laws of motion that will be done in the numerical simulation must be similar to the movement recommended for rehabilitation prescribed by the physiotherapist.

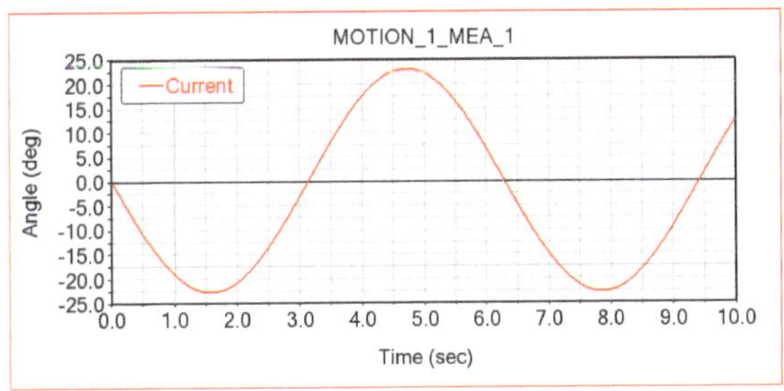

Figure 7. Law of motion imposed in ADAMS on robot motor joints.

A dynamic simulation was performed for the situation when the robotic system is used by the virtual patient dummy and all three modules of the ASPIRE robot are active simultaneously (flexion/extension, abduction/adduction, pronation/supination). The simulation was performed for a period of 20 s. Figure 8 shows the extreme operating positions of the ASPIRE robot corresponding to the situation when the flexion/extension and abduction/adduction movements are active, indicating a combined movement. The trajectories described by the elbow and shoulder joints are observed, which are spatial (spherical) circular portions.

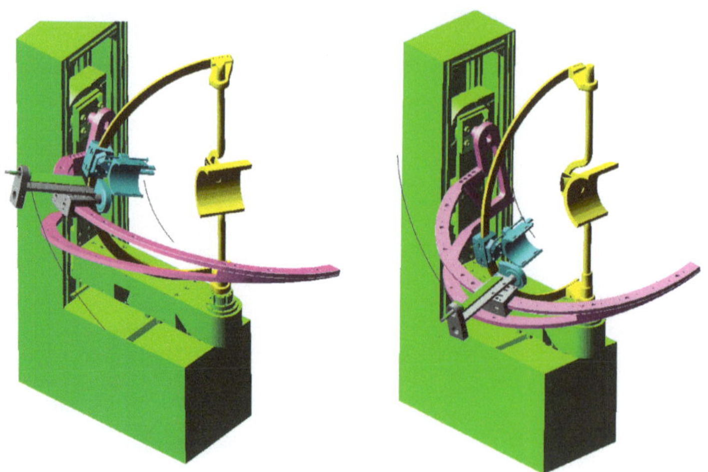

Figure 8. Extreme operating positions of the ASPIRE robot corresponding to a combined flexion-extension and abduction-adduction movement.

In the first phase, a dynamic simulation with the specified laws of motion, without taking into account the friction in the kinematic joints of the robot, was performed. The results obtained for this simulation, meaning the variations of the motor moment calculated by numerical simulation in ADAMS when simultaneously performing those three types of movement—flexion-extension, adduction-abduction and pronation-supination of the arm—are presented in Figure 9.

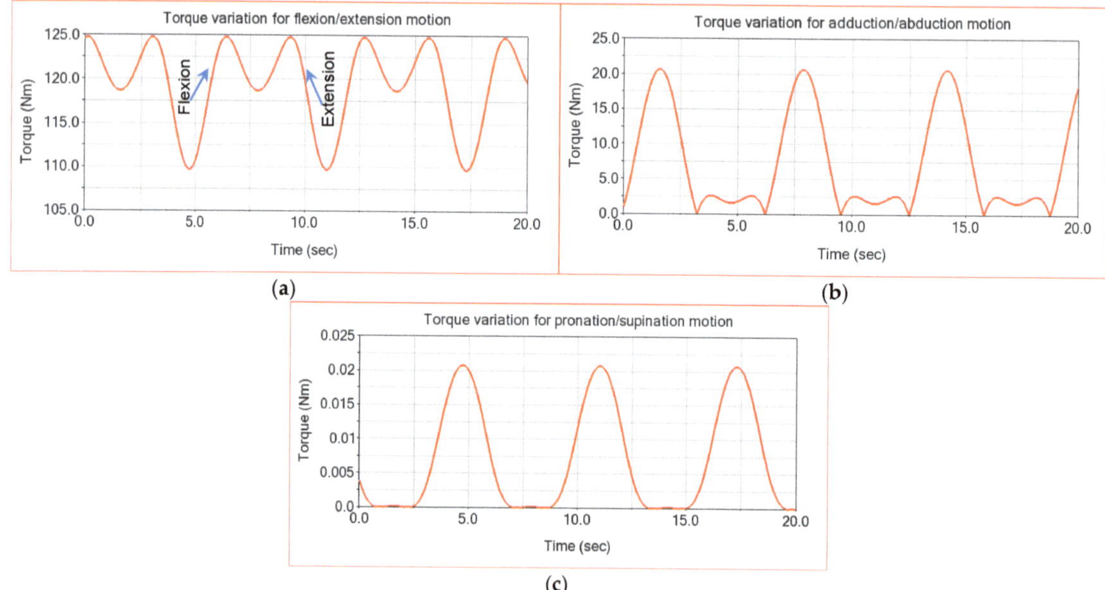

Figure 9. ADAMS simulation computed torque for: (**a**) the flexion-extension module; (**b**) the adduction-abduction module; (**c**) the pronation-supination module.

It is observed that the maximum value of the actuation moment of the flexion/extension module reaches a maximum value of 125 Nm, and the minimum is 110 Nm. It is important to note that these results are obtained when all three modules are active simultaneously.

For the adduction-abduction movement, the motor moment obtained by numerical simulation in ADAMS is presented in Figure 9a. It is observed that its amplitude is around 22 Nm. For the pronation-supination movement, the value calculated by numerical simulation in ADAMS is equal to 0.023 Nm (Figure 9c).

The results of the dynamic simulation conclude that, for the movement performed by the flexion-extension motion module, a very high actuation moment is required, namely 125 Nm (Figure 9a). In this case, it is necessary to use worm gearboxes to amplify the torque of the electric drive motor. The gear ratio of the planetary gearbox must be chosen in accordance with the size of the resistive torque but also with the angular velocity required at the output. For the flexion-extension module mechanism, a worm gearbox gear ratio I = 1:45 was used. For the adduction-abduction module mechanism a planetary gearbox with gear ratio I = 1:100 is implemented, and for the pronation-supination mechanism a planetary gearbox with gear ration I = 1:11 is required.

As the flexion/extension module requires the greatest moment of actuation, it is clear that this module will be the most structurally stressed. For the flexion/extension module, as well as the adduction/abduction one, the connecting forces calculated from the couplings B and C, because these forces will be used later to perform the structural analysis with finite elements, are presented.

For a good understanding and interpretation of the obtained results, in the case of the connection forces from the kinematic couplings of the robot's modules, Figure 10 presents the orientation of the local reference axis systems attached to the kinematic couplings of the ASPIRE robot in relation to the calculated components of the connecting forces.

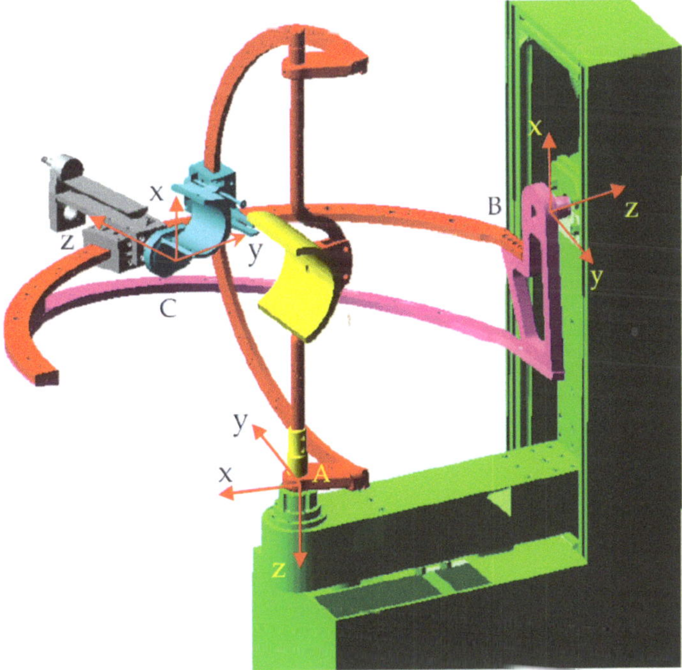

Figure 10. Positioning of the axes of the kinematic couplings of the ASPIRE robot.

Figure 11a–c shows the time variation graphs for the components of the connecting force in the coupling A in relation to the local reference system attached to this coupling, oriented as in Figure 10. Through this rotation coupling, the module that performs the adduction/abduction movement is connected to the base of the robot with the help of a bearing. By means of an electric motor and a transmission with a worm gearbox, this module is operated and controlled. As previously detailed, the mass of this module is 24.3 kg, to which is added a part of the mass of the lower limb. The analysis of these graphs shows that the maximum value recorded along the X axis of the local reference system in coupling A is −34 N (Figure 11a). The component along the Y axis of the connecting force in coupling A reaches a maximum value of 23 N, and the minimum value is equal to 2.5 N (Figure 11b). Since the weight of 243 N of the adduction/abduction module is oriented along the Z axis, the component along this axis of the connecting force has the highest value, reaching −271 N (Figure 11c). The variation in the negative range of this component is explained by the orientation of the Z axis of the local reference system, which is oriented downwards, and the reaction transmitted to the base is oriented in the opposite direction. The connecting force in coupling A along the Z axis has an additional value of 31 N; this is as compared to the weight of this module of 243 N, which comes from the load given by the support of the patient's arm and from the interaction with the flexion/extension module. The resulting value of the connecting force in this coupling has a maximum of 274 N and a minimum of 263 N, according to Figure 11d.

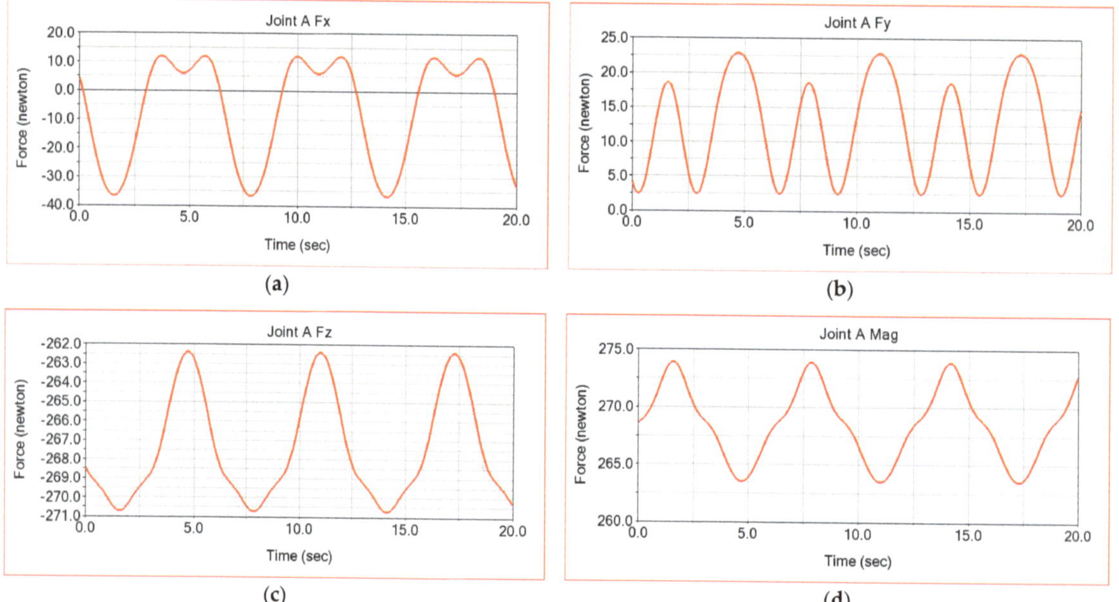

Figure 11. Connecting forces in the rotation coupling A connected to the base of the adduction/abduction module: (**a**) in the x direction; (**b**) in the y direction; (**c**) in the z direction; and (**d**) the resultant.

Figure 12 shows the variation laws of the connecting forces in the rotation coupling B, which is defined in ADAMS, between the flexion-extension module and the base. Quite high values are observed of 347 N for the vertical component of the connecting force, which is the one along the X axis of the local reference system from the coupling B (Figure 12a). The reaction along the Y axis has much lower values, reaching a maximum of 57 N (Figure 12b). The connecting force along the Z axis has much lower values, with a maximum of 10 N. This occurs due to the friction interaction between the flexion/extension module and slide 3, as well as due to the opposite resistance of the patient, in this case only by the force of upper limb weight. The fact that the highest value is recorded by the component along the X axis of the connecting force in the coupling B is explicable, because of the weight of the flexion/extension module, equal to 175 N, to which is added both the weight of the upper limb and the interaction between the patient's upper arm and the robot module (opposite resistance of the upper limb), such that it acts in the opposite direction of the X axis of the local reference system. The fact that the values of the resulting reaction along the X axis are negative is again natural, because the component of the connecting force along the X axis acts downwards and has the opposite direction to the axis of the local reference system attached in the rotation torque B.

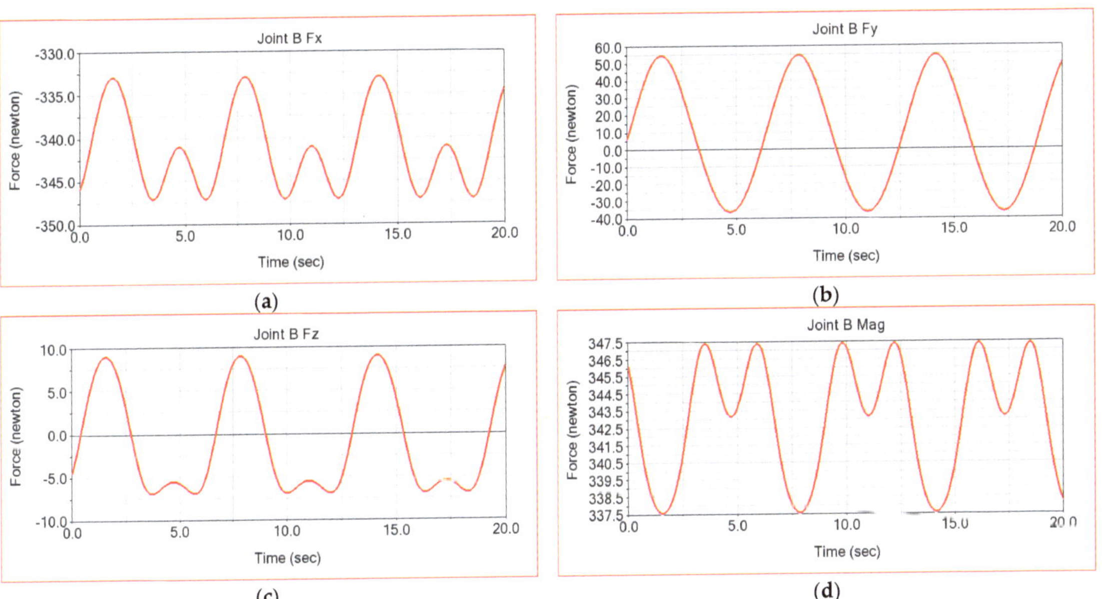

Figure 12. Connecting forces in the rotating coupling B, connected to the base, of the flexion/extension module: (**a**) on x direction; (**b**) on y direction; (**c**) on z direction; (**d**) resultant.

In conclusion, according to the results obtained for the connecting forces in couplings A and B, it results that the flexion/extension module is subjected to the highest mechanical stresses. It is therefore necessary to verify its structural integrity with the finite element method, a verification that will be performed later in ANSYS.

4.2. Dynamic Analysis in MSC.ADAMS of the Robot, Considering the Rigid Solid Kinematic Elements and the Friction in the Couplings

The previously presented results for the actuation moments and the connection forces from the kinematic couplings are obtained by numerical simulation in MSC.ADAMS in the simplifying hypothesis, by which the frictions from the kinematic couplings were not taken into account. In reality, the presence of friction in the couplings imprints another character on the dynamics of the ASPIRE robot. For these reasons, in the second phase, we will consider the frictions that appear in the kinematic couplings A, B and C. The parameters necessary for defining the frictions in the kinematic couplings are defined according to the data from Figure 13, for the definition of which the existing data in the literature were taken into account [54]. According to the existing data in the specialized literature, we adopted for the static friction coefficient the value of 0.3, and for the dynamic friction coefficient, we specified the value 0.2.

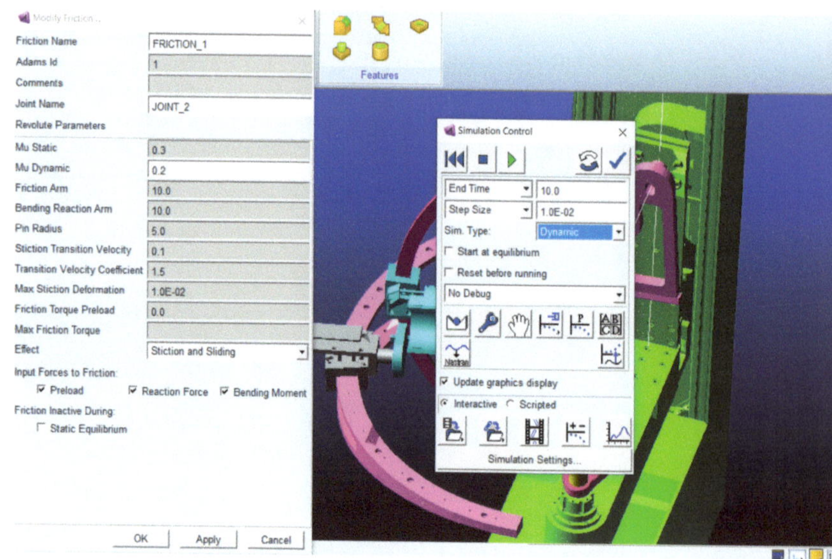

Figure 13. Parameters defining the friction force in the ASPIRE robot couplings.

The results obtained by numerical simulation in ADAMS, where the friction in the kinematic couplings of the ASPIRE robot is considered, are presented below. The laws of variation of the moment necessary to operate the flexion/extension module, as well as the abduction/adduction module, are presented in Figure 14.

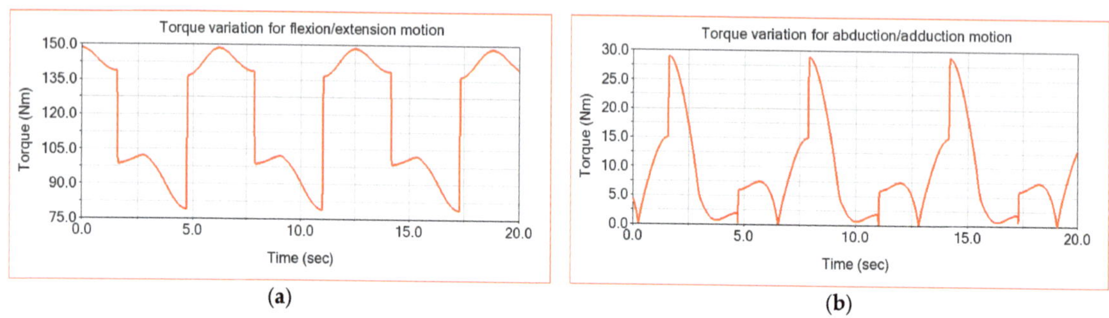

(a)

(b)

Figure 14. Second setup ADAMS simulation computed torque for: (**a**) the flexion-extension module; (**b**) the adduction-abduction module.

There is an increase in the motor torque required to operate the modules in this case, as well as a steeper variation, similar to that determined experimentally [43].

The results obtained for the connection forces from the kinematic couplings in the hypothesis of considering the friction from the robot joints are presented below. Figure 15 shows the graphs obtained for the variation laws of the connection forces for the coupling A, calculated in the simulation from ADAMS. If we compare these results with those shown in Figure 11, in which case we did not consider the friction in the kinematic couplings, it is found that we have an increase in the connecting forces. Thus, the component along the X axis of the connecting force shows an increase from the value of −36 N to a maximum value of −70 N (Figure 15a). In the case of the connecting force component along the Y axis, the increase occurs from 23 N to 70 N (Figure 15b). There is also an increase in amplitude from 20 N to 100 N. The component of the reaction along the Z axis shows an increase from -271 N to -285 N (Figure 15c). The maximum value of the reaction resulting from the coupling A increases from the value of 274 N to 284 N. It is also found that the presence of friction in the couplings imposes a more dynamic character of the movement, an aspect which is highlighted by the steeper variation of the variation graphs.

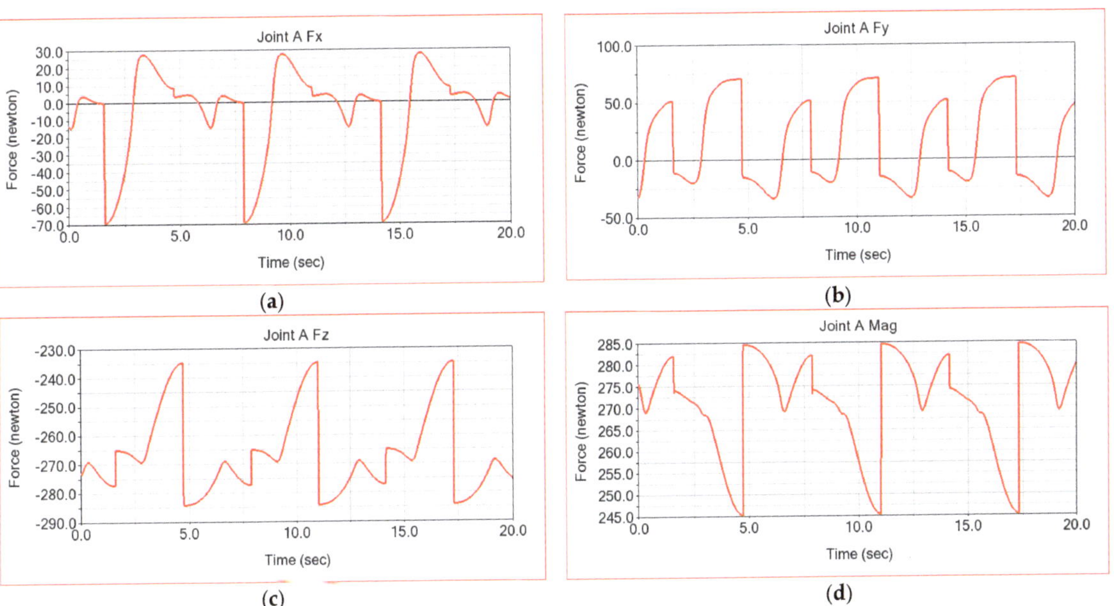

Figure 15. Connecting forces in the rotation coupling A of the adduction/abduction module, calculated assuming friction: (**a**) in the x direction; (**b**) in the y direction; (**c**) in the z direction; and (**d**) the resultant.

For coupling B, the connecting forces obtained in Figure 16 are presented in the case of considering the friction in couplings. There is also an increase in the maximum values obtained, assuming the friction of the couplings. Thus, the reaction resulting from coupling B shows an increase from the maximum value of 347 N to the value of 370 N, meaning an increase of 6.62%. The steeper change of the parameter's variation laws is noticed at the moment of the transition from the flexion movement to the extension for the module that realizes this movement of the arm from adduction to abduction.

4.3. Dynamic Analysis in MSC.ADAMS of the Robot, Considering the Kinematic Elements as Deformable Solids and Joint Friction

In the third stage of the simulation, the deformability of the kinematic elements of the ASPIRE robot are also considered (flexibility for the two modules of the ASPIRE robot, the flexion/extension and the adduction/abduction modules). The procedure for considering the kinematic elements as deformable solids involves their discretization in finite elements, the calculation of their own modes of vibrations, as well as the calculation of specific deformations and mechanical stresses. The second stage consists of establishing the connection points of the discretized element through kinematic connections with the other elements. Figure 17 shows a construction aspect of a deformable kinematic element in MSC.ADAMS.

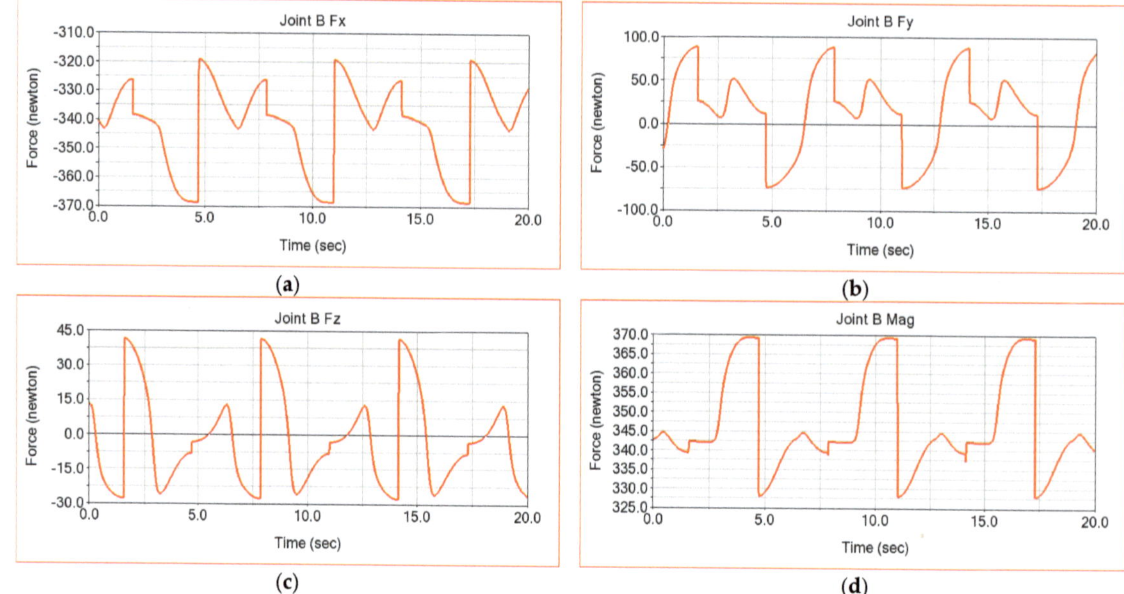

Figure 16. Connecting forces in coupling B for the rotation of the flexion/extension module calculated assuming friction: (**a**) in the x direction; (**b**) in the y direction; (**c**) in the z direction; and (**d**) the resultant.

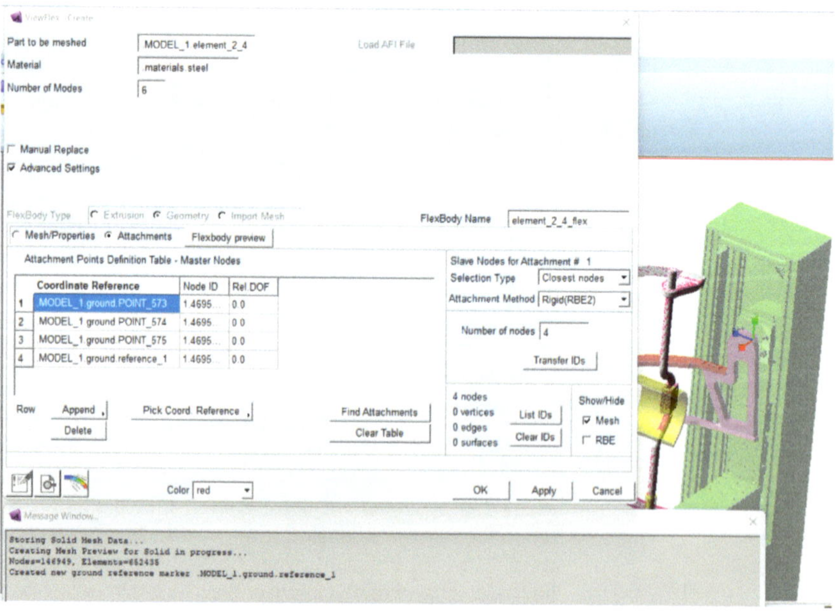

Figure 17. Defining a deformable element in MSC.ADAMS.

The purpose of this simulation is to obtain results of the dynamic operating parameters of the ASPIRE robot in real operating conditions. Through the dynamic analysis, considering also the flexibility of the kinematic elements, the friction from the kinematic couplings, as well as the external loads, results regarding the real variation of the dynamic parameters have been obtained. Thus, in this stage results will be presented, such as the specific deformation of the robot elements, deformation velocities and accelerations, own frequencies and normal modes of vibration, connection forces in the kinematic couplings and actuation moments.

A first category of results of interest is represented by the moments of actuation of the flexion/extension module, as well as the adduction/abduction module. For this simulation, obviously we used the same loads, given by the weight of the upper limb, as well as the same laws of motion implemented in the motor torques. The dynamic simulation is performed only for a period of 10 s due to the complexity of the model. Thus, for this phase of the dynamic simulation, in Figure 18, the actuation moments corresponding to the flexion/extension and abduction adduction modules are presented. There are very abrupt variations of the dynamic parameters at the beginning of the robot's movement, which disappear after a second of operation. These aspects are also manifested in the real operation only in the initial phase of the robot's operation, so it does not affect the proper functioning of the robot and especially the patient.

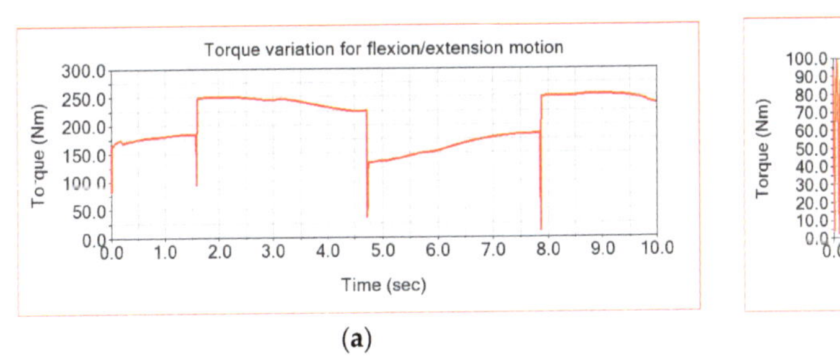
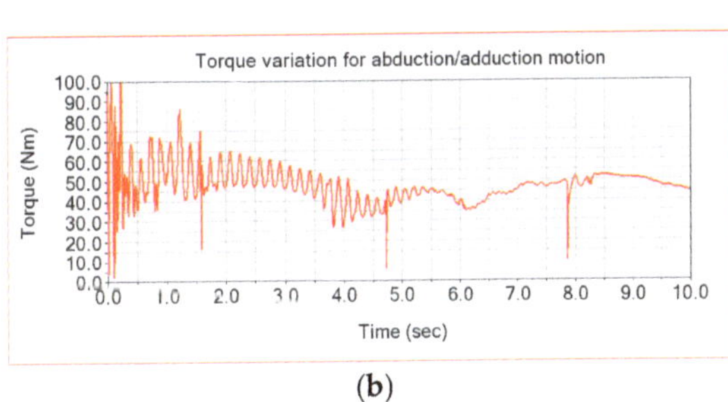

(a) **(b)**

Figure 18. Third setup ADAMS simulation computed torque for: (**a**) the flexion-extension module; (**b**) the adduction-abduction module.

The other parameters of interest for the simulation in the dynamic regime, considering the deformability of the elements, are the translational deformations of some characteristic points of the ASPIRE robot kinematic elements. The variation graphs of the translational deformations for the centre of mass of the flexion/extension module are shown in Figure 18. From these graphs, it is found that the centre of mass of the flexion/extension module registers larger deformations in the first two seconds of operation. As expected, the largest deformation is recorded along the X and Z axes, because the loading produced by the movement of the robot takes place along these axes. The maximum value of the translational deformation of the marker detailed in Figure 19 reaches 0.08 mm.

By simulating in dynamic regime and considering the deformability of the kinematic elements of the robot, an overview of the elastic displacements of the elements has been obtained. Thus, Figure 20 presents the distribution maps of the elastic displacements for robot modules. It is observed that the maximum displacement reaches the value of 2.51 mm at the unsupported end of the flexion/extension module

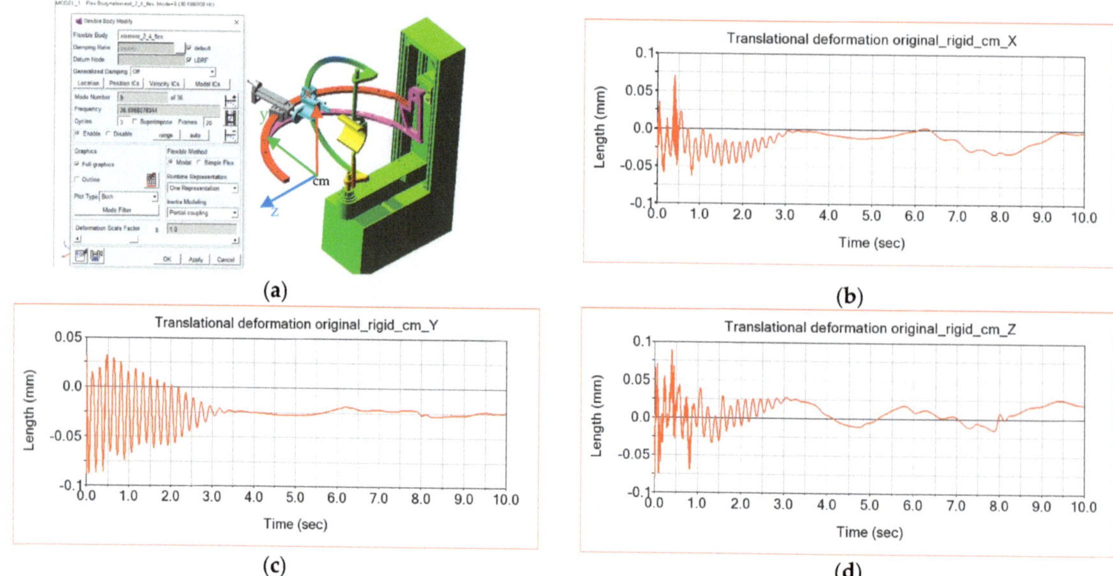

Figure 19. Translational deformations of the centre of mass of the flexion/extension module in relation to the X, Y and Z axes: (**a**) marker attached to flexion extension module mass centre; (**b**) X axis translational deformation; (**c**) Y axis translational deformation; (**d**) Z axis translational deformation.

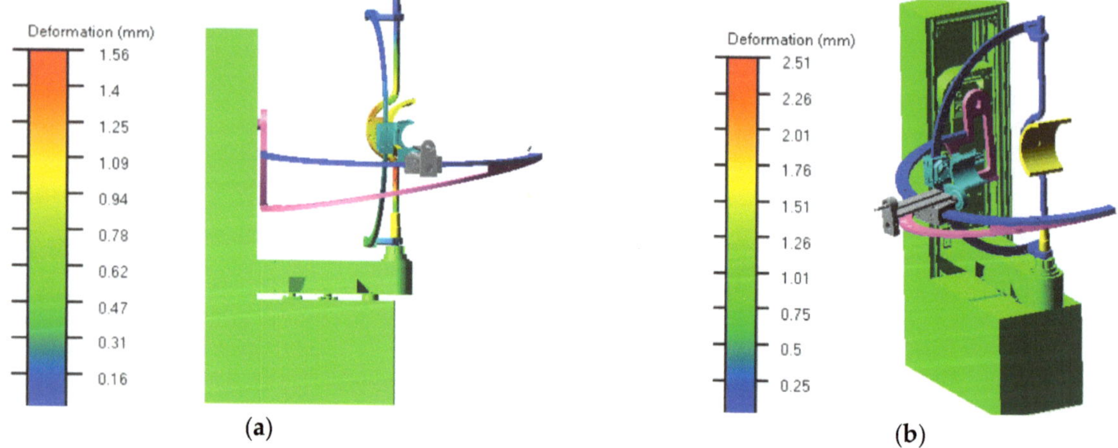

Figure 20. ASPIRE robot total deformation distribution map in two different moments: (**a**) at the moment t = 0.001; (**b**) at the moment t = 0.008 s.

5. Finite Element Analysis in ANSYS of the ASPIRE Robot Flexion/Extension Module

In this section, the results of a finite element analysis in ANSYS for the ASPIRE robot flexion/extension module will be presented. The objectives of this study are as follows:

- Determine the stress concentrators that occur in the structure;
- Optimize the design by eliminating these stress concentrators.

The steps taken to achieve these goals are described below. The verification of interferences and discontinuities was made (lack of contact between elements for the virtual assembly of the flexion-extension module). The existence of interferences prevents the discretization in finite elements for structural analysis. Discontinuities lead to erroneous results because the connecting forces between the components of the subassembly are no longer transmitted. At the same time, the elimination of stress concentrators was made by making a fillet radius in the area of sharp edges or jumps in diameter between circular sections.

The optimized model was transferred to the ANSYS software in order to obtain maps of equivalent stresses based on the finite element method analysis, which represents a modern and very powerful method which is useful for understanding the behavior of the robotic structures, rehabilitation devices and human biomechanics [22,27,55]. It is observed that the analyzed element is composed of six distinct parts which are assembled by threaded joints. The combination of all these pieces to obtain a single element (part) with the help of ANSYS Design Modeler is performed. This aspect is shown in Figure 21. The discretization of the whole set of bodies in finite elements (Figure 21) is achieved using a size of the finite element of 2 mm. Additionally, in the areas where fillet radius appears, a refinement of discretization was performed by using the option "mesh refinement".

The next step was to fix the cylindrical surface of the coupling B shaft. The load of the element comes from the contact pressure, which has a value of 0.47 MPa. This results from the dynamic model made in MSC.ADAMS, which appears between the guidance surface and the circular portion.

By finite element analysis of the optimized structure (which was obtained by changing the geometry and introducing the fillet radius), equivalent stresses with lower values were obtained compared to those registered at the initial structure, not optimized, when equivalent stresses were 123 MPa. Due to the bending of the circular portions, the contact edge, visible in Figure 22c), produces stress with high values. Filleting these edges reduces the values of these maximum mechanical stresses. The maximum von Mises stress obtained in this case is 64.271 MPa and is recorded at the contact of the joint surfaces.

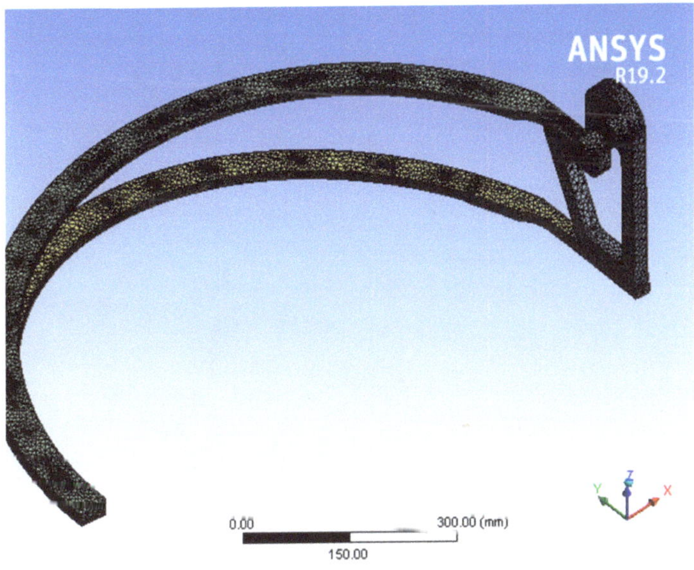

Figure 21. *Cont.*

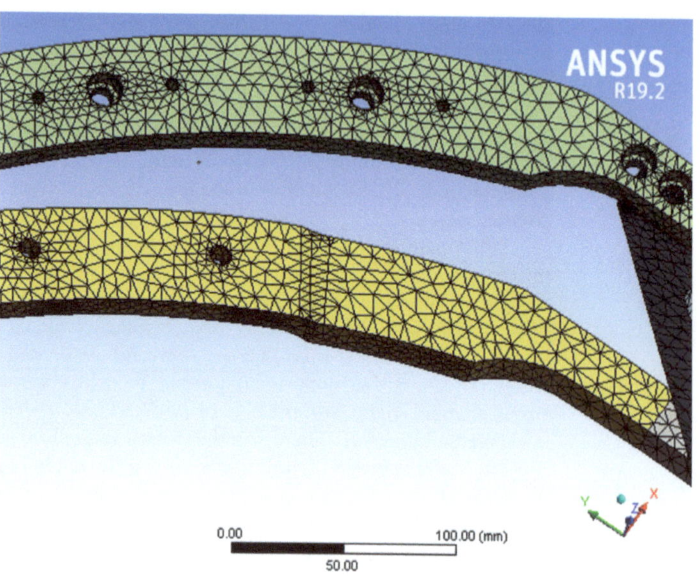

Figure 21. Discretization in finite elements of the flexion/extension module.

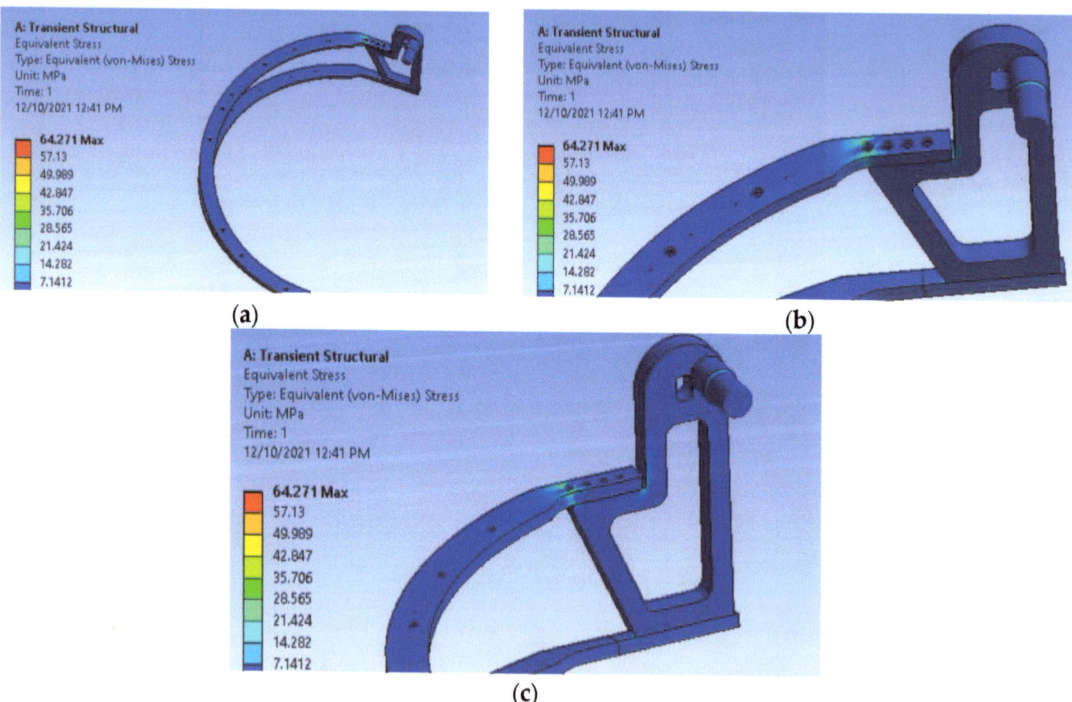

Figure 22. Distribution of equivalent stresses in the optimized model of the structure: (**a**) flexion/extension module view; (**b**) upper corner assembly detail; (**c**) lower corner detail view.

The maps with the distribution of the displacements recorded along the X and Y axes of the coordinate system are presented in Figure 23. The maximum values recorded are relatively small, at 0.40 mm along the X axis and 0.34 mm along the Y axis, and do not prevent the proper functioning of the module made of aluminum alloy Al 1060. The vertical movement of the module free end reaches 2.6 mm, keeping it within acceptable limits. It is noted that the value of the total elastic displacement obtained in ANSYS is close in value to the result obtained in MSC.ADAMS, presented in Figure 20, where the maximum values are 2.51 mm.

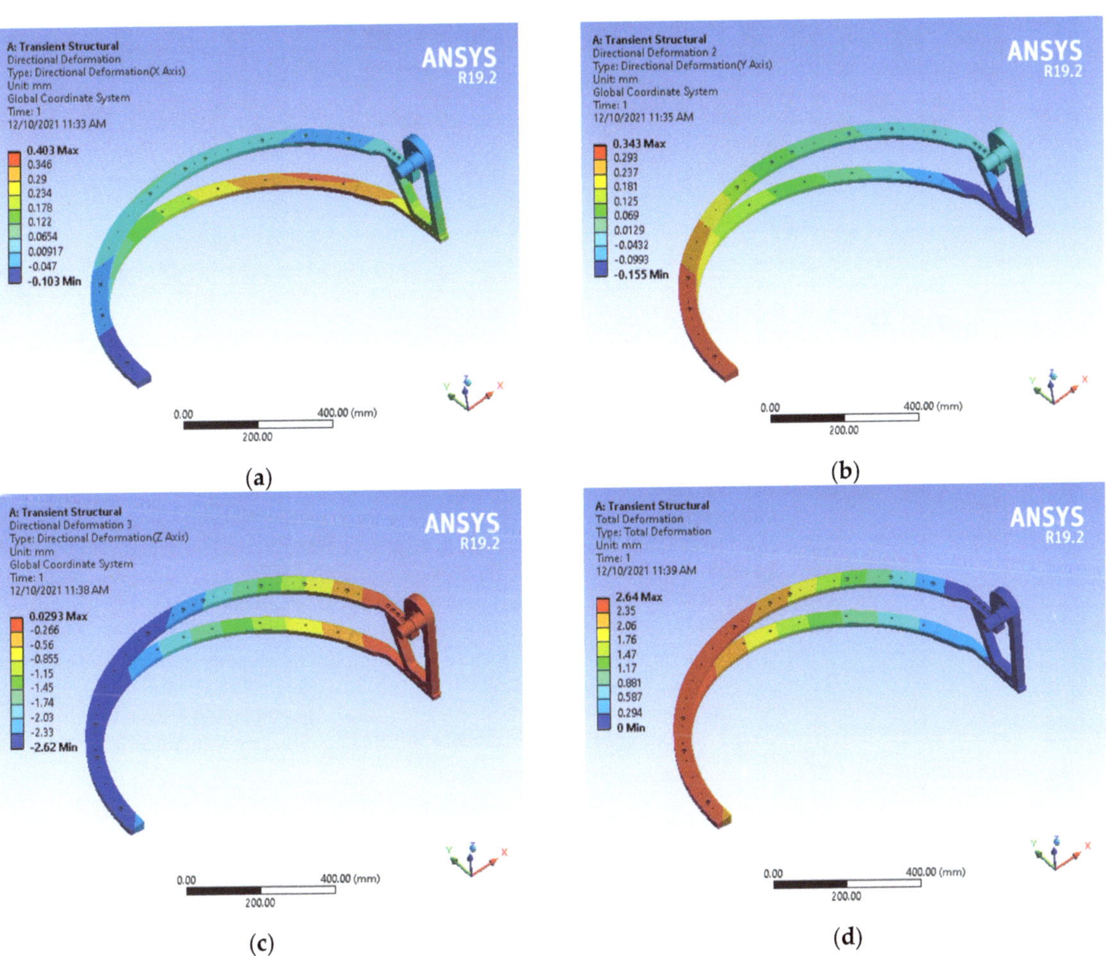

Figure 23. Distribution of displacements along the axes: (**a**) X; (**b**) Y; (**c**) Z and (**d**) total displacement.

6. Experimental Determination of the ASPIRE Robot Operating Motor Torques

The results obtained by numerical simulation in ADAMS must also be validated by those obtained experimentally. Thus, in Figure 24, the results obtained experimentally for the actuation moments of the ASPIRE robot modules have been shown [43].

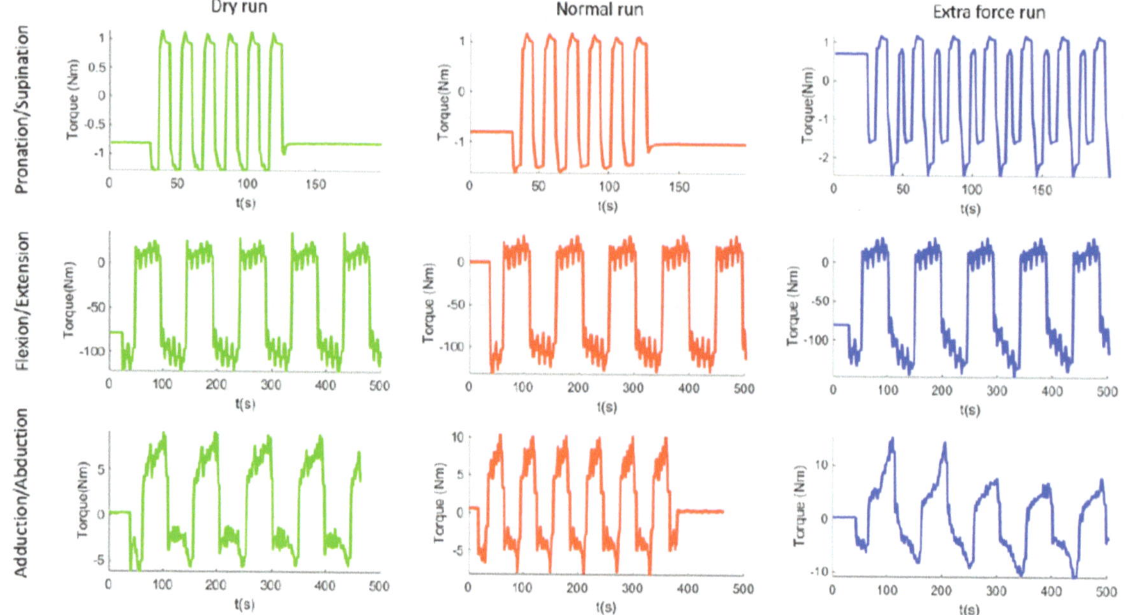

Figure 24. The torques required to operate the ASPIRE robot, measured experimentally [43].

7. Discussion

A comparison between the experimental results and the results obtained by numerical simulation in MSC.ADAMS, presented in Figure 14, has been made. Therefore, it was observed that the maximum value of the moment necessary to operate the flexion/extension module is 147.37 Nm experimentally, and by numerical simulation in ADAMS we obtained the value of 149 Nm. The negative value recorded in the experimental torque determination comes from the calibration of the torque-recording transducer, with the maximum value being recorded in the negative direction of the axis. MSC.ADAMS shows the absolute, and therefore positive, value of the torque. As a result, in absolute terms, the difference between the maximum experimental value and that obtained by simulation is 1.63 Nm, i.e., less than 1%.

For the moment necessary to operate the adduction/abduction module, we obtained experimentally the maximum moment value of 14.4 Nm, and by numerical simulation in MSC.ADAMS we obtained the value of 25 Nm. It is observed, however, in this case that the allure of the variation curve of the moment obtained by simulation is similar to that obtained experimentally.

A similar approach and the obtained results are presented in previous papers [45] with reference to a dynamic study of a robotic system for lower limb rehabilitation.

In a preliminary study with finite elements of the robot modules, the conclusion is that high stresses appear in the assembly parts of the elements. The optimization of these assemblies was realized by introducing the fillet radius and chamfering the sharp edges. In this way, the maximum stresses decreased from 123 MPa to 64.271 MPa.

Thus, the structural strength of the flexion/extension module was checked in ANSYS FEA software. The kinematic and dynamic parameters of the optimized version of the ASPIRE robot mounted on a virtual mannequin that respects the anthropometric data of a user are studied, considering the component elements as being deformable.

The numerical simulation in MSC.ADAMS should consider the real aspects of the operation, such as friction and deformability of the elements, to obtain results as close as possible to the real situation. To obtain an improved variant of the robot, studies are performed on optimizing the shape and constructive structure of the robot that improves the mechanical strength and dynamic behaviour and leads to the increased stability of the movement of the robot system—the human arm.

The difference between the experimentally recorded maximum values and those obtained by simulation comes from the parameters introduced in the MSC.ADAMS simulation, which may differ from the actual operating conditions. However, the similar allure of variation of the torques obtained experimentally and those obtained by numerical simulation in MSC.ADAMS can be noticed. The results are similar when comparing between the torques measured experimentally and those obtained by numerical simulation in MSC.ADAMS in the case of considering the deformability of the elements.

Additionally, the differences that appear between the experimental results [43] and those obtained by numerical simulation in MSC.ADAMS are explained by the manufacturing tolerances of the parts, as well as the assembly tolerances, which are responsible for the different elasticity of the real robot modules compared to the one considered in the simulation. Moreover, in the simulation, the modules were considered as a unitary assembly, united by fixed couplings. In reality, these elements are joined by removable assemblies, which offer a slightly different elasticity. Nonetheless, there is a clear similarity of the mode of variation for the operating moments of the modules obtained by numerical simulation and experimental determination.

8. Conclusions

The paper presents theoretical research, through virtual and experimental simulation on the dynamics of the ASPIRE robot, used to rehabilitate patients suffering from brachial monoparesis. A model of the ASPIRE robot which includes modules with optimized geometry in terms of how to assemble the components of the modules has been made in SolidWorks.

Another novelty that this study brings is that the numerical simulation in ADAMS, in the situation when the ASPIRE robot is used by a virtual mannequin, elaborated as a 3D Solid Works model based on an average human subject, has been performed. The challenge of the dynamic simulation was to find a way to define the translation couplings between the circular and sliding guides. The option was to define contact as point on the curve. Results for the laws of variation of the actuation moments as well as for the connection forces from the kinematic couplings in three situations have been obtained:

The original elements of this article consist of the dynamic study in MSC.ADAMS of the ASPIRE robot. Results are obtained in three distinct situations:

(a) The kinematic elements were considered as rigid bodies and the friction in the kinematic couplings was not considered;
(b) The kinematic elements were considered as rigid bodies and the friction in the robot couplings was considered;
(c) Both the flexibility of the kinematic elements and the friction in the couplings were considered.

In each of these situations, the results obtained are presented and commented on. Thus, by numerical simulation in MSC.ADAMS, we obtained the connecting forces from the kinematic torques as well as the actuation moments of the robot modules. We obtained these data in the first phase without considering the friction in the couplings; the results are presented in Figures 9, 11 and 12. In the next phase, we also considered the friction from the kinematic torques of the robot, with the results obtained for the actuation moments being presented in Figure 14 and for the connecting forces in Figures 15 and 16. It is noted that if we introduce in the simulation the real operating conditions, the results obtained are much closer to the real ones obtained experimentally.

It is found that the results closest to the experimental ones are those obtained for the most complex situation, namely situation (c). The results obtained by numerical simulation are validated by the experimental results. Moreover, the structural integrity check is performed for the most solicited module, the flexion/extension module. In this way, the stable dynamic behavior of the robot is highlighted, and its safety good functioning is validated for the rehabilitation of the brachial monoparesis.

A methodology for transforming rigid, complex-shaped bodies, such as ASPIRE robot modules, into deformable solids in MSC.ADAMS has been developed. An optimal way to define the contact translational joints in the structure of the ASPIRE robot has been established. In this situation, it is important to correctly define the translational joints, to consider the flexibility of the elements, as well as to define the friction functions, which should consider the stick-slip phenomenon.

The virtual simulation in MSC.ADAMS by comparing the actuation moments determined by simulation with those obtained experimentally has been validated. Thus, the data on the load forces of the elements was used for a structural analysis with finite elements in ANSYS.

Author Contributions: Conceptualization, I.D.G., D.T., G.C. and D.P.; methodology, I.D.G., D.T. and D.P.; software, I.D.G. and D.T.; validation, P.T., D.P. and D.N.T.; formal analysis, I.D.G., P.T. and A.B.; investigation, A.B and M.G.; resources, D.T.; writing—original draft preparation, I.D.G. and A.B.; writing—review and editing, D.T., D.P. and G.C.; visualization, D.N.T. and M.G.; supervision, D.T., D.P. and G.C.; project administration, D.P. All authors have read and agreed to the published version of the manuscript.

Funding: This research was funded by a grant from the Romanian Ministry of Research and Innovation, CCCDI—UEFISCDI, project number PN-III-P2-2.1-PED-2019-3022/546PED/2020 (NeuroAssist) within PNCDI III, and by the project POCU/380/6/13/123927-ANTREDOC, "Entrepreneurial competencies and excellence research in doctoral and postdoctoral studies programs", co-funded from the European Social Fund through the Human Capital Operational Program 2014–2020.

Institutional Review Board Statement: Not applicable.

Informed Consent Statement: Not applicable.

Data Availability Statement: The data presented in this study are openly available in reference number [30,31,45].

Conflicts of Interest: The authors declare no conflict of interest.

References

1. Bureau, U.S.C. *The Nation's Older Population Is Still Growing*; Census Bureau: Houtland, MD, USA, 2017.
2. NSCISC. *National Spinal Cord Injury Statistical Center, Facts and Figures at a Glance*; NSCISC: Birmingham, UK, 2017.
3. SPARC. *Robots Association. Multi Annual Roadmap for Robotics in Europe*; Robots Association: Bruxelles, Belgium, 2017.
4. Feigin, V.L.; Norrving, B.; Mensah, G.A. Global Burden of Stroke. *Circ. Res.* **2017**, *120*, 439–448. [CrossRef] [PubMed]
5. Lanctôt, K.L.; Lindsay, M.P.; Smith, E.E.; Sahlas, D.J.; Foley, N.; Gubitz, G.; Austin, M.; Ball, K.; Bhogal, S.; Blake, T.; et al. Canadian Stroke Best Practice Recommendations: Mood, Cognition and Fatigue Following Stroke, 6th Edition Update 2019. *Int. J. Stroke Off. J. Int. Stroke Soc.* **2020**, *15*, 668–688. [CrossRef]
6. Skilbeck, C.E.; Wade, D.T.; Hewer, R.L.; Wood, V.A. Recovery after Stroke. *J. Neurol. Neurosurg. Psychiatry* **1983**, *46*, 5–8. [CrossRef]
7. Kwakkel, G.; Kollen, B.J.; van der Grond, J.; Prevo, A.J.H. Probability of Regaining Dexterity in the Flaccid Upper Limb: Impact of Severity of Paresis and Time since Onset in Acute Stroke. *Stroke* **2003**, *34*, 2181–2186. [CrossRef] [PubMed]
8. Iqbal, J.; Bruno, A.; Berger, M. Stroke Causing Pure Brachial Monoparesis. *J. Stroke Cerebrovasc. Dis.Off. J. Natl. Stroke Assoc.* **1995**, *5*, 88–90. [CrossRef]
9. Song, I.-U.; Kim, J.-S.; Kim, Y.-I.; Lee, K.-S. Brachial Monoparesis as an Isolated Manifestation of a Paramedian Pontine Ischemic Lesion. *Clin. Neurol. Neurosurg.* **2007**, *109*, 376–378. [CrossRef] [PubMed]
10. Sivan, M.; Gallagher, J.; Makower, S.; Keeling, D.; Bhakta, B.; O'Connor, R.; Levesley, M. Home-Based Computer Assisted Arm Rehabilitation (HCAAR) Robotic Device for Upper Limb Exercise after Stroke: Results of a Feasibility Study in Home Setting. *J. Neuroeng. Rehabil.* **2014**, *11*, 163. [CrossRef] [PubMed]
11. Bayona, N.A.; Bitensky, J.; Salter, K.; Teasell, R. The Role of Task-Specific Training in Rehabilitation Therapies. *Top. Stroke Rehabil.* **2005**, *12*, 58–65. [CrossRef] [PubMed]

12. Turchetti, G.; Vitiello, N.; Trieste, L.; Romiti, S.; Geisler, E.; Micera, S. Why Effectiveness of Robot-Mediated Neurorehabilitation Does Not Necessarily Influence Its Adoption. *IEEE Rev. Biomed. Eng.* **2014**, *7*, 143–153. [CrossRef]

13. Gorelick, P.B. The Global Burden of Stroke: Persistent and Disabling. *Lancet Neurol.* **2019**, *18*, 417–418. [CrossRef]

14. UK Stroke Association. *State of the Nation: Stroke Statistics*; UK Stroke Association: London, UK; Available online: https: //www.stroke.org.uk/resources/state-nation-stroke-statistics (accessed on 20 August 2021).

15. ReNeuron Group plc. *CTX Cells for Stroke Disability 15.%09*; ReNeuron Group PlC: Bridgend, UK, 2021.

16. Manna, S.K.; Dubey, V.N. Comparative Study of Actuation Systems for Portable Upper Limb Exoskeletons. *Med. Eng. Phys.* **2018**, *60*, 1–13. [CrossRef]

17. Murie-Fernández, M.; Irimia, P.; Martínez-Vila, E.; John Meyer, M.; Teasell, R. Neuro-rehabilitation after stroke. *Neurologia* **2010**, *25*, 189–196. [CrossRef]

18. Clarke, D.J.; Burton, L.-J.; Tyson, S.F.; Rodgers, H.; Drummond, A.; Palmer, R.; Hoffman, A.; Prescott, M.; Tyrrell, P.; Brkic, L.; et al. Why Do Stroke Survivors Not Receive Recommended Amounts of Active Therapy? Findings from the ReAcT Study, a Mixed-Methods Case-Study Evaluation in Eight Stroke Units. *Clin. Rehabil.* **2018**, *32*, 1119–1132. [CrossRef] [PubMed]

19. Norouzi-Gheidari, N.; Archambault, P.S.; Fung, J. Effects of Robot-Assisted Therapy on Stroke Rehabilitation in Upper Limbs: Systematic Review and Meta-Analysis of the Literature. *J. Rehabil. Res. Dev.* **2012**, *49*, 479–496. [CrossRef] [PubMed]

20. Huang, J.; Tu, X.; He, J. Design and Evaluation of the RUPERT Wearable Upper Extremity Exoskeleton Robot for Clinical and In-Home Therapies. *IEEE Trans. Syst. Man Cybern. Syst.* **2015**, *46*, 1–10. [CrossRef]

21. Al-Fahaam, H.; Davis, S.; Nefti-Meziani, S. Wrist Rehabilitation Exoskeleton Robot Based on Pneumatic Soft Actuators. In Proceedings of the 2016 International Conference for Students on Applied Engineering (ICSAE), Newcastle upon Tyne, UK, 20–21 October 2016; pp. 491–496. [CrossRef]

22. Tarnita, D.; Pisla, D.; Geonea, I.; Vaida, C.; Catana, M.; Tarnita, D.N. Static and Dynamic Analysis of Osteoarthritic and Orthotic Human Knee. *J. Bionic Eng.* **2019**, *16*, 514–525. [CrossRef]

23. Vaida, C.; Plitea, N.; Carbone, G.; Birlescu, I.; Ulinici, I.; Pisla, A.; Pisla, D. Innovative Development of a Spherical Parallel Robot for Upper Limb Rehabilitation. *Int. J. Mech. Robot. Syst.* **2018**, *4*, 256. [CrossRef]

24. Carbone, G.; Gherman, B.; Ulinici, I.; Vaida, C.; Pisla, D. Design Issues for an Inherently Safe Robotic Rehabilitation Device. In *Advances in Service and Industrial Robotics, Proceedings of the International Conference on Robotics in Alpe-Adria Danube Region, Patras, Greece, 6–8 June 2018*; Springer: Cham, Switzerland, 2017; pp. 1025–1032. [CrossRef]

25. Görgülü, İ.; Carbone, G.; Dede, M. Time Efficient Stiffness Model Computation for a Parallel Haptic Mechanism via the Virtual Joint Method. *Mech. Mach. Theory* **2020**, *143*, 103614. [CrossRef]

26. Vaida, C.; Birlescu, I.; Pisla, A.; Ulinici, I.; Tarnita, D.; Carbone, G.; Pisla, D. Systematic Design of a Parallel Robotic System for Lower Limb Rehabilitation. *IEEE Access* **2020**, *8*, 34522–34537. [CrossRef]

27. Tarnita, D.; Marius, M.C.; Tarnita, D. Contributions on the Modeling and Simulation of the Human Knee Joint with Applications to the Robotic Structures. In *Mechanisms and Machine Science*; Springer: Cham, Switzerland, 2014; Volume 20, pp. 283–297. [CrossRef]

28. Gherman, B.; Birlescu, I.; Plitea, N.; Carbone, G.; Tarni, D.; Pisla, D.; Tarniţă, D.; Tarni, D.; Pisla, D. On The Singularity-Free Workspace of a Parallel Robot For Lower-Limb Rehabilitation. In Proceedings of the Romanian Academy, Series A; The Publishing House of the Romanian Academy: Bucharest, Romania, 2020; pp. 383–391.

29. Tarnita, D.; Marghitu, D.B. Analysis of a Hand Arm System. *Robot. Comput. Manuf.* **2013**, *29*, 493–501. [CrossRef]

30. Tucan, P.; Vaida, C.; Plitea, N.; Pisla, A.; Carbone, G.; Pisla, D. Risk-Based Assessment Engineering of a Parallel Robot Used in Post-Stroke Upper Limb Rehabilitation. *Sustainability* **2019**, *11*, 2893. [CrossRef]

31. Tucan, P.; Vaida, C.; Ulinici, I.; Banica, A.; Burz, A.; Pop, N.; Birlescu, I.; Gherman, B.; Plitea, N.; Antal, T.; et al. Optimization of the ASPIRE Spherical Parallel Rehabilitation Robot Based on Its Clinical Evaluation. *Int. J. Environ. Res. Public Health* **2021**, *18*, 3281. [CrossRef] [PubMed]

32. Gull, M.A.; Bai, S.; Bak, T. A Review on Design of Upper Limb Exoskeletons. *Robotics* **2020**, *9*, 16. [CrossRef]

33. Valevicius, A.M.; Jun, P.Y.; Hebert, J.S.; Vette, A.H. Use of Optical Motion Capture for the Analysis of Normative Upper Body Kinematics during Functional Upper Limb Tasks: A Systematic Review. *Off. J. Electromyogr. Kinesiol. J. Int. Soc. Electrophysiol. Kinesiol.* **2018**, *40*, 1–15. [CrossRef]

34. Li, J.; Cao, Q.; Dong, M.; Zhang, C. Compatibility Evaluation of a 4-DOF Ergonomic Exoskeleton for Upper Limb Rehabilitation. *Mech. Mach. Theory* **2021**, *156*, 104146. [CrossRef]

35. Culmer, P.R.; Jackson, A.E.; Makower, S.G.; Cozens, J.A.; Levesley, M.C.; Mon-Williams, M.; Bhakta, B. A Novel Robotic System for Quantifying Arm Kinematics and Kinetics: Description and Evaluation in Therapist-Assisted Passive Arm Movements Post-Stroke. *J. Neurosci. Methods* **2011**, *197*, 259–269. [CrossRef] [PubMed]

36. Shahar, N.; Schwartz, I.; Portnoy, S. Differences in Muscle Activity and Fatigue of the Upper Limb between Task-Specific Training and Robot Assisted Training among Individuals Post Stroke. *J. Biomech.* **2019**, *89*, 28–33. [CrossRef]

37. Byl, N.N.; Abrams, G.M.; Pitsch, E.; Fedulow, I.; Kim, H.; Simkins, M.; Nagarajan, S.; Rosen, J. Chronic Stroke Survivors Achieve Comparable Outcomes Following Virtual Task Specific Repetitive Training Guided by a Wearable Robotic Orthosis (UL-EXO7) and Actual Task Specific Repetitive Training Guided by a Physical Therapist. *J. Hand Ther.* **2013**, *26*, 343–352. [CrossRef] [PubMed]

38. Sanjuan, J.; Castillo, A.D.; Padilla Ramirez, M.; Quintero, M.C.; Gutierrez, E.E.; Sampayo, I.; Hernandez, J.R.; Rahman, M. Cable Driven Exoskeleton for Upper-Limb Rehabilitation: A Design Review. *Rob. Auton. Syst.* **2020**, *126*, 103445. [CrossRef]

39. Ben Hamida, I.; Laribi, M.A.; Mlika, A.; Romdhane, L.; Zeghloul, S.; Carbone, G. Multi-Objective Optimal Design of a Cable Driven Parallel Robot for Rehabilitation Tasks. *Mech. Mach. Theory* **2021**, *156*, 104141. [CrossRef]

40. Nelson, C.; Nouaille, L.; Poisson, G. A Redundant Rehabilitation Robot with a Variable Stiffness Mechanism. *Mech. Mach. Theory* **2020**, *150*, 103862. [CrossRef]

41. Major, Z.Z.; Vaida, C.; Major, K.A.; Tucan, P.; Simori, G.; Banica, A.; Brusturean, E.; Burz, A.; Craciunas, R.; Ulinici, I.; et al. The Impact of Robotic Rehabilitation on the Motor System in Neurological Diseases. A Multimodal Neurophysiological Approach. *Int. J. Environ. Res. Public Health* **2020**, *17*, 6557. [CrossRef] [PubMed]

42. Vaida, C.; Plitea, N.; Pisla, D.; Carbone, G.; Gherman, B.; Ulinici, I. Spherical Robot for Medical Recovery of Upper Limb Proximal Area. Patent NO. RO132233 3, 30 March 2020.

43. Pisla, D.; Tarnita, D.; Tucan, P.; Tohanean, N.; Vaida, C.; Geonea, I.D.; Bogdan, G.; Abrudan, C.; Carbone, G.; Plitea, N.C. A Parallel Robot with Torque Monitoring for Brachial Monoparesis Rehabilitation Tasks. *Appl. Sci.* **2021**, *11*, 9932. [CrossRef]

44. Available online: https://www.mscsoftware.com/product/adams (accessed on 25 May 2021).

45. Geonea, I.; Tarnita, D. Design and Evaluation of a New Exoskeleton for Gait Rehabilitation. *Mech. Sci.* **2017**, *8*, 307–321. [CrossRef]

46. Tarnita, D.; Geonea, I.; Petcu, A. Experimental Human Walking and Virtual Simulation of Rehabilitation on Plane and Inclined Treadmill. In *Acoustics and Vibration of Mechanical Structures—AVMS-2017*; Springer Proceedings in Physics; Springer: Cham, Switzerland, 2018; pp. 149–155. [CrossRef]

47. Nair, A.; Deenadayalan, E. Performance Analysis of Super Twisting Sliding Mode Controller by ADAMS–MATLAB Co-Simulation in Lower Extremity Exoskeleton. *Int. J. Precis. Eng. Manuf. Technol.* **2020**, *7*, 743–754. [CrossRef]

48. Soliman, A.F.; Sendur, P.; Ugurlu, B. 3-D Dynamic Walking Trajectory Generation for a Bipedal Exoskeleton with Underactuated Legs: A Proof of Concept. In Proceedings of the 2019 IEEE 16th International Conference on Rehabilitation Robotics (ICORR), Toronto, ON, Canada, 24–28 June 2019; pp. 599–604. [CrossRef]

49. Lee, Y.; Kim, Y.-J.; Lee, J.; Lee, M.; Choi, B.; Kim, J.; Park, Y.J.; Choi, J. Biomechanical Design of a Novel Flexible Exoskeleton for Lower Extremities. *IEEE/ASME Trans. Mechatron.* **2017**, *22*, 2058–2069. [CrossRef]

50. Berceanu, C.; Tarnita, D.; Filip, D. About an experimental approach used to determine the kinematics of a human. *Solid State Phenom.* **2010**, *166–167*, 45–50. [CrossRef]

51. Menga, G.; Ghirardi, M. Lower Limb Exoskeleton for Rehabilitation with Improved Postural Equilibrium. *Robotics* **2018**, *7*, 28. [CrossRef]

52. Lin, P.-Y.; Shieh, W.-B.; Chen, D.-Z. A Theoretical Study of Weight-Balanced Mechanisms for Design of Spring Assistive Mobile Arm Support (MAS). *Mech. Mach. Theory* **2013**, *61*, 156–167. [CrossRef]

53. Craig, R.R.; Bampton, M.C.C. Coupling of substructures for dynamics analyses. *AIAA J.* **1968**, *6*, 1313–1319. [CrossRef]

54. Pennestri, E.; Rossi, V.; Salvini, P.; Valentini, P.P. Review and Comparison of Dry Friction Force Models. *Nonlinear Dyn.* **2016**, *83*, 1785–1801. [CrossRef]

55. Tarnita, D.; Marius, M.; Dumitru, N.; Tarnita, D.N. Design and Simulation of an Orthotic Device for Patients with Osteoarthritis. In *New Trends in Medical and Service Robots*; Mechanisms and Machine Science; Springer: Cham, Switzerland, 2016; Volume 38, pp. 61–77. [CrossRef]

Article

Preliminary Biomechanical Evaluation of a Novel Exoskeleton Robotic System to Assist Stair Climbing

Max Böhme [1,*], Hans-Peter Köhler [2], Robert Thiel [1], Jens Jäkel [1], Johannes Zentner [1] and Maren Witt [2]

1 Faculty of Engineering, Leipzig University of Applied Sciences, Karl-Liebknecht-Str. 132, 04277 Leipzig, Germany
2 Department of Biomechanics in Sports, Faculty of Sport Science, University Leipzig, Jahnallee 59, 04109 Leipzig, Germany
* Correspondence: max.boehme@htwk-leipzig.de

Abstract: A novel exoskeleton robotic system was developed to assist stair climbing. This active demonstrator consists of a motor with a cable system, various sensors, and a control system with a power supply. The objective of this preliminary study is a biomechanical evaluation of the novel system to determine its effectiveness in use. For this purpose, three test persons were biomechanically investigated, who performed stair ascents and descents with and without the exoskeleton. Kinematics, kinetics, and muscle activity of the knee extensors were measured. The measured data were biomechanically simulated in order to evaluate the characteristics of joint angles, moments, and reaction forces. The results show that the new exoskeleton assists both the ascent and the descent according to the measured surface electromyography (sEMG) signals, as the knee extensors are relieved by an average of 19.3%. In addition, differences in the interaction between the test persons and the system were found. This could be due to a slightly different operation of the assisting force or to the different influence of the system on the kinematics of the users.

Keywords: biomechanics; exoskeleton; stair climbing; knee extensors; sEMG

Citation: Böhme, M.; Köhler, H.-P.; Thiel, R.; Jäkel, J.; Zentner, J.; Witt, M. Preliminary Biomechanical Evaluation of a Novel Exoskeleton Robotic System to Assist Stair Climbing. *Appl. Sci.* **2022**, *12*, 8835. https://doi.org/10.3390/app12178835

Academic Editors: Ionuţ Daniel Geonea and Cristian Copilusi Petre

Received: 17 August 2022
Accepted: 31 August 2022
Published: 2 September 2022

Publisher's Note: MDPI stays neutral with regard to jurisdictional claims in published maps and institutional affiliations.

1. Introduction

An exoskeleton is a wearable device with rigid and/or soft components, that augments, assists, enables, and/or enhances physical activity through mechanical interaction with the body [1]. Furthermore, an exoskeleton robotic system is an exoskeleton with all of the associated components such as a power supply, equipment, software, and communications that are necessary to make it fully functional. In recent years, more and more exoskeletons have been developed and tested for a wide range of applications [2]. Thus, a number of systems were also developed to assist people, especially elderly people, exclusively during stair climbing [3,4]. For example, an active system has been developed by [5] that consists of five parts and weighs about 13.5 kg. In this system, additional torque is applied to the knee and ankle joints of both legs by tension elements, which are actuated by two DC motors that are located on the back with a power of 140 W each [5]. Another example is a lower limb exoskeleton that is described in [6]. This carbon fibre bilateral system assists the extension and flexion moment of the knee joint by two pneumatic actuators each. The actuators are controlled by five non-invasive surface electromyographs (sEMG) that have to be placed on the user's thigh muscles in advance [6].

All systems of the state of the art devices for the exclusive assistance of elderly people during stair climbing have different actuator configurations. However, biomechanical studies to evaluate the need for assistance have shown that only knee extension should be assisted during ascent and descent [3]. In this way, the natural muscle weakness that occurs with age can be compensated [7]. The current systems do not meet this requirement and assist or actuate more than is actually needed.

Furthermore, the effectiveness of these exoskeletons has not been evaluated or has only been evaluated in very simple biomechanical studies. For example, trials were conducted with the active system by [5] and it was described that the system helps the user to climb stairs. However, this statement is only based on the experience and sense of the user and has not been evaluated by measurements. The lower limb exoskeleton was tested with only one subject, but for a period of 10 days [6]. From the recordings of knee angle, sEMG of knee extensors and flexors, and metabolic costs, a reduction in both flexor muscles activity by up to 61% could be concluded [6]. Also, only one subject (age 37 years, height 1.70 m, weight 69 kg) was examined in the preliminary walking experiments for the test of the robotic hip exoskeleton [8]. Although the test was carried out on a standardized staircase (step height 0.18 m, tread 0.30 m) according to [9], the aim of the test was not the biomechanical evaluation of the exoskeleton. A more comprehensive investigation was carried out on the Hiteexosuit with three test persons [10]. Here, the tensile force of the actuator and the knee joint angular displacements were measured during stair climbing [10]. The lower extremity exoskeleton achieved a knee joint moment of 0.8 Nm/kg in stair climbing experiments, which were not further described [11]. It was concluded that the design and hybrid control strategy are functional and improve mobility during stair climbing. Furthermore, trials were conducted with Beihang University's lower limb exo on various indoor and outdoor staircases [12]. However, only the kinematics were measured to evaluate a gait algorithm, but not to investigate the biomechanical effectiveness of the system itself [12]. The passive exoskeleton PKAExo was only mechanically verified in one experiment [13,14]. Nevertheless, the authors noted that this experiment does not sufficiently prove that the passive system neither reduces muscle activity during weight-bearing ascending locomotion, nor that it improves human metabolic productivity [14]. They already mentioned that sEMG signals, metabolic costs, and also external forces should be measured in order to evaluate the system [14]. This was done, for example, for an evaluation of an exoskeleton to assist sit-to-stand movements [15].

In recent and more extensive studies on the effectiveness of exoskeletons, kinematics, external forces, sEMG signals, and metabolic costs are measured in several subjects during exercises with and without the exoskeleton [16,17]. In addition, biomechanical simulations are increasingly used for the evaluation of results [17]. The advantage is that internal forces that cannot be measured directly, such as joint reaction forces, can be determined and analyzed for an evaluation of the system.

Considering the state of the art devices, there are no studies which quantify the complex kinetics and kinematics during stair climbing with an exoskeleton. Therefore, the objective of this study is to present and biomechanically evaluate the novel robotic exoskeleton for stair climbing. In detail, the change of kinematics in the sagittal and frontal plane during ascent and descent due to the influence of the exoskeleton will be investigated. Furthermore, it will be examined whether the required joint moments of the user can be reduced by the active system, thus providing assistance. This is also to be evaluated by the muscle activities of the knee extensors with and without the use of the exoskeleton. Finally, a potential change in the internal knee joint reaction forces should be evaluated.

2. Materials and Methods

2.1. Exoskeleton

The novel exoskeleton robotic system to assist stair climbing, shown in Figure 1, is according to ISO 13482 a restraint type physical assistant robot, since the system is fastened to the human during use. The intended purpose of the exoskeletal movement assisting system (**eBUS**; German abbreviation for exoskelettales **B**ewegungsunterstützungssystem) is to assist people with a natural lack of strength, such as elderly people. In previous biomechanical studies, it was found that an exoskeleton for elderly people should assist them in knee extension during ascent and descent as well [3]. Therefore, the active exoskeleton eBUS was designed that assists only the knee joint actively in extension in order to maintain minimal actuation. Furthermore, the eBUS is characterized by a unilateral and rigid structure and is symmetrically designed for the left and right leg. The system works independently for each leg and weighs 6.1 kg per leg. The components that are attached to the thigh have a total mass of 4.4 kg (72%), those on the shank 1.5 kg (24%), and on the foot 0.2 kg (4%). Thus, it was possible to consider the fact that the low mass at the distal segment (foot) is advantageous for the acceleration of the foot.

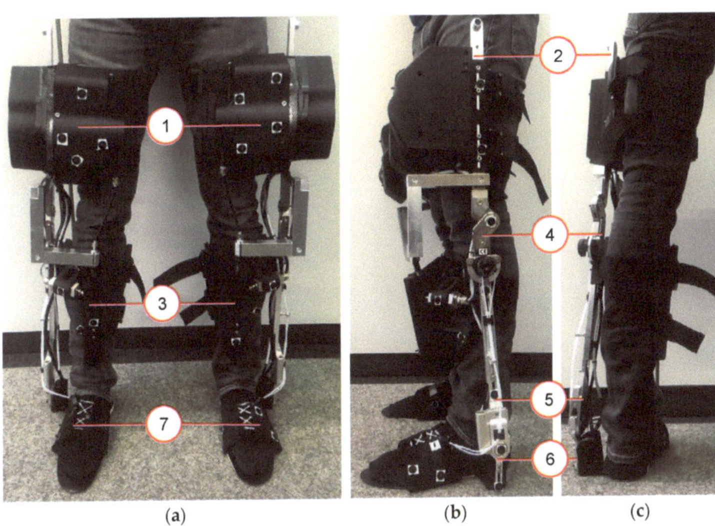

(a)   (b)   (c)

Figure 1. Novel exoskeleton robotic system to assist stair climbing called **eBUS** [4]: (**a**) attached to both legs in the frontal plane and (**b**) in the sagittal plane; (**c**) attached to the left leg in posterior view; (1) thigh segment, (2) thigh rail, (3) shank segment, (4) shank upper rail, (5) shank lower rail, (6) foot rail, and (7) foot segment.

The eBUS consists of seven parts per leg with two rotational joints corresponding to flexion/extension of the knee and dorsiflexion/plantarflexion of the ankle (Figure 1). Three of the seven segments are attached to the user's thigh, shank, and foot and represent the interface. These three segments are connected to each other through four sliding rails. The four rails create the two rotational joints with one degree of freedom, that should ideally be coaxial with the human knee and ankle joints, respectively. The sliding rails enable the eBUS to be adjustable for different anthropometries of users, as shown in Figure 2.

The foot segment of the exoskeleton surrounds the human foot and consists of a compliant material to allow the foot to roll off. This design is intended to take into account the second DOF in the ankle joint so that lateral movements in the frontal plane are possible to increase the user comfort. All the three interface attachment components are adjustable to consider individual anthropometry to increase comfort.

The active force to assist stair climbing is implemented through a Bowden cable, which is tensioned between the thigh segment and the shank segment of the eBUS and spans the human knee anteriorly. The drive system from a Maxon motor ag (Sachseln, Switzerland) consists of a brushless DC motor (EC 60 flat, 150 W) with integrated encoder (Encoder MILE 4096 steps per turn, two channels with Line Driver) and a gearhead (GP 52 C) that is fixed to the exoskeleton thigh. The energy storage unit is also attached to the thigh segment and consists of seven lithium-ion cells (Samsung INR18650-29E) of type 7S1P (FRIWO Gerätebau GmbH, Ostbevern, Germany). The driver system pulls the cable and creates a pulling force. The motor is controlled by a servo controller (ESCON Module 50/5, Maxon motor ag, Sachseln, Switzerland) in combination with an Arduino® Mega 2560 (Arduino SA, Lugano, Switzerland). The force in the bowden cable is measured at any time by a ZAD 500-T tension force sensor (HAEHNE Elektronische Messgeräte GmbH, Erkrath, Germany). The force vector direction is independent of the knee angle as the cable has a fixed distance with 0.1 m from the patella of the knee. The pulling force results in an external moment through the rigid structure of the exoskeleton to assist knee extension. This external force is active only if the corresponding leg is in the stance phase of the movement, which is detected by four pressure sensors (FSR 402, Interlink Electronics, Camarillo, CA, USA) that are integrated in the sole of the foot. In addition, inertial measurement units (BNO055, Bosch Sensortec GmbH, Reutlingen, Germany) are attached to the thigh, shank, and foot segments. These allow to determine the joint angle positions during use in order to distinguish the ascent and descent as well as their phases from each other.

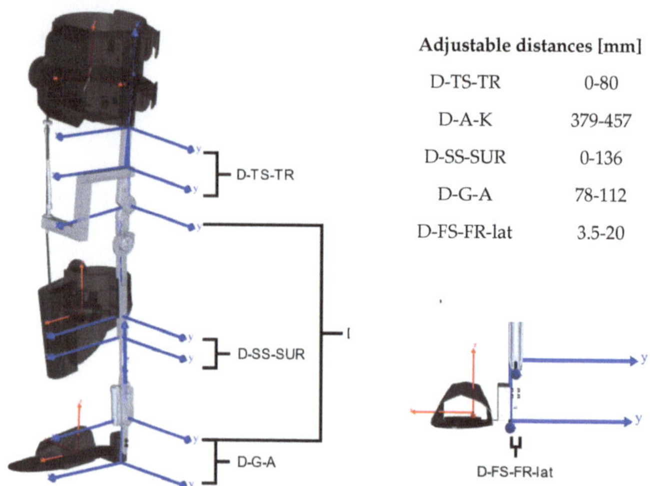

Figure 2. Adjustment possibilities on the eBUS exemplary for the left side (**left**) with a detailed view in the frontal plane (**bottom right**) and the adjustable distances (**top right**) [4]; D-A-K (Distance-Ankle-Knee); D-G-A (Distance-Ground-Ankle); D-TS-TR (Distance-ThighSegment-ThighRail); D-SS-SUR (Distance-ShankSegment-ShankUpperRail); D-FS-FR-lat (Distance-FootSegment-FootRail-lateral).

Further details about the exoskeleton, such as design requirements, specifications, motor selection, or computer architecture design, are described in [4].

2.2. Participants, Setup and Procedure

There were three subjects (1 m; 2 f; 26.0 ± 3.6 years; 24–31 years; 1.72 ± 0.03 m; 70.2 ± 6.0 kg) that participated in the preliminary biomechanical study. According to their own statements, each subject was healthy and had no orthopaedic impairments. The relevant anthropometric data of each subject were measured first and is shown in Table 1. The hip joint centres were estimated to determine the hip width [18,19].

Table 1. Relevant anthropometric data of the subjects with SID = subject identification, A = age in years, S = sex (m = male, f = female), bm = body mass [kg], bh = body height [m], hw = hip width [m], mtl = morphological thigh length [m], sl = shank length [m], fl = foot length [m].

SID	A	S	bm	bh	hw	Right Leg			Left Leg		
						mtl	sl	fl	mtl	sl	fl
S1	31	m	76.5	1.76	0.171	0.486	0.426	0.257	0.494	0.412	0.260
S2	24	f	62.2	1.72	0.176	0.428	0.412	0.265	0.442	0.400	0.267
S3	23	f	71.8	1.69	0.186	0.458	0.386	0.248	0.445	0.387	0.248

In order to investigate the biomechanical effectiveness of the novel exoskeleton system eBUS, all of the participants performed ascents and descents both without and with eBUS.

For this purpose, a four-step staircase (Figure 3) with a step height of 0.16 m, tread of 0.28 m, and a resulting inclination angle of 30° was used. In the second step of the staircase is a force plate (MiniDyn type 9119AA2, Kistler Instrumente AG, Winterthur, Switzerland) is implemented to record the ground reaction force (GRF) at 1500 Hz. A marker-based motion capture system (Qualisys AB, Goteburg, Sweden) with twelve active infrared cameras recorded the kinematics at 100 Hz. For recording and synchronization, the software Qualisys Track Manager 2021 was used. The tensile assistive force in the Bowden cable was recorded through the tension force sensor, which is mentioned above. This force data, together with the data of the inertial measurement units for position determination and the pressure sensors for stands detection, were stored with a time stamp on the internal memories of the eBUS. Furthermore, sEMG signals of the m. vastus medialis (VAM) and m. vastus lateralis (VAL) were measured on both legs using dual electrodes and a wireless measurement system (Noraxon Inc., Scottsdale, AZ, USA).

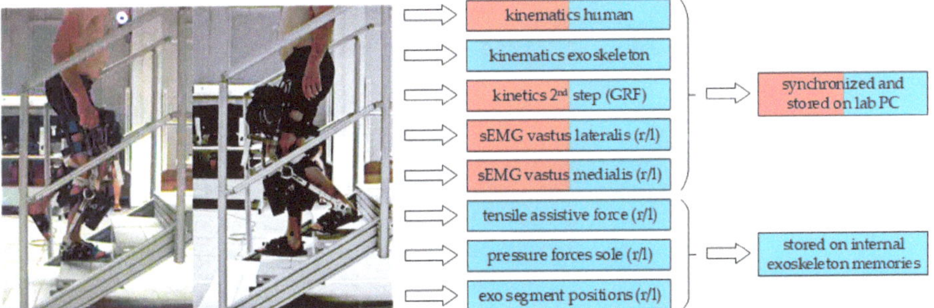

Figure 3. Measured and stored data during stair ascents and descents with the novel exoskeleton system eBUS (bright blue) The red bares indicate data that are measured during stair ascent and descent without eBUS as a reference. r/l stands for right and left leg.

Firstly, the relevant body parameters such as body weight, body height, and segment lengths were measured. Subsequently, the electrodes for the sEMG measurement were placed on the subject according to the seniam project [20]. After that, the subjects performed an isolated knee extension within a flexion angle of 90° to 70° to determine the maximum voluntary contraction (MVC) of the quadriceps femoris muscles according to [21]. Based on the CAST-model [22] and the placement of two markers on the shoulder, a total of 36 passive infrared markers were applied. Furthermore, eight additional markers were placed medially and laterally on the knee and ankle joints of both legs. The participants then performed six ascents and descents without the exoskeleton, barefoot, at a self-selected speed and climbing step-over-step.

In the second phase, the exoskeleton was adjusted according to the anthropometry of the participants (Table 2). Subsequently, the marker setup was adapted so that a total of 26 markers were attached to the human body and 18 markers were attached to each exoskeleton leg (Figure 4).

Table 2. Adjusted lengths of the exoskeleton to consider the anthropometry of each subject.

Distances [mm] According to Figure 2	Subject 1		Subject 2		Subject 3	
	Left	Right	Left	Right	Left	Right
D-TS-TR	20	20	0	0	0	0
D-A-K	416	424	400	414	400	400
D-SS-SUR	136	136	133	135	133	135
D-G-A	78	78	78	78	80	79
D-FS-FR-lat	5	5	5	5	5	5

The eBUS, which was adjusted specifically for the subject, was attached to the subject's legs under supervision and with the help of laboratory staff. Before the tests were carried out, the subject was able to familiarize him/herself with the attached eBUS and climbed the stairs several times without recording any data. The subject then performed a series of six ascents and descents at a self-selected speed, step-over-step, and without using the handrail.

2.3. Data Preparation, Modelling, and Calculations

After all data were recorded, a check of completeness and errors was carried out. In the first step, gaps in the trajectories of up to ten frames that were caused by the partial occlusion of markers by the test bed, were filled by polynomial interpolation using Qualisys Track Manager Version 2021. For markers belonging to a cluster, relational gap filling was used.

A moving average filter was used to improve the signal-to-noise ratio. The sEMG data with eBUS were generally noisier, so a filter length of 333 ms and 16.7 ms without the eBUS were used. Since MVC-normalized EMG-data showed values above 100% in two subjects, amplitude normalization to the within-trial peak value was performed for each subject and each movement task (ascent/descent; without and with eBUS). Subsequently, each sEMG dataset was time normalized in the stance phase between the start and end points that were marked in the dataset. To compare the subjects in each stance phase, we calculated the average of all of the sEMG datasets per subject. As we only had a limited number of datasets and there were some outliers, it was best to calculate the average using the median method. To compare the outliers per subject, the standard deviation was calculated and scaled and shown as a shaded area around the median. The scaling was necessary because the standard deviation was out of bounds in some datasets and was done using the square root function.

All of the movements without the eBUS were simulated using the Anybody Modelling System (AMS) (version 7.3.4, AnyBody Technology A/S, Aalborg, Denmark) [23]. A musculoskeletal model from the AnyBody Managed Model Repository (AMMR) (version 2.3.0) [24] without arms was used. The segment lengths were adjusted manually by using the specific subject data that were obtained from Table 1. In addition, the simplest muscle model in AMS was used. The measured data were then implemented, filtered with a low-pass filter (2nd order, 5 Hz Butterworth), and kinematically optimized [25] to calculate the joint angle curves. It was assumed that all of the markers have the same influence. The joint moments were calculated according to [26]. The calculation was automated using Python version 3.7 and the AnyPyTools package [27].

Subsequently, the data that were stored on the exoskeleton memories was synchronized with the data from the lab PC. For this purpose, the knee angle curve of both datasets was compared and synchronized. The force data that were measured with the tensile force sensor was then processed and used for the subsequent simulation with the eBUS. For this simulation, the CAD model of the eBUS was translated into an AMS script using the plug-in AnyExp4SOLIDWORKS version 1.2.0. The masses and inertias of the seven individual components of the eBUS per leg were taken into account accordingly. The exoskeleton model was attached to the human model that was described above with the help of dummy segments according to [28] and a contact model with virtual force plates [29]. The Bowden cable was modelled with endpoints and via points between the thigh and the shank of the exoskeleton to implement the assistive force through separate force files. The kinematics and kinetics were calculated in the same way as in the simulation without exoskeleton. The whole simulation model is shown in Figure 4 during the inverse calculation of a stair ascent.

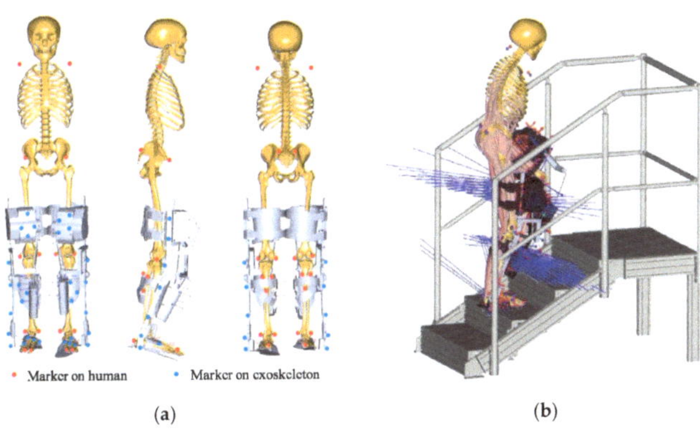

Figure 4. (**a**) Marker placement on the human and exoskeleton [29]. Markers on the human: two at the shoulders, four at the pelvis, four at the knees, six at the shanks, four at the ankles, and six at the feet. Markers on each exoskeleton leg: four at the thigh segment, three at the thigh rail, three at the shank segment, three at the shank lower rail, one at the foot rail, and four at the foot segment; (**b**) Human-exoskeleton-model during the inverse calculation of a stair ascent [4].

2.4. Analyses

For the evaluation of exoskeletons, muscle activities, joint moments, and joint reaction forces are often analyzed as simulation outputs [30,31]. A comprehensive evaluation of the exoskeleton should consider the effects of the system on the entire human body. For this reason, all of the angular curves of the lower extremities in sagittal and frontal planes, the moments of the hip, knee and foot, as well as the assistive moments that were applied by the exoskeleton in the knee are evaluated on a subject specific basis. Since the eBUS is designed to assist knee extension, the measured sEMG signals from VAL and VAM during the stance phase were also analyzed. The muscle activity is indicated as a percentage of the within-subject maximum. Furthermore, the joint reaction forces in the knee with and without eBUS were compared. The joint moments and the joint reaction force were normalized to the respective body weights.

All curves of angles, moments, and joint reaction forces are presented as a cycle for ascent and descent on a per-subject basis. The start of the cycle (0% cycle time) is defined by the first foot contact on the force plate with a threshold value of 20 N. The stance phase ends when the contact is released, and the swing phase begins at the same time. The end of the swing phase (100% cycle time) was determined by the marker trajectories. Each trial was trimmed, resampled, and linearly normalized to the duration of the stance respectively swing phase. The biomechanical results of each trial setting were averaged, standard deviated and plotted on the same graph.

The data that are summarized in Table 3 were simulated successfully and could be used for the analysis. Due to large gaps in the trajectories and incomplete data from the exoskeleton, not all six trials could be simulated for subjects 5 and 6 and were excluded from the analysis.

Table 3. Analyzable data per subject (S1, S2, S3), movement (ascent, descent), and setting (with/without eBUS).

Movement	Subject 1		Subject 2		Subject 3	
	With eBUS	Without eBUS	With eBUS	Without eBUS	With eBUS	Without eBUS
Ascent	6	6	5	6	4	6
Descent	6	6	5	6	3	6

3. Results

The evaluation method that was used here can generate various results, but only a few relevant ones will be analyzed. As the exoskeleton actively assists the knee joint, we first present the assistive moment in the knee joint, that is the additional torque that is provided by the exoskeleton during the movements. Furthermore, the system has an influence on the kinematics of the subject during use, especially on the lower extremities. For this reason, the kinematic patterns of all three joints (hip, knee, foot) for sagittal and frontal planes were selected for evaluation. Since the eBUS influences not only the kinematics but also the kinetics, we also present the results of the joint moments in the sagittal plane. As the advantage of the applied analysis method is to calculate also the internal forces during the movement, the effect of the system on the knee joint reaction force was investigated. Finally, to determine the effect of the system, the measured sEMG signals of VAM and VAL were evaluated.

All of the curves are available as the mean with one standard deviation. The curves without the exoskeleton are displayed in red together with the curves with the exoskeleton in blue in the same graph.

3.1. Assistive Moment in Knee Joint

The curves of the assistive moment for knee extension are shown for ascent and descent in Figure 5 for the tests with the eBUS as the mean (black) and with one standard deviation (grey). The mean curve of the eBUS moments is characterized by one peak for both movements. During the ascent, this peak occurs during the forward motion while during the descent it also occurs at the end of the stance phase during the subphase of controlled lowering. In 60 ± 30 Nm the first subject, the eBUS generates a maximum torque of during ascent and 45 ± 20 Nm during descent. For the other two subjects, these values are 45 ± 10 Nm each during the ascent and 60 ± 20 Nm during the descent.

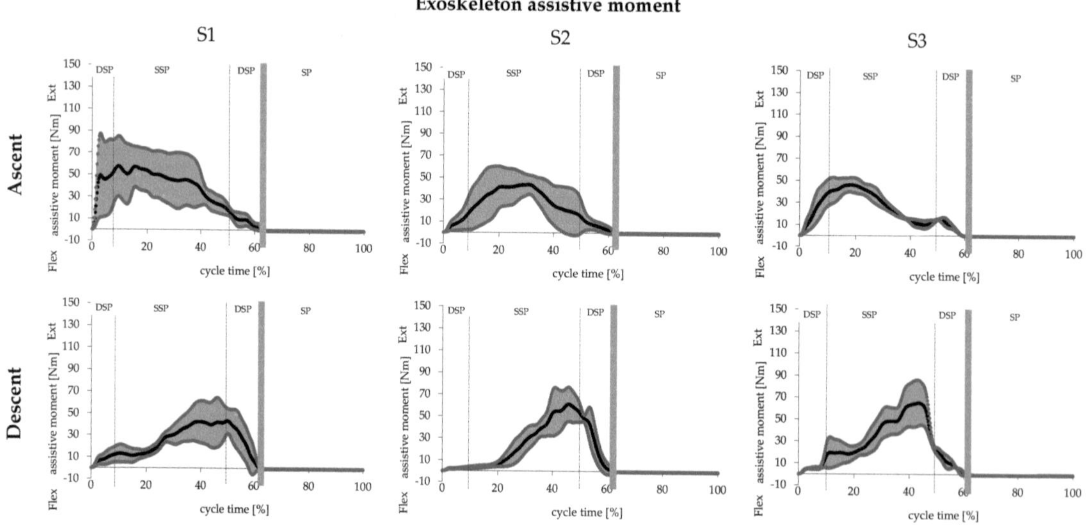

Figure 5. Exoskeleton assistive moment in the knee joint to assist knee extension for Subject 1 (**left**), Subject 2 (**middle**), and Subject 3 (**right**); DSP = double stance phase, SSP = single stance phase, SP = swing phase.

3.2. Joint Angles in Sagittal Plane

The kinematic curves of the foot, knee, and hip angles in the sagittal plane are shown for ascent and descent in Figure 6 for the tests with the eBUS (blue) and without the eBUS (red). In general, we observe a high degree of similarity in the curves of the test persons. With the exoskeleton, the standard deviation is significantly greater, whereby individual differences become obvious.

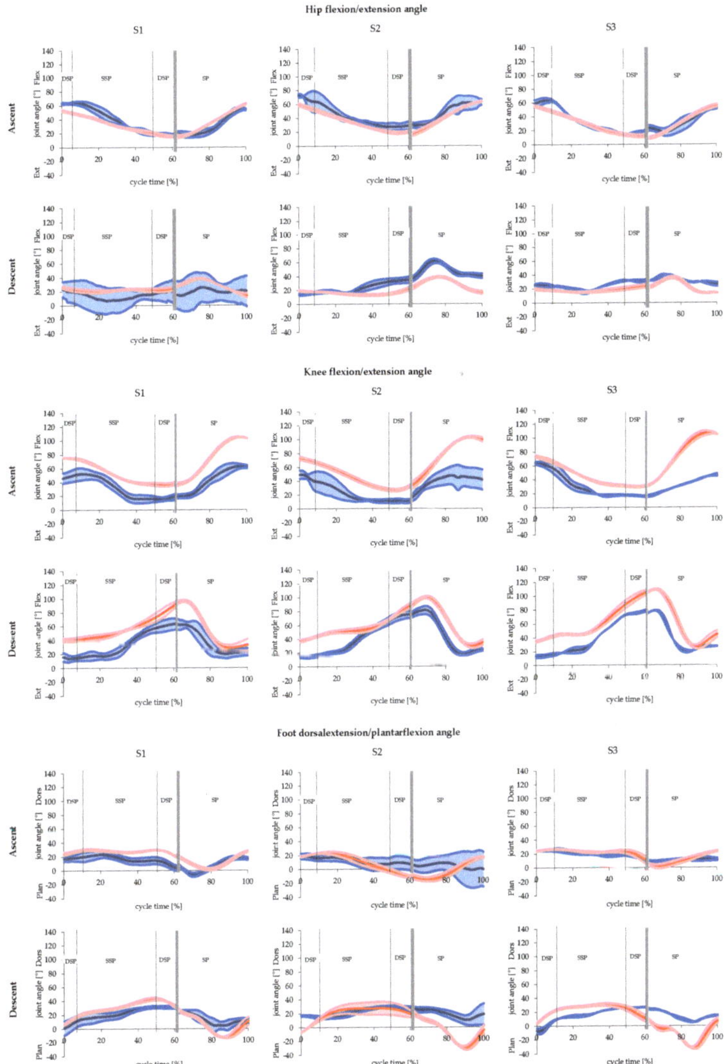

Figure 6. Foot, knee, and hip angle curves in the sagittal plane for ascent and descent without (red) and with (blue) the exoskeleton for Subject 1 (**left**), Subject 2 (**middle**), and Subject 3 (**right**); DSP = double stance phase, SSP = single stance phase, SP = swing phase.

Subjects 2 and 3 reduce the movement amplitude in the ankle joint by using the eBUS in ascent and descent. For the first subject, these angle curves are qualitatively similar for both movements, but still differ quantitatively. The largest average deviation of the curves with and without the eBUS occurs for Subject 2 during the descent in the swing phase with a value of 35°. However, the qualitative curves without the eBUS are intersubjectively similar for all three subjects. In both movements, the dorsal and plantar peaks of the curves without the eBUS are not achieved by the curves with the eBUS.

The knee angle curves with and without the eBUS are qualitatively similar for both movements of all three subjects. Quantitatively, the knee is flexed about 20° less with the eBUS during ascent over the entire cycle than without the eBUS. During descent, this effect occurs for all three subjects only in the first two phases until the controlled lowering at about 35% cycle time. In the swing phase, the knee is less flexed with the eBUS than without. The maximum flexion angles of the curves without the eBUS are not achieved in the trials with the eBUS for both movements in all three subjects.

The qualitative hip angle curves with and without the eBUS are similar for both ascent and descent. The exception to this is the phase of controlled lowering from 30 to 60% cycle time for the second subject during descent. In this phase, the second subject flexes his/her hip during the trials with the eBUS, so that the mean value is also beyond the standard deviation of the curve without the eBUS. For the first subject, the largest standard deviation of 20° occurs during the descent. The maximum flexion angle of 75° on average occurs in the second subject during the ascent with the eBUS and thus exceeds the maximum without the eBUS.

3.3. Joint Angles in Frontal Plane

The kinematic curves of the foot and hip angles in the frontal plane are shown for the ascent and descent in Figure 7 for the tests with eBUS (blue) and without the eBUS (red).

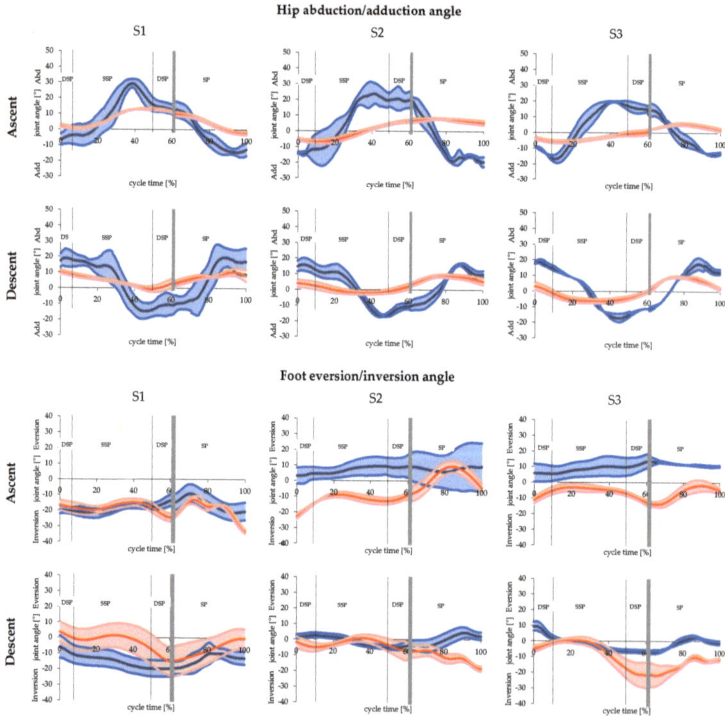

Figure 7. Foot and hip angle curves in the frontal plane for ascent and descent without (red) and with (blue) the exoskeleton for Subject 1 (**left**), Subject 2 (**middle**), and Subject 3 (**right**); DSP = double stance phase, SSP = single stance phase, SP = swing phase.

The angular curves of the subtalar eversion or inversion of the trials with the eBUS already differ qualitatively in all three subjects from the curves without the eBUS in both movements during most of the phases. However, all curves without the eBUS also differ from each other in an intersubjective comparison both qualitatively and quantitatively during ascent and descent. The same applies to the curves with the eBUS in the intersubjective comparison.

Both during ascent and descent, the adduction and abduction angle curves of the hip with the eBUS differ qualitatively and quantitatively from the curves without the eBUS in all three subjects. This is particularly noticeable in the phase of pulling up and forward as well as in the swing phase. It should also be noted that the qualitative curves without the eBUS are similar in the intersubjective comparison for the ascent, which also applies to the descent. The qualitative progressions for both movements with the eBUS are also similar in the intersubjective comparison.

3.4. Joint Moments in Sagittal Plane

The net joint moments of the foot, knee, and hip in the sagittal plane are shown for ascent and descent in Figure 8 for the tests with the eBUS (blue) and without the eBUS (red).

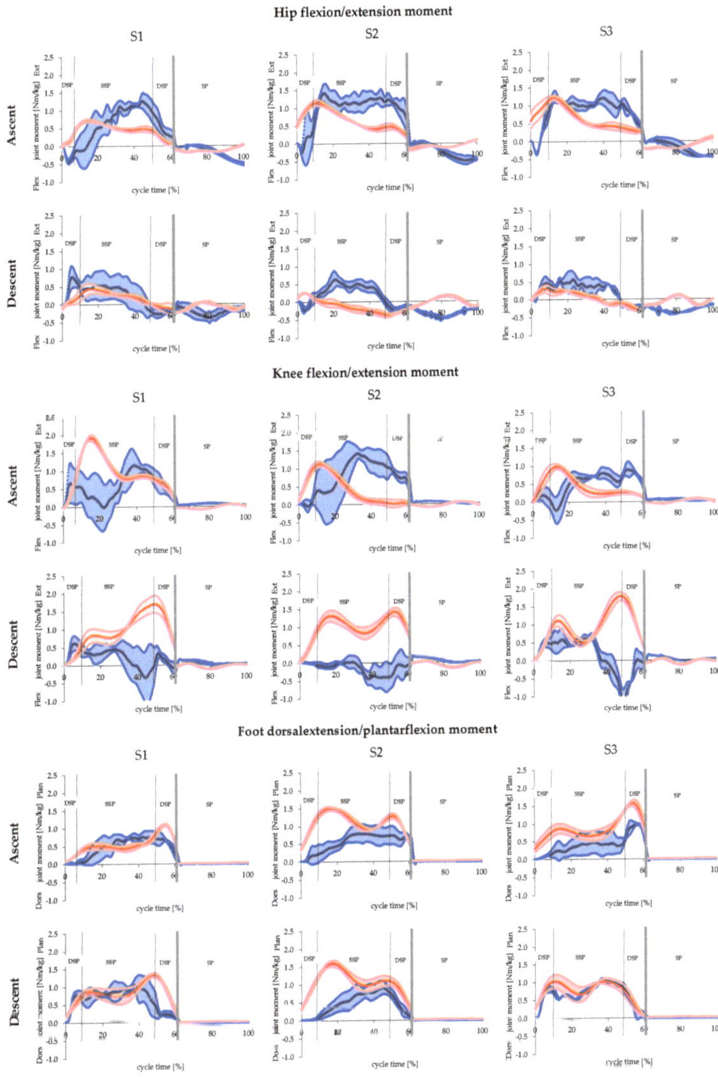

Figure 8. Foot, knee, and hip moment curves in the sagittal plane for ascent and descent without (red) and with (blue) the exoskeleton for Subject 1 (**left**), Subject 2 (**middle**), and Subject 3 (**right**); DSP = double stance phase, SSP = single stance phase, SP = swing phase.

The curves of the foot or ankle moments without the eBUS are characterized by two peaks for both movements. The foot moment curves with the eBUS already differ qualitatively from those without the eBUS, since only one peak occurs here, except for the descent in Subject 3. Quantitatively, the foot joint moments that are applied without the eBUS, with mean maximum values of up to 1.6 Nm/kg, are significantly higher than the mean maximum values of 1.0 Nm/kg with the eBUS.

The results of the knee moment without the eBUS already show qualitative differences in the intersubjective comparison during the ascent. During descent, these curves without the eBUS are again similar. On the other hand, the knee moment curves with the eBUS are qualitatively different for each subject, which applies to the ascent and the descent. It is obvious during the descent in the trials with the eBUS that even flexion moments occur in all three subjects during the phase of controlled descent. In the trials without the eBUS, an extension moment always occurs in this phase. During descent, the quantitative maxima with the eBUS are significantly lower compared to the maxima without the eBUS. During the ascent, this again varies from subject to subject.

During the ascent, the curves of the hip moments coincide qualitatively and quantitatively in the phase of weight acceptance. In the subsequent phases, when the eBUS is used, the extension moment continues to increase for Subject 1 up to a cycle time of 45%, until a maximum of 1.3 ± 0.2 Nm/kg is achieved. For Subjects 2 and 3, a plateau is maintained here until the second double stance phase. Subsequently, the curve with the eBUS decreases more rapidly compared to the curve without the eBUS. The swing phase of the ascent with the eBUS is characterized by a lower flexion moment up to 80% cycle time. Subsequently, the flexion moment increases when using the eBUS until a maximum value of 0.5 ± 0.1 Nm/kg is achieved at the end of the cycle for all three subjects. During descent, the two trajectories differ both qualitatively and quantitatively in the most frequent phases for all three subjects. In the swing phase, the hip moment without the eBUS increases between 75 and 85% cycle time and then decreases again. In the swing phase, the hip moment curve with the eBUS is exactly the opposite of the curve without the eBUS.

3.5. Knee Joint Reaction Force

In the calculation of the knee joint reaction forces, the muscle forces that are involved in the movement, the masses and inertias of the human body and the exoskeleton, the measured ground reaction force, and the assisting force of the eBUS were taken into account besides the kinematics. The curves of the calculated joint reaction forces in the knee normalized to the body weight are shown accordingly in Figure 9 for the tests with the eBUS (blue) and without the eBUS (red).

The two peaks in the stance phase are characteristic for the ascent and descent, as it can be found in the curves without the eBUS. Compared to the stance phase, the swing phase is characterized by very low values with a maximum of 5 N/kg. Compared to the stance phase, the knee joint reaction forces during the swing phase are negligible.

However, when using the eBUS, the characteristic progression in the stance phase is not achieved in neither the ascent nor the descent. For both movements, only one peak occurs in the stance phase with values of 68 ± 10 N/kg during ascent and 35 ± 10 N/kg during descent. In comparison, the maximum values of Subjects 1 and 2 without the eBUS are about half as large during ascent. However, these values are comparatively similar during descent, and even 15 N/kg lower for Subject 2.

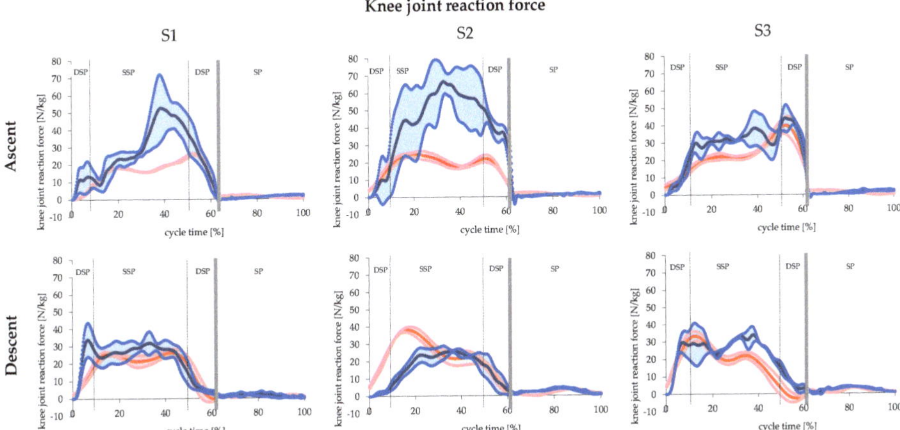

Figure 9. Knee joint reaction force (proximal-distal) for ascent and descent without (red) and with (blue) the exoskeleton for Subject 1 (**left**), Subject 2 (**middle**), and Subject 3 (**right**); DSP = double stance phase, SSP = single stance phase, SP = swing phase.

3.6. sEMG of Knee Extensor Muscles

The curves of the muscle activity of the knee extensor muscles (VAM and VAL) have a high degree of similarity in the contraction pattern as shown in Figure 10 for the tests with the eBUS (blue) and without the eBUS (red), which varies individually. The muscle activity varies greatly in individual cases with the eBUS, especially in Subject 1.

The measured muscle activities of VAM and VAL are characterized by two peaks in Subjects 2 and 3 during both movements without using the eBUS. In Subject 1, only one peak occurs at the beginning of the stance phase. When using the eBUS, the curve of both muscle activities during ascent is characterized by only one peak at the beginning of the stance phase. During descent, this peak occurs in the phase of controlled lowering with the use of the eBUS. Quantitatively, for all the subjects the maximum of the measured muscle activity during the trials with the eBUS is up to 50% lower than the maximum during the trials without. It should also be noted that for Subject 2 during descent in the phase of controlled lowering, the maximum values of both muscle activities with the eBUS are higher than the maximum values of the activities without the eBUS. The integrals of each sEMG curve are shown in Table 4 to highlight differences in the net effort of both muscles by using the eBUS. It becomes obvious that the values of all the subjects are lower when using the system, both for ascent and descent.

Table 4. Integrals of the normalized median sEMG curves of m. vastus medialis (VAM) and m. vastus lateralis (VAL) in time normalized stance phase with and without the exoskeleton per subject (S1, S2, S3) and movement (ascent, descent).

Movement	Subject 1		Subject 2		Subject 3	
	With eBUS	Without eBUS	With eBUS	Without eBUS	With eBUS	Without eBUS
m. vastus medialis (VAM)						
Ascent	35.40	53.70	15.76	41.04	18.05	52.40
Descent	27.52	46.99	33.91	42.22	19.38	52.32
m. vastus lateralis (VAL)						
Ascent	20.08	39.07	17.77	31.69	17.54	31.12
Descent	26.44	58.72	33.55	35.89	26.80	38.66

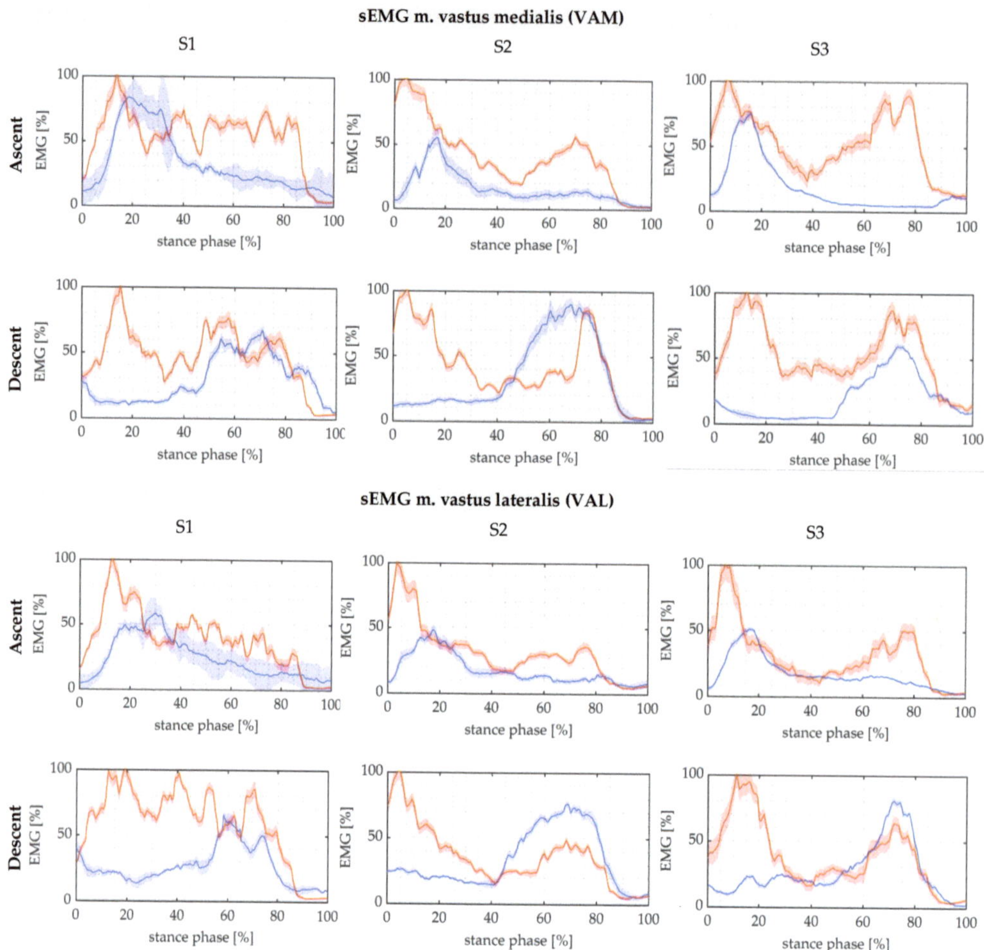

Figure 10. sEMG curves of m. vastus medialis (VAM) and m. vastus lateralis (VAL) in stance phase for ascent and descent without (red) and with (blue) the exoskeleton for Subject 1 (**left**), Subject 2 (**middle**), and Subject 3 (**right**).

4. Discussion

The results that were obtained in the preliminary evaluation will now be discussed. In general, it can be postulated that the eBUS can significantly influence the climbing of stairs in three ways. Firstly, the additional masses of the eBUS have an influence on the dynamic human-machine system, which could result in a change in the kinematics and kinetics of human movements. The second influence relates to the restriction of the range of motion of the human locomotion system by the eBUS. Finally third, the tensile force transmitted from the eBUS to the human will most likely influence the human's movement pattern, in an assistive manner in the desirable case. Within this discussion, only assumptions can be made about these influences, since the causal research of a potential change in movement was not the aim of this investigation. Rather, the results are to be interpreted in such a way that recommendations for further development of the evaluated eBUS can be derived and a general statement on the effect of the exoskeleton can be made.

4.1. Assistive Moment in Knee Joint

The assistive knee extension moment is applied by the eBUS during ascent in the phase of forward progression and during descent in the phase of controlled lowering. Therefore, in both ascent and descent, the phases with a power deficit that were identified by [3] could be assisted by the eBUS, as shown by the qualitative progression of the applied knee extension moment. Since the eBUS applied a maximum moment of 60 ± 30 Nm for the 76.5 kg Subject 1 (0.8 ± 0.4 Nm/kg) during the ascent, the exoskeleton that was presented here can compensate for the power deficit of 0.8 Nm/kg that was determined in [3]. The required assistive knee extension moments of 0.4 Nm/kg during the descent were also applied by the eBUS for all three subjects. It should be mentioned here that the selected drive can apply a cable force of 1500 N and thus a moment of 1.5 Nm/kg, so the maximum performance was not yet achieved during these tests.

4.2. Kinematics in Sagittal and Frontal Plane

The changes in the kinematics in the sagittal plane are significant in both movements due to the influence of the eBUS, because the angular curves with the eBUS are not within the standard deviation of the curves without the eBUS over the entire cycle in either the foot, knee, or hip joint. The qualitative trajectory can be maintained for the most part in the knee and hip joints. However, during descent in the controlled lowering phase, the hip becomes much more flexed when using the eBUS, as seen in Subjects 2 and 3. This could be interpreted as a compensatory movement for the lower dorsiflexion in the ankle. Since the freedom of movement in the hip joint is not affected by the eBUS, larger angular amplitudes can be achieved here, which favours an easier movement compensation via the hip. On the other hand, it can be noted that in the foot and knee joint the angular amplitudes are not achieved with the eBUS. This could be caused by the restriction of the freedom of movement by the eBUS, which is amplified by the potential misalignment of the joint axes between the human and the eBUS.

Since in the frontal plane all the mean angular curves of the foot are qualitatively altered by the eBUS in both movements and all three subjects, it could be postulated that this is due to a restriction of movement by the eBUS. In the hip joint, greater abduction occurs in the phase of pulling up and forward, especially during the ascent. During this phase, compensatory movements of the upper body were observed, which can presumably be attributed to the additional mass of the eBUS that is acting on the swing leg. During the stance phase, the subjects leaned laterally with their upper body over their own stance leg to facilitate the pulling up of the swing leg that was loaded with additional mass. Due to the larger abduction angle during the stance phase in the ascent, a smaller adduction angle is consequently required in the swing phase with the same experimental setup. This phenomenon can be confirmed by the measurement results. It can be concluded that the change in kinematics with the eBUS in the frontal plane is also significant.

4.3. Human Joint Moments

Human joint moments are influenced by the exoskeleton in various ways. On the one hand, the kinematics of the human exoskeleton system changes significantly, which is also reflected in longer stance times. On the other hand, the assistive moment of the eBUS influences at least the net knee moment of the human. Furthermore, the additional mass of the exoskeleton and the changed inertias of the lower musculoskeletal system influence the joint moments that must be generated.

For instance, the assisting extension moment in the knee also influences the foot moment that has to be generated by the user. On one hand, this is shown by only one peak in the curves with the eBUS and, on the other hand, by the reduction of the applied foot moment itself. Here, it can already be determined that the eBUS affects the subjects in different ways or that the subjects have different movement strategies when using the eBUS.

This assumption can be further confirmed by the curves of the applied knee moments, since the subjects with the eBUS also show different curves here. However, it can also be seen that the subjects have different knee moment curves even without the use of the exoskeleton, which suggests different movement strategies from the outset. Particularly noticeable is the change during the descent in all three subjects in exactly the phase in which the knee moment is assisted by the exoskeleton. In this phase, the required knee extension moment is even reduced to such an extent that the subjects generate a flexion moment to counteract the movement, as can be seen for Subject 3 especially. This could already indicate too much assistance by the system.

The hip moments to be applied by the human show a large discrepancy between the two test scenarios during the ascent in the phase of pulling up and forward. In these phases, the maximum extension moment is 1.4 ± 0.3 Nm/kg and could possibly be critical for the elderly. Presumably, this phenomenon has many causes. Firstly, an additional hip extension moment could be required to contribute potential energy to the additional mass of the eBUS during ascent. Secondly, the increased hip extension moment could be an indication of a redistribution of moments in favor of the hip in order to relieve the knee joint even more. Another cause for the change in hip moment could be the increased forward lean of the upper body during the trials with the eBUS. The exact causes of this phenomenon cannot be definitively answered at this point, as the changes in the kinematics due to the use of the eBUS must also be considered. For the hip moment that has to be generated in the swing phase, it can be stated that due to the additional mass of the eBUS, it is significantly higher compared to the tests without the exoskeleton, both for ascent and descent.

4.4. Knee Joint Reaction Forces

It was assumed that the joint reaction forces in the knee would decrease significantly through the use of the eBUS. This can neither be confirmed for the ascent nor for the descent by the results. In the descent, the mean value of the peak of the curve with the use of the eBUS is below the mean value without the eBUS only for Subject 2. During the ascent, the knee joint reaction force is even greater in all three subjects due to the use of the eBUS. The reason for this significant increase during ascent could be the misalignment of the knee joint axes between the human and the exoskeleton. These mismatches can cause additional shear loads in the knee joint and have to be compensated by the locomotion system [32].

Since the external moment of the eBUS is introduced from 20% to about 60% of the cycle time and at the same time the knee joint reaction force during ascent is significantly larger, it could be suspected that the additional moment of the eBUS influences the knee joint reaction force directly and even increases it. However, Stoltze and co-authors [33] figured out for level walking that the first peak of the knee joint reaction force is significantly influenced by the knee moment. On the other hand, the second peak is influenced by the foot moment, or gastrocnemius muscle, during the transition from single to double stance phase. In a pilot study, the authors were also able to demonstrate that an externally applied knee moment leads to a reduction of the knee joint reaction force in the first peak during level walking [34]. If these findings are also valid for stair climbing, only the first peak is lowered using the eBUS, as it is the case for the descent of the second subject. However, the qualitative curve of the knee joint reaction force with the eBUS is not characterized by two peaks, which is due to an acyclic movement.

4.5. Muscle Activity of Knee Extensors

Since all maxima of the muscle activities of the VAM and VAL were decreased using the eBUS, it could be concluded that the exoskeleton contributes to the relief of the knee extensors during ascent and descent. In addition to the maxima, the integrals of all the test persons were also reduced by using the eBUS according to Table 4. It can be concluded that the net effort of the test persons is lower when using the eBUS. Based on the integrals, the eBUS reduces the effort by an average of 19.3%. Thus, the performance deficit that is evaluated in [3] would be compensated by the exoskeleton. However, it should be noted that the presumed redistribution of joint moments in favour of the knee can also be the cause of the release of the knee extensors. However, this can only be postulated in this study, since no muscle activities of the gastrocnemius muscle nor of the hip flexors or extensors were measured.

A contradiction can be observed in the curves of the muscle activities compared to the knee moment curves and the assisting moment. The sEMG curves show a decrease in the muscle activity in the unassisted phases, while the activity in the assisted phases tends to remain the same. One reason for this could be the kinematics that have changed due to the eBUS. When using the eBUS, there might be a stronger backlift than leglift. Another cause could be passive assistance by the system itself in the phases in which it does not provide active assistance, although the measured muscle activities are decreased.

Furthermore, the measured muscle activities show that the subjects react individually to wearing and assistance by the exoskeleton. From this it can be concluded that not only an individual adjustment of the hardware of the exoskeleton is required, but also a control-related adjustment regarding the time of providing the assistance and the magnitude of the assisting moments.

5. Conclusions

In this pilot study it could be shown that the developed eBUS assists the ascent and descent of stairs. On one hand, this is indicated by the applied knee extension moment of the eBUS and, on the other hand, by the reduced muscle activities of the knee joint extensors. However, it also became evident that exoskeletons can affect subjects individually and should, therefore, be designed or adjusted to be subject-specific. This concerns, for example, the shape of the eBUS or its settings regarding segment lengths. Since the additional mass of the eBUS leads to compensatory movement of the upper body and the kinematics are significantly changed during both movements, the mass of the eBUS should be further reduced.

Since stair climbing is a comparatively slow movement with low angular velocities but large joint angles, assistance could be obtained for all three subjects. The results of this pilot study clearly show that a more individual application of force as well as an individual amount are required to provide adequate assistance. Furthermore, the goal of further optimizations of the system that is presented here should be to achieve shorter cycle times and a more cyclic movement.

Since the results in this pilot study are based only on the movement patterns of the subjects and the eBUS as well as on externally measured forces, additional sEMG measurements of all relevant muscles of the lower musculoskeletal system should be performed for future studies. This would also allow for the detection of subject-specific changes in movement strategies. Furthermore, an analysis of the joint power is useful in order to find out in which phases the human has to generate or absorb energy. A study with a larger number of subjects and trials would increase the statistical power. In subsequent studies, elderly people should also be tested, since they are the actual target group of the eBUS. In any case, further studies are necessary to investigate and ideally exclude potential damage to humans by the eBUS. For this purpose, it should be investigated, in particular, whether a redistribution of joint moments takes place as well as what effects the increase in the knee joint reaction force that is determined during the ascent has on humans.

All in all, it has been shown that the complex methods that are used here are suitable for evaluating the biomechanical function of exoskeletons. Other types of exoskeletons could also be evaluated with the methods that were used here.

Author Contributions: Conceptualization, J.J. and J.Z.; methodology, M.B.; formal analysis, M.B. and R.T.; investigation, M.B., H.-P.K. and M.W.; resources, M.W., J.J. and J.Z.; data curation, M.B. and H.-P.K.; writing—original draft preparation, M.B.; writing—review and editing, M.B., R.T., H.-P.K., J.J. and M.W.; visualization, M.B. and R.T.; supervision, M.W., J.J. and J.Z.; project administration, J.J. and J.Z.; funding acquisition, J.J. and J.Z. All authors have read and agreed to the published version of the manuscript.

Funding: This research was funded by Sächsische AufbauBank and the European Regional Development Fund (ERDF), grant number 3000917185. We acknowledge support by the Open Access Publication Funds of the HTWK Leipzig.

Institutional Review Board Statement: The study was conducted in accordance with the Declaration of Helsinki and approved by the Ethics Committee of University Leipzig (approved on 05 August 2021 with protocol code 2021.07.13_eb_107).

Informed Consent Statement: Informed consent was obtained from all subjects that were involved in the study.

Data Availability Statement: The data that are presented in this study are available on request from the corresponding author. The data are not publicly available due to privacy restrictions.

Acknowledgments: We would like to thank Felix Weiske, who developed the control system for the exoskeleton and played a key role in the data recording. We would also like to thank Julius Hannig, a member of the biomechanics laboratory at the University of Leipzig, who provided technical support during the study and helped with the data processing afterwards. Further thanks go to all study participants.

Conflicts of Interest: The authors declare no conflict of interest. The funders had no role in the design of the study; in the collection, analyses, or interpretation of data; in the writing of the manuscript, or in the decision to publish the results.

References

1. ASTM F48 Committee. *Terminology for Exoskeletons and Exosuits*; ASTM International: West Conshohocken, PA, USA, 2020.
2. Young, A.J.; Ferris, D.P. State of the Art and Future Directions for Lower Limb Robotic Exoskeletons. *IEEE Trans. Neural Syst. Rehabil. Eng.* **2017**, *25*, 171–182. [CrossRef] [PubMed]
3. Böhme, M.; Weiske, F.; Jäkel, J.; Zentner, J.; Witt, M. Evaluation of the power deficit of elderly people during stair negotiation: Which joints should be assisted at least by an exoskeleton and with what amount? *Wearable Technol.* **2022**, *3*, E4. [CrossRef]
4. Böhme, M. Untersuchungen zur Biomechanik und seniorengerechten Gestaltung von exoskelettalen Bewegungsunterstützungssystemen zur Überwindung von Treppen. Doctoral Thesis, Technische Universität Berlin, Berlin, Germany, 2022.
5. Joudzadeh, P.; Hadi, A.; Alipour, K.; Tarvirdizadeh, B. Design and implementation of a cable driven lower limb exoskeleton for stair climbing. In Proceedings of the 5th RSI International Conference on Robotics and Mechatronics (IcRoM 2017), Tehran, Iran, 25–27 October 2017; IEEE: Piscataway, NJ, USA, 2018; pp. 76–81.
6. Chandrapal, M.; Chen, X.; Wang, W. Preliminary evaluation of a lower-limb exoskeleton-stair climbing. In Proceedings of the 2013 IEEE/ASME International Conference on Advanced Intelligent Mechatronics (AIM 2013), Wollongong, NSW, Australia, 9–12 July 2013; IEEE: Piscataway, NJ, USA, 2013; pp. 1458–1463.
7. Grimmer, M.; Riener, R.; Walsh, C.J.; Seyfarth, A. Mobility related physical and functional losses due to aging and disease—A motivation for lower limb exoskeletons. *J. Neuroeng. Rehabil.* **2019**, *16*, 2. [CrossRef] [PubMed]
8. Jang, J.; Kim, K.; Lee, J.; Lim, B.; Shim, Y. Assistance strategy for stair ascent with a robotic hip exoskeleton. In Proceedings of the IROS 2016: 2016 IEEE/RSJ International Conference on Intelligent Robots and Systems, Daejeon, Korea, 9–14 October 2016; IEEE: Piscataway, NJ, USA, 2016; pp. 5658–5663.
9. Deutsches Institut für Normung e.V. *Gebäudetreppen–Begriffe, Messregeln, Hauptmaße*; Beuth Verlag GmbH: Berlin, Germany, 2020.
10. Zhao, S.; Yang, Y.; Gao, Y.; Zhang, Z.; Zheng, T.; Zhu, Y. Development of a soft knee exosuit with twisted string actuators for stair climbing assistance. In Proceedings of the IEEE International Conference on Robotics and Biomimetics, Dali, China, 6–8 December 2019; IEEE: Piscataway, NJ, USA, 2019; pp. 2541–2546.
11. Zhang, Z.; Zhu, Y.; Zheng, T.; Zhao, S.; Ma, S.; Fan, J.; Zhao, J. Lower extremity exoskeleton for stair climbing augmentation. In Proceedings of the IEEE ICARM 2018—2018 3rd IEEE International Conference on Advanced Robotics and Mechatronics (ICARM), Sigapore, 18–20 July 2018; IEEE: Piscataway, NJ, USA, 2018; pp. 762–768.
12. Zhao, X.; Chen, W.-H.; Li, B.; Wu, X.; Wang, J. An adaptive stair-ascending gait generation approach based on depth camera for lower limb exoskeleton. *Rev. Sci. Instrum.* **2019**, *90*, 125112. [CrossRef] [PubMed]
13. Li, B.; Yuan, B.; Tang, S.; Mao, Y.; Zhang, D.; Huang, C.; Tan, B. Biomechanical design analysis and experiments evaluation of a passive knee-assisting exoskeleton for weight-climbing. *Ind. Robot. Int. J.* **2018**, *45*, 436–445. [CrossRef]
14. Yuan, B.; Li, B.; Chen, Y.; Tan, B.; Jiang, M.; Tang, S.; Wei, Y.; Wang, Z.; Ma, B.; Huang, J. Designing of a Passive Knee-Assisting Exoskeleton for Weight-Bearing. In *Intelligent Robotics and Applications*; Huang, Y., Wu, H., Liu, H., Yin, Z., Eds.; Springer International Publishing: Cham, Switzerland, 2017; Volume 10463, pp. 273–285.
15. Pott, P.P.; Wolf, S.I.; Block, J.; van Drongelen, S.; Grün, M.; Heitzmann, D.W.; Hielscher, J.; Horn, A.; Müller, R.; Rettig, O.; et al. Knee-ankle-foot orthosis with powered knee for support in the elderly. *Proc. Inst. Mech. Eng. Part H J. Eng. Med.* **2017**, *231*, 715–727. [CrossRef] [PubMed]

16. Schmalz, T.; Colienne, A.; Bywater, E.; Fritzsche, L.; Gärtner, C.; Bellmann, M.; Reimer, S.; Ernst, M. A Passive Back-Support Exoskeleton for Manual Materials Handling: Reduction of Low Back Loading and Metabolic Effort during Repetitive Lifting. *IISE Trans. Occup. Ergon. Hum. Factors* **2022**, *10*, 7–20. [CrossRef] [PubMed]
17. Fritzsche, L.; Galibarov, P.E.; Gärtner, C.; Bornmann, J.; Damsgaard, M.; Wall, R.; Schirrmeister, B.; Gonzalez-Vargas, J.; Pucci, D.; Maurice, P.; et al. Assessing the efficiency of exoskeletons in physical strain reduction by biomechanical simulation with AnyBody Modeling System. *Wearable Technol.* **2021**, *2*, E6. [CrossRef]
18. Bell, A.L.; Brand, R.A.; Pedersen, D.R. Prediction of hip joint centre location from external landmarks. *Hum. Mov. Sci.* **1989**, *8*, 3–16. [CrossRef]
19. C-Motion. Coda Pelvis: Visual3D Wiki Documentation. Available online: https://www.c-motion.com/v3dwiki/index.php?title=Coda_Pelvis (accessed on 4 July 2022).
20. Hermens, H.J. *European Recommendations for Surface ElectroMyoGraphy: Results of the SENIAM Project*; Roessingh Research and Development: Enschede, The Netherlands, 1999.
21. Konrad, P. EMG-Fibel: Eine Praxisorientierte Einführung in Die Kinesiologische Elektromyographie. Doctoral Thesis, Technische Universität Berlin, Berlin, Germany, 2011.
22. Cappozzo, A.; Catani, F.; Della Croce, U.; Leardini, A. Position and orientation in space of bones during movement: Anatomical frame definition and determination. *Clin. Biomech.* **1995**, *10*, 171–178. [CrossRef]
23. Damsgaard, M.; Rasmussen, J.; Christensen, S.T.; Surma, E.; de Zee, M. Analysis of musculoskeletal systems in the AnyBody Modeling System. *Simul. Model. Pract. Theory* **2006**, *14*, 1100–1111. [CrossRef]
24. Lund, M.E.; Tørholm, S.; Jensen, B.K.; Galibarov, P.E.; Dzialo, C.M.; Iversen, K.; Sarivan, M.; Marra, M.A.; Simonsen, S.T. The AnyBody Managed Model Repository (AMMR). *Zenodo* **2020**. [CrossRef]
25. Andersen, M.S.; Damsgaard, M.; MacWilliams, B.; Rasmussen, J. A computationally efficient optimisation-based method for parameter identification of kinematically determinate and over-determinate biomechanical systems. *Comput. Methods Biomech. Biomed. Eng.* **2010**, *13*, 171–183. [CrossRef] [PubMed]
26. Rasmussen, J.; Damsgaard, M.; Voigt, M. Muscle recruitment by the min/max criterion—A comparative numerical study. *J. Biomech.* **2001**, *34*, 409–415. [CrossRef]
27. Lund, M.; Rasmussen, J.; Andersen, M. AnyPyTools: A Python package for reproducible research with the AnyBody Modeling System. *J. Open Source Softw.* **2019**, *4*, 1108. [CrossRef]
28. Chander, D.S.; Böhme, M.; Andersen, M.S.; Rasmussen, J.; Cavatorta, M.P. Simulating the Dynamics of a Human-Exoskeleton System Using Kinematic Data with Misalignment Between the Human and Exoskeleton Joints. In *Computer Methods, Imaging and Visualization in Biomechanics and Biomedical Engineering II: Selected Papers from the 17th International Symposium CMBBE and 5th Conference on Imaging and Visualization, 7–9 September 2021, 1st ed.*; Tavares, J.M.R.S., Bourauel, C., Geris, L., Vander Slote, J., Eds.; Springer International Publishing: Cham, Switzerland, 2021; Volume 38, pp. 65–73.
29. Chander, D.S.; Böhme, M.; Andersen, M.S.; Rasmussen, J.; Zentner, J.; Cavatorta, M.P. A comparison of different methods for modelling the physical human-exoskeleton interface. *Int. J. Hum. Factors Model. Simul.* **2021**, *7*, 204. [CrossRef]
30. Tröster, M.; Wagner, D.; Müller-Graf, F.; Maufroy, C.; Schneider, U.; Bauernhansl, T. Biomechanical Model-Based Development of an Active Occupational Upper-Limb Exoskeleton to Support Healthcare Workers in the Surgery Waiting Room. *Int. J. Environ. Res. Public Health* **2020**, *17*, 5140. [CrossRef] [PubMed]
31. Zhou, L.; Li, Y.; Bai, S. A human-centered design optimization approach for robotic exoskeletons through biomechanical simulation. *Robot. Auton. Syst.* **2017**, *91*, 337–347. [CrossRef]
32. Plegge, C. Bedeutung und kinematische Untersuchung der Passform eines aktiven Exoskeletts für die untere Extremität. In *Technische Unterstützungssysteme, DIE Die Menschen Wirklich Wollen: Zweite Transdisziplinäre Konferenz: Hamburg 2016*; Weidner, R., Ed.; Laboratorium Fertigungstechnik smartASSIST Helmut Schmidt Universität: Hamburg, Germany, 2016; pp. 13–20.
33. Stoltze, J.S.; Rasmussen, J.; Andersen, M.S. On the biomechanical relationship between applied hip, knee and ankle joint moments and the internal knee compressive forces. *Int. Biomech.* **2018**, *5*, 63–74. [CrossRef]
34. Stoltze, J.S.; Pallari, J.; Eskandari, B.; Oliveira, A.S.C.; Pirscoveanu, C.I.; Rasmussen, J.; Andersen, M.S. Development and Functional Testing of an Unloading Concept for Knee Osteoarthritis Patients: A Pilot Study. *J. Biomech. Eng.* **2022**, *144*, 011007. [CrossRef] [PubMed]

applied sciences

Article

The Energy Consumption and Robust Case Torque Control of a Rehabilitation Hip Exoskeleton

Rabé Andersson * and Niclas Björsell

Department of Electrical Engineering, Mathematics and Science, University of Gävle, 801 76 Gävle, Sweden; niclas.bjorsell@hig.se
* Correspondence: rabe.andersson@hig.se; Tel.: +46-026-648944

Abstract: Gait disorders, muscle weakness, spinal cord injuries (SCIs) and other work-related disorders have increased the need for rehabilitation exoskeletons—specifically, for the hip because a huge percentage of mechanical power comes from the hip joint. However, realising a lightweight rehabilitation hip exoskeleton for mobility and at-home use with reliable control is challenging. The devices developed are restricted by a joint actuator and energy source design and tend to have various uncertainties. Thus, this study tested the robustness of four optimal controller cases in a simulation-based environment. We sought to determine whether the most robust optimal controller consumed less energy and demonstrated better performance in tracking the desired signal. The robustness of the optimal cases was tested with the hip torque signals of healthy subjects. The number of sit-to-stand (STS) instances and the walking distance at various speeds were calculated. The results showed that the most robust case controller was more energy efficient for STS, but not for walking activity. Furthermore, this study provides compelling evidence that various optimal controllers have different degrees of robustness and effects on energy consumption.

Keywords: exoskeleton; robust controller; energy consumption; hip rehabilitation exoskeleton; series elastic actuator (SEA); LQR control; Luenberger state observer; torque control

Citation: Andersson, R.; Björsell, N. The Energy Consumption and Robust Case Torque Control of a Rehabilitation Hip Exoskeleton. *Appl. Sci.* **2022**, *12*, 11104. https://doi.org/10.3390/app122111104

Academic Editors: Ionuţ Daniel Geonea and Cristian Copilusi Petre

Received: 25 September 2022
Accepted: 28 October 2022
Published: 2 November 2022

Publisher's Note: MDPI stays neutral with regard to jurisdictional claims in published maps and institutional affiliations.

1. Introduction

An urgent need for robotic exoskeletons has been widely observed in the military, industry, medicine and hazardous situations such as disasters and pandemics over the last few decades [1,2]. Some exoskeleton robots augment human mobility or endurance, while others are used for rehabilitation and ambulatory assistance [3]. In addition, some are used for paediatric recovery, while others serve the elderly, soldiers or athletes [4–6].

According to a 2019 United Nations' report [7,8], the number of strokes and the percentage of mobility accidents are the highest among people older than 65 years and the 250,000–500,000 patients suffering from spinal cord injury (SCI) every year, increase research interest in robotic exoskeletons [9]. Additionally, more than 17% of work-related disorders are associated with lower limb activity, which also increased the need for robotic exoskeletons for mobility assistance and rehabilitation [9,10].

Thus, a lower limb exoskeleton, specifically, the rehabilitation hip exoskeleton, is the focus of this paper because 45% of the mechanical energy of lower limb movements comes from the hip joint [8,11]. Robotic exoskeletons can contribute in repetitive training of human extremities to enable patients to pursue rehabilitation sessions at home and enable therapists to focus on more specialised rehabilitation sessions [12]. Among the different rehabilitation exoskeletons are those used for sit-to-stand and walking movements. Therefore, we focus more on these activities because of their importance in daily life [13].

The rehabilitation hip exoskeleton is a mechanical structure worn to enhance gait mobility but can also be used, for instance, after hip replacement surgery or a stroke [14]. Since, in the very early stages of rehabilitation sessions, patients' limbs are usually trained passively to track predefined trajectory motions, which vary according to the patients' needs or limb disorders [15], different controller strategies have been used to track and control the trajectories because human extremities are intolerant of any unexpected harmful movement errors; in particular, rehabilitation exoskeletons serve people who already have some sort of limb disorder or mobility deficiency [16,17].

Some control strategies have used fuzzy hybrid controller, while others utilised Bayesian optimisation to minimise the metabolic cost of walking by adjusting various control parameters for different individuals [18,19]. Some applications were required to apply different parameters based on assistance onset timing, assistance magnitude and duration, as used in [20]. However, building a seamless controller integrated with all applications and human biological systems is still challenging. Therefore, some researchers used reinforcement learning for optimal controllers or to perform a normative range of motion and gait patterns during locomotion [21,22]. While others applied real-time optimisation control strategies based on measurements of the user, machine learning to implement various dynamic controllers, EMG-based controllers or observer-based robust controllers [23–26]. However, the most used controller is the proportional integral derivative (PID), which showed its insufficient performance in tracking a predefined trajectory and overcoming changes to model parameters and uncertainties [16,27,28]; therefore, an optimal controller, a linear quadratic regulator (LQR), is proposed to be used in this work.

However, the hip exoskeleton, like other active exoskeleton devices, contains sensors, actuators, links, batteries and electronic circuits that operate in close contact with patients; for this reason, it should operate safely and reliably. For a safe hip rehabilitation exoskeleton, a series elastic actuator (SEA) is chosen for this research, as it presents a compliant interface, accurate and stable torque control over time, low impedance and shock tolerance (elasticity), all of which afford highly confident force control and comfortable wearability and it extends the operational life time of the joint [8,16,28–34]. An SEA was used at the hip joint to replicate the hip torque signals of a healthy joint as desired movements in the controller.

However, the functions of the controller depend on the mathematical representation and modelling of the system. This can be a challenge because it is an approximation to a real system, which can also be affected by parametric uncertainties, variations or short-term and long-term stochastic noise disturbances (ageing) [27,35]. To the best of our knowledge, no research has validated the LQR for robustness and energy consumption. Therefore, the goal of this study is to validate the LQR controller of a rehabilitation hip exoskeleton for such variations and uncertainties and to establish the foundation of safe human–robot interactions for future research. Hence, the basic design for the exoskeleton and the SEA was introduced in [8,28,36]. Estimating the states of the system was essential for the controller; therefore, the Luenberger observer (LO) was used. The implementation and design of the LO are elaborated in [8].

Not only did the controllers attract the attention of researchers and innovators, but also the weight of such devices, as rehabilitation exoskeletons are usually attached to deficit limbs. Therefore, the energy consumption is worth studying to avoid the extra weight of additional batteries, especially when the powered hip rehabilitation exoskeleton is used outdoors and at home. The contributions of this paper are threefold: an examination of the robustness of four LQR controller cases, validation of the performance of such controllers and investigation of the energy consumption of our rehabilitation hip exoskeleton performing STS and walking movements at each control case and at various speeds. The experiments were conducted in a simulated environment as it provided insight into how an exoskeletal system behaved in a virtual environment that mimicked a real one. They assessed the performance of the model by testing various parameters that were difficult, expensive and time-consuming to reproduce using the physical model [37]. This was especially true of safety, which is a critical factor in experimental designs for human–robot interactions.

This paper is organised as follows: Section 2 presents the importance of the STS and walking activities trained by the rehabilitation hip exoskeleton, LQR strategy, concepts of robustness, performance of the controller in tracking a trajectory signal measured by root mean square errors (RMSEs), energy consumption and STS and walking power calculations. In Section 3, the methods used to test the robustness of four optimal control cases, the RMSE and a simulation energy consumption of a proposed rechargeable battery during the STS and walking movements are analysed. Finally, the results and conclusion are provided in Sections 4 and 5, respectively.

2. Theory

The worldwide demand for rehabilitation exoskeletons is increasing along with the number of people who have partial or complete mobility disorders and SCI. Gait disorders also include people with cerebral palsy, osteoarthritis and joint replacements who need repetitive rehabilitation sessions and, usually, long-term hospitalisation.

A mobility disorder does not affect physical movement but can lead to medical complications such as osteoporosis, high blood pressure, muscle spasticity and pressure ulcers [38]. Thus, the sit-to-stand and walking activity was the main means by which our rehabilitation hip exoskeleton could provide mobility by overcoming the aforementioned clinical conditions.

2.1. Sit-to-Stand and Walking Activity

Rising from a seated position is a primary step towards pursuing gait mobility for daily activities and improving the quality of life. The STS is a crucial link between a static position and a dynamic activity, which is still a challenging aspect after a stroke [39]. In addition, the STS stimulates the circulatory and respiratory systems, reduces spasticity and increases bone density [40].

Therefore, for rehabilitation purposes, repetitive STS activities at various torque levels representing five angular speed movements were performed and analysed, followed by an analysis of walking movements at five torque levels equivalent to walking on a treadmill at various linear speeds (0.5–2.6 m/s). To control the rehabilitation hip exoskeleton, an admittance control strategy was used in which the hip torque of a healthy person, τ_{HH}, at various torque levels ($TL1$ *to* $TL5$) was used for reference control signals. The torque signal of a healthy hip performing STS was used to control the STS activity, $\tau_{HH\text{-}STS}$, whereas the torque signal of a healthy hip performing walking movements, $\tau_{HH\text{-}W}$, was applied to the walking control strategy. The torque control signals for both STS and walking are further discussed in [8,41,42] and are also shown in Figures 1 and 2.

Because higher movement speeds require a higher torque, a lower torque level was selected for low-speed movement or when the patient's joint required a small amount of torque [43]. In addition, people have different disorders and rehabilitative needs; therefore, five torque levels were used in the control strategy to fulfil the various rehabilitation sessions. Figures 1 and 2 show that, if the exoskeleton was designed to train patients at a faster speed, the maximum hip torque signal of a healthy person was set at the fifth level as a reference control signal. On the other hand, the torque was set to the first level if the minimum walking speed (0.5 m/s) was needed.

Lastly, the rehabilitation hip exoskeleton aimed to restore the impaired hip torque signal to that of a healthy one by the end of the treatment program. Healthy torque signals can vary with speed, which can be assigned and modified by a physiotherapist to tackle different conditions and needs. For this reason, the hip torque signal at various speeds was controlled using an LQR. Then, the control behaviour was analysed for the robustness; RMSEs, which show the differences between the desired and measured torque signals; and the energy needed to perform repetitive sessions, which will be illustrated later.

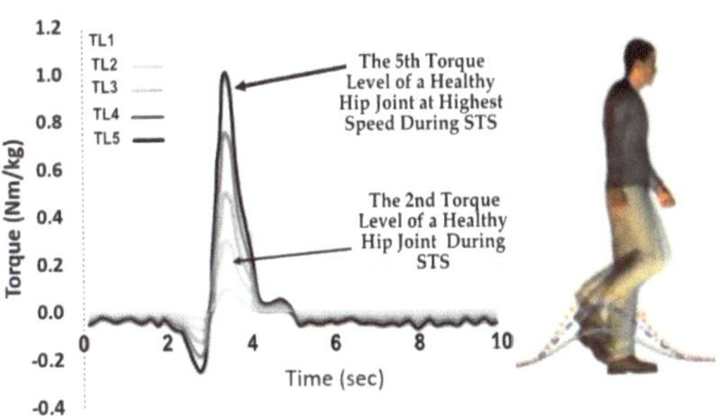

Figure 1. The hip torque during STS activity [8,44].

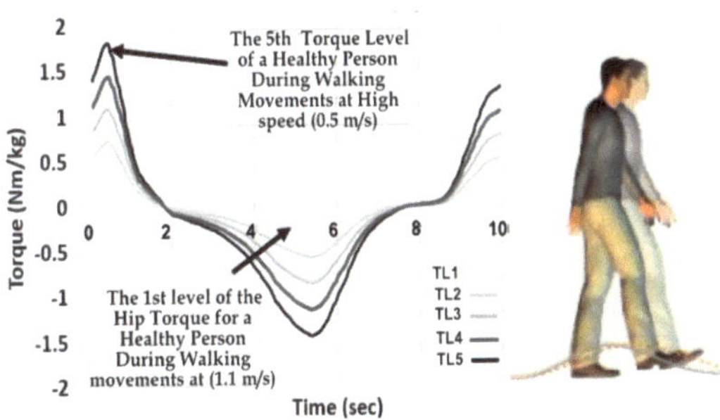

Figure 2. The hip torque during walking activity [8,44].

2.2. The Linear Quadratic Regulator

Trajectory tracking with predefined movements is commonly used in rehabilitation exoskeletons for the lower extremities because patients can be trained passively to follow desired reference motions [8,10]. Therefore, an admittance controller based on the torque of a healthy hip joint τ_{HH} was used for the rehabilitation exoskeleton to train the patient's weak muscles. The controller was used with an SEA and LQR that were chosen for the reasons set out in Section 1. The SEA was used to generate the proper current for the desired torque $\tau_{reference}$ and applications. Thus, an LQR with a Luenberger observer was used, but a mathematical model expressing the rehabilitation hip exoskeleton was needed, as illustrated in [8] and depicted based on the literature [28,36,45].

To use the LQR, the state space model of the exoskeleton was represented in the following form

$$\dot{x} = Ax + Bu$$
$$y = Cx + Du,$$

(1)

where $\dot{x}$ is a vector matrix representing the system states; A, B, C and D are the system's dynamic matrices; u is the system input signal; and y is the output signal. However, the states of the system are not always accessible for sensors; thus, an observer is commonly used. Many observers are used through control systems, but the one that can increase the control system's bandwidth and improve the disturbance response indirectly is the Luenberger observer [46], which estimates the states $\dot{x}$ by using the inputs u and the outputs y of the system. Finally, it subtracts the output of the observer, $\dot{y}$, from the output of the system, y.

On the LQR controller side, the optimal placement of the system poles was based on the weights of the Q and R matrices by minimising the quadratic cost function J and finding the minimum steady-state error and the best feedback gain [47].

$$J = E\left\{ \lim_{T \to \infty} \frac{1}{T} \int_0^T \left[x^T Q x + u^T R u \right] dt \right\}, \tag{2}$$

where Q and R are the semi-definite and positively definite design matrices, respectively. By tuning Q and R, the placement of the poles was changed, whereas the feedback gain, K, was determined from the Hamilton–Jacobi–Bellman method and solved by using the Riccati equation [8,48] that ends with

$$u = -Kx. \tag{3}$$

The feedback gain, K, increased or decreased depending on the values selected for Q and R. A larger weight value in R set a higher penalty on the input signal, u, which decreased the feedback gain, K and slowed the steady state and the energy needed. Conversely, the large weight values of Q meant a higher penalty on the state of the system, x, when finding the minimum cost function, J, according to (2), which increased the value of the feedback gain, K, in (3) and accelerated the system's response accordingly.

The system states, $\hat{x}$, were necessary for the LQR algorithm, which can be inaccessible in many real systems or expensive to measure with certain sensors; thus, an LO was a suitable solution for deriving the system states [8].

The system states, $\hat{x}$, can be estimated by putting the input, u, and the system output, y, into LO. The procedures for obtaining the states and the LO diagram are discussed in more detail elsewhere [8]. Although the states are critical to finding the mathematical model of the system, which needs to be optimally controlled using the LQR, the model of the system will nevertheless include errors and imperfect model representation is more common compared to real systems. Therefore, the system robustness for the various Q and R is analysed and the energy consumption was studied regarding the sensitivity of the system to the model uncertainty.

2.3. Robustness

The control strategy depended on modelling the system, yet the mathematical model representation was based on physical components that are quite often simplified [49]. Additionally, a system may also have disturbances, stochastic noise resources or a dynamic behaviour that is hard to represent. Consequently, the mathematical representation or transfer function that describes a system is only an approximation of the real system. This means that the control system may not work in exactly the same manner on the real hardware as it does in the mathematical model. For this reason, the stability and robustness of the system in the presence of model uncertainty were simply analysed using a disk-based uncertainty (disk margin, DM) for gain (DGM) and phase (DPM) perturbation representation [50].

The DM visualises the stability of a closed-loop system for a combination of gain and phase variation. The margin shows how much uncertainty the controller can tolerate before the system becomes unstable. Another method for investigating system stability is using Nyquist's stability criterion, which is elaborated in [51]. The stability of the system is described by how far the Nyquist curve can move away from the point (-1) before the system collapses.

After studying the Nyquist curve for all the optimal controller cases to identify the most robust optimal controller for a certain activity, the controller behaviour of the optimal cases was then studied at various speeds. The performance of the optimal controllers with various degrees of robustness was then analysed to determine the relationships among robustness, different speeds and energy consumption for different activities. The performance of a certain controller was measured by the RMSEs, comparing the torque signals with the desired torque reference signals.

2.4. The Performance and the Root Mean Square Error

The performance of the LQR case strategy was evaluated according to the RMSE between the measured torque signal, τ_{actual}, of the LQR and the desired torque reference signal, $\tau_{\text{reference}}$, as deduced by

$$RMSE = \sqrt{\frac{\sum_{i=0}^{n}(\tau_{\text{reference}} - \tau_{\text{actual}})^2}{n}},\tag{4}$$

where n is the number of observation points over time. The $\tau_{\text{reference}}$ represented in this work is the $\tau_{\text{HH-STS}}$ for STS and $\tau_{\text{HH-W}}$ for walking at different speeds.

2.5. Energy Consumption and Power Calculation

The energy consumption of an SEA is derived from the power needed to rotate the motor with torque τ and angular velocity $\dot{\theta}$, so the power, P, is

$$P = \tau \cdot \dot{\theta},\tag{5}$$

where $\dot{\theta}$ is the derivative of the motor angular position θ and the angular velocity is found by the simulation program, but can also be calculated based on the literature [42,52]. It is worth mentioning that we assumed the motor required the same energy for thrusting, P_{Th} and damping, P_{Da}. Thus, the energy, E, for each step in STS and walking achieved by the rehabilitation hip exoskeleton can be calculated as

$$E = E_{Th} + |E_{Da}| = \int P_{Th}(t)dt + \int P_{Da}(t)dt.\tag{6}$$

Then, the number of steps of the STS or walking activity can be calculated as the total amount of energy stored in a battery divided by the amount of energy consumed in each step movement.

3. Methods

The simulation and design of the rehabilitation hip exoskeleton were modelled on the Denavit–Hartenberg (DH) parameters as shown in Figure 3 using the MATLAB and Autodesk Inventor software. The actuator selection was an SEA for the active hip joint. The hip rehabilitation consisted of a pelvis, four links for both legs and four joints that were active, except the knees, which were passive, ending up with a four-degree-of-freedom (DoF) prototype moving in the sagittal plane. The limitations and the range of motion (RoM) for each joint were based on the data in Table 1 [8]. The trunk and exoskeleton masses were connected to a body-weight-supported treadmill similar to the one used with the LokomatPro exoskeleton [53].

The control framework was based on a linear quadratic regulator of an SEA for a hip rehabilitation exoskeleton in which the torque control signals were the reference signals in the admittance control strategy. The control signals were five levels of torque ($TL1$ to $TL5$) of STS and the walking activities for healthy subjects, as shown in Figures 1 and 2 and in articles [8,41,42].

Table 1. The hip and knee joints with their range of motions (RoMs) [8].

Joint	DoF's Description	Type	RoM
Hip	Flexion–Extension	Active	120° to −122°
Knee	Flexion–Extension	Passive	120° to 0°

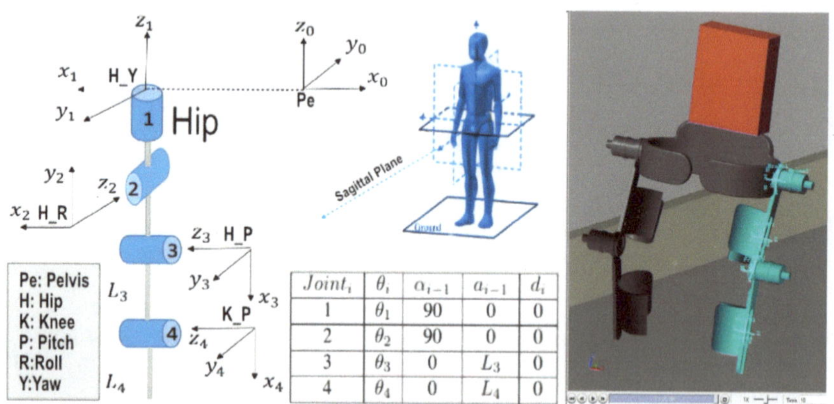

$Joint_i$	θ_i	a_{i-1}	a_{i-1}	d_i
1	θ_1	90	0	0
2	θ_2	90	0	0
3	θ_3	0	L_3	0
4	θ_4	0	L_4	0

Figure 3. The DH parameters and the rehabilitation hip exoskeleton model [8].

The kinematic and kinetic formulation, the system mathematical representation and the admittance control are thoroughly discussed in [8], where the modelling was based on the average height and weight of male subjects. The length of the thigh link was 52 and the shank link was 39 cm, as mentioned in [54]. The SEA was modelled based on [55–58] and the parameters used are shown in Table 2, where J_m, J_s and J_l are the inertia of the motor, spring and load link, respectively. B_m, B_s and B_l are the damping coefficients for the motor, spring and load, respectively, whereas K_s and N are the spring stiffness and gear ratios in the SEA. Therefore, these parameters with the system links concluded with the transfer function $G(S)$.

$$G(s) = \frac{17.54\, s^3 + 0.02\, s^2 + 4.04 \cdot 10^8 s + 6.69 \cdot 10^{10}}{s^3 + 0.02\, s^2 + 1.11 \cdot 10^4\, s + 2.19 \cdot 10^5}, \tag{7}$$

Table 2. The simulation parameters of the SEA [56].

Parameter	Value
J_m	2.7×10^{-4} kg·m^2
J_s	45×10^{-5} kg·m^2
J_l	7×10^{-2} kg·m^2
B_m	1.5×10^{-3} Nm·s/rad
B_s	2×10^{-2} Nm·s/rad
B_l	5×10^{-3} Nm·s/rad
K_s	5×10^3 Nm/rad
N	1×10^2

The LQR control strategy was considered to operate in a linear time-invariant system for this study. The LQR was examined with four different values of the Q and R matrices (four cases), which were chosen empirically with different weights on Q and R to check the speed and robustness with different penalties on the input signal and states as shown in Table 3, which were the basis for finding the minimum quadratic cost function, J, regarding (2) and, consequently, the optimal controller.

Moreover, the result of using the regulator to follow the trajectory torque signal was investigated in the form of the RMSE. Finally, the energy consumption and distance for each step activity when using the hip exoskeleton powered by a charged battery (lithium–iron 48 V, 150 Wh, with 30 LiFePO$_4$ cells mounted in a battery tray) for each hip joint were simulated and studied.

Table 3. Four controller cases with different Q and R values.

Case 1	Case 2	Case 3	Case 4
$Q = \begin{bmatrix} 0.025 & 100 & 100 \\ 0 & 0.001 & 0 \\ 0 & 0 & 0.001 \end{bmatrix}$	$Q = \begin{bmatrix} 0.025 & 100 & 100 \\ 0 & 0.001 & 0 \\ 0 & 0 & 0.001 \end{bmatrix}$	$Q = \begin{bmatrix} 0.025 & 100 & 100 \\ 0 & 0.001 & 0 \\ 0 & 0 & 0.001 \end{bmatrix}$	$Q = \begin{bmatrix} 100 & 0 & 0 \\ 0 & 100 & 0 \\ 0 & 0 & 100 \end{bmatrix}$
$R = 1 \times 10^{-6}$	$R = 1 \times 10^{-5}$	$R = 1 \times 10^{-3}$	$R = 1 \times 10^{-2}$

4. Results and Discussion

For the robustness analyses of the system, we checked the gain (GM) and phase margins (PM), as the margin is one way to specify how much uncertainty the controller design can handle before the system becomes unstable [59]. Therefore, we checked the pure gain margin/classical gain margin (CGM) and pure phase margin/classical phase margin (CPM) for all controller cases that examined the controller's sensitivity to different uncertainties. Table 4 shows, for instance, that the controller Case 1 was still stable if the gain variations in a real system were up to 33.2 dB higher than those in our designed model. The same was valid for stability if the phase variations in the real system were up to 178.9 deg higher than those in the designed model. However, this variation was valid only if there was a CGM or CPM only, which is uncommon in real scenarios; thus, we analysed the controller cases with a combination of both GM and PM called DM (see Table 4). The GM & $PM - C$ showed that the border of DM in Case 1 was within (0.2–4.5) around the nominal point 1 of the Nyquist plane, while the tolerance for phase variation was within ± 64.9 deg. Another way to analyse the robustness of a system to uncertainty and variation was studying the Nyquist plot (see Figure 4) for each controller case.

Table 4. Gain and phase disk margins for four different controller cases.

Controller	CGM [dB]	CPM [deg]	GM & PM-C	DM
Case 1	33.2	178.9	$GM : \begin{bmatrix} 0.2 & 4.5 \end{bmatrix}$ $PM : \begin{bmatrix} -64.9 & 64.9 \end{bmatrix}$	1.3
Case 2	12	179.3	$GM : \begin{bmatrix} 0.3 & 2.6 \end{bmatrix}$ $PM : \begin{bmatrix} -47.8 & 47.8 \end{bmatrix}$	0.9
Case 3	12	21.4	$GM : \begin{bmatrix} 0.8 & 1.3 \end{bmatrix}$ $PM : \begin{bmatrix} -13.1 & 13.1 \end{bmatrix}$	0.2
Case 4	27	179.3	$GM : \begin{bmatrix} 0.1 & 6.1 \end{bmatrix}$ $PM : \begin{bmatrix} -71.6 & 71.6 \end{bmatrix}$	1.4

CGM = Classical Gain Margin, CPM = Classical Phase Margin, GM & PM-C = Gain and Phase Margin Combined, DM = Disk Margin.

As shown in Figure 4 and Table 4, the most robust controller was Case 4 with ($DM = 1.4$), which tolerated more uncertainties based not only on the Nyquist plot but also on the combined gain and phase disk margin ($GM : [0.1 \quad 6.1] \Rightarrow DGM = 16$ dB) and disk phase margin ($PM : [-71.6 \quad 71.6] \Rightarrow DPM = 72$ deg) with respect to the other three cases as also shown in Figure 5.

We then examined how each controller case tracked the hip torque reference signal for a healthy person in STS and walking activities at five torque levels, which represented various angular velocities for the joint's movements. The energy consumption per step activity with the four controller cases and five torque levels ($TL1$ *to* $TL5$) have also been studied.

Lastly, the number of times for the rehabilitation hip exoskeleton in STS was compared with the numbers of times for different controller cases and torques. The distances reached with a fully charged battery for each hip joint in the walking activity were also calculated. The results for STS and walking activity—based on a simulation—were as follows.

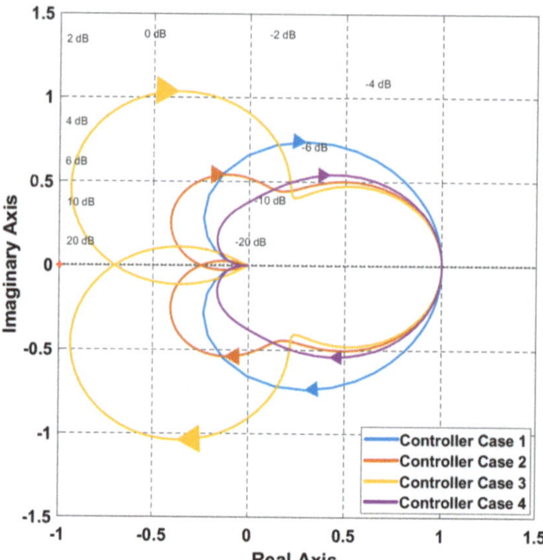

Figure 4. The Nyquist plots of the four controller cases.

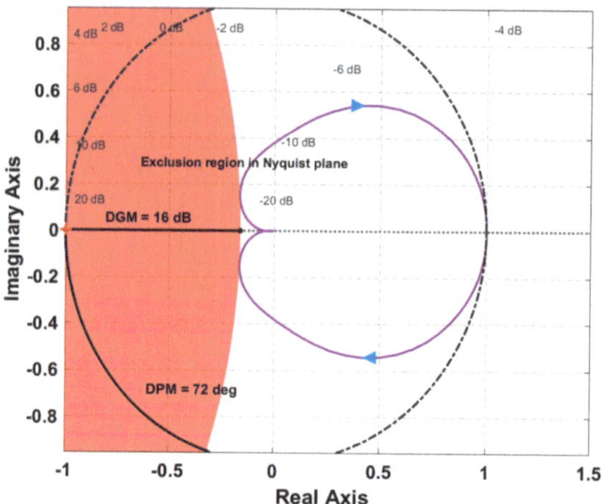

Figure 5. The disk margin of the fourth case controller.

4.1. Sit-to-Stand Activity

The RMSE and energy consumption for each controller case with various torques are shown in Tables 5 and 6, while the numbers of times that each case and torque in STS could be reached are shown in Table 7. As the function of the LQR controller is based on the system states and the control signal combined at the same time, it is interesting to study the effects on each of the two criteria, but we chose the RMSE as a base of our evaluation of the control performance.

Table 5. The RMSE for sit to stand ($\times 10^{-6}$ Nm/kg).

Torque Level	Case 1 *	Case 2	Case 3	Case 4
TL 1	26.5	44.7	43.3	68.8
TL 2	106.3	178.9	173.5	275.2
TL 3	239.1	402.5	390.5	619.3
TL 4	425.1	715.5	694.2	1100.1
TL 5	664.1	1100.1	1100.2	1700.1

(*) The case with lowest RMSEs.

Table 6. The energy consumption for sit to stand (milliwatts).

Torque Level	Case 1	Case 2	Case 3	Case 4
TL 1	1.4	0.4	0.4	0.4
TL 2	5.1	2.2	2.4	2.1
TL 3	10.9	6.6	6.9	6.5
TL 4	19.2	15.6	16.1	15.4
TL 5	30.0	30.1	31.2	29.9

Table 7. The number of sit-to-stand movements with a fully charged battery (step).

Torque Level	Case 1	Case 2	Case 3	Case 4
TL 1	108430	350970	342710	356800
TL 2	30052	69720	67194	70621
TL 3	13727	22575	21824	22929
TL 4	7804	9630	9379	9747
TL 5	5005	4981	4813	5014

4.2. Walking Activity

The RMSE, the energy consumption for the walking activity and the distance when using each controller at different torque levels are illustrated in Tables 8–10, respectively.

Despite the fact that the regulator in Case 4 was the most robust controller, the RMSEs for STS and walking activity at the various torques were still not the minimum values compared with the other cases. The RMSEs were the highest for STS and walking in Case 4 at all torque levels compared with those for other cases, while the RMSEs for Case 1 were the lowest for STS and walking movements. The reason controller Case 1 had the lowest RMSE values was the longer distance of the poles on the left half of complex planes to the origin. The faster response of the controller occurred when the poles lay far away from the origin or the controller had a large input signal [60]. For this reason, controller Case 1 provided a faster response to the dynamics of the system because its poles lay on ($P_{1,2} = -49.4 \pm 46.3i$, $P_3 = -47.8$) and its $R = 1 \times 10^{-6}$, compared with controller Case 4, which had its poles on ($P_{1,2} = -50.1 \pm 93.6i$, $P_3 = -19.4$) with a minimum input signal due to the high penalty value of $R = 1 \times 10^{-2}$. Although $P_{1,2}$ Case 4 was farther away from the origin than Case 1, it still had its dominant pole at ($P_3 = -19.4$), which was closer to the origin than the dominant pole of Case 1 at ($P_3 = -47.8$). Consequently, the system slowed in Case 4, which affected the system's behaviour and speed [61].

Table 8. The RMSE of walking ($\times 10^{-6}$ Nm/kg).

Torque Level	Case 1 *	Case 2	Case 3	Case 4
TL 1	153.3	209.6	204.3	300.6
TL 2	613.3	838.6	817.2	1200.1
TL 3	1400.1	1900.1	1800.1	2700.1
TL 4	2500.1	3400.1	3300.1	4800.2
TL 5	3800.2	5200.2	5100.1	7500.2

(*) The case with the lowest root mean square errors.

Table 9. The energy consumption per walking step (milliwatts).

Torque Level	Case 1	Case 2	Case 3	Case 4
TL 1	4.4	4.6	5.4	4.5
TL 2	12.3	11.3	13.3	11.4
TL 3	27.8	23.1	25.4	23.9
TL 4	53.7	52.1	55.7	53.4
TL 5	88.5	112.1	121.6	114.1

Table 10. The distance of walking (kilometres).

Torque Level	Case 1	Case 2	Case 3	Case 4
TL 1	25.7	25.1	21.1	25.2
TL 2	9.3	10.2	8.6	10.1
TL 3	4.1	4.9	4.5	4.8
TL 4	2.1	2.2	2.1	2.1
TL 5	1.3	1.1	0.9	1.0

Additionally, the higher the torque/angular speed of the STS and walking activity, the higher the RMSEs. However, in analysing the energy consumption required for each step in STS and walking, we noticed that the controller Case 4, considered to be the most robust regulator, demanded less energy in STS compared with other controller cases (see Figure 6 and Table 6). Therefore, we found the highest number of times by pursuing STS using controller Case 4 at all torques.

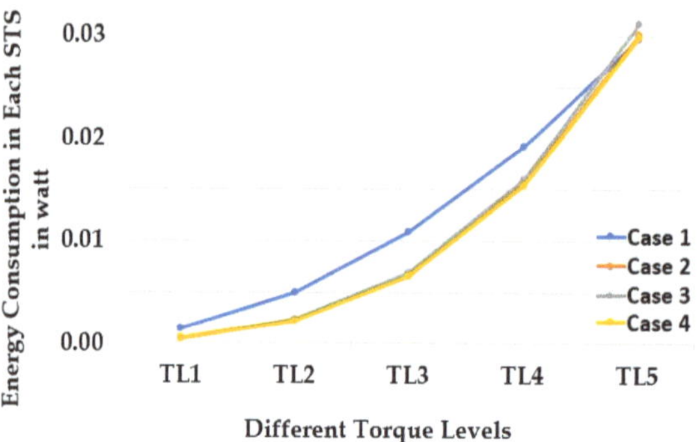

Figure 6. The energy consumption for sit to stand with different Q and R matrices and torque levels.

Interestingly, controller Case 4 in the walking activity did not have the lowest consumed energy at all torque levels as it showed in STS. The differences are obvious in Figure 7 and Table 9; in particular, the controller Case 1 and Case 4 had the lowest energy consumption at the speed of 0.5 m/s = $TL1$) and the second-lowest energy consumption at a speed of 2.6 m/s = $TL5$ for Case 4; still, the controller Case 1 had the lowest energy consumption at speeds of 0.5 m/s = $TL1$ and 2.6 m/s = $TL5$, respectively. Therefore, we obtained the longest distance for the walking activity in Case 1 at speeds of 0.5 m/s = $TL1$ and 2.6 m/s = $TL5$, as shown in Table 10, compared to in other cases.

Another notable feature of the walking activity was that the most extended distances were reached in Case 2 at speeds of 1.1–2.1 m/s = $TL2$–$TL4$ compared with the other cases for these speeds. The benefit of obtaining the longest distances was caused by the minimal energy consumed in Case 2 at these speeds (see Tables 9 and 10). This finding supported the hypothesis that the highest penalty of the R matrix in the quadratic cost function led to instability in walking activity at certain speeds. By contrast, the STS refuted this hypothesis by the choice of the highest $R = 0.01$ for controller Case 4, which was also the most robust controller case.

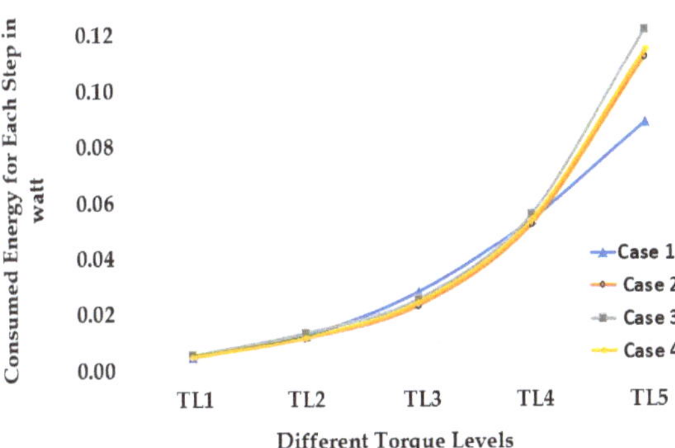

Figure 7. The energy consumption of walking with different Q and R matrices and torque levels.

Thus, a possible solution is having a different mode in the operating system of the rehabilitation hip exoskeleton if the economic control strategy with the longest distance is the final target. However, this could be a solution for walking activity for certain speeds but not for STS, as the controller Case 4 in STS showed its efficient energy consumption and the number of times for achieving STS. For the walking activity, however, choosing the more energy-efficient controller cases such as Case 1 or 2 could lead to a penalty in robustness against uncertainties, as these cases were not robust cases compared with Case 4, as shown in Figure 4 and Table 4.

Subsequently, as Case 4, the most robust controller, gave the second-longest walking distance at all speeds and the highest number of times for STS, it would be the more reliable regulator choice for our rehabilitation hip exoskeleton. With the Case 4 controller, we obtain the optimal robust controller with high performance and energy efficiency.

It is worth mentioning that the controller-related energy consumption comparison between the sit-to-stand and walking activities will have differences compared to the real world as well as between the real walking on a treadmill or overground. The differences between the simulation and real world are caused by many reasons—firstly, as they depend on the modelling process, which in most cases are approximations of the real systems. Secondly, the error caused in experimental data could be affected by instruments' errors and disturbances, which can be illuminated or processed in simulation environments. Thirdly, the simulation models are usually bounded by some assumptions. However, the simulation process can be used to verify if the errors or certain behaviour lie within the accepted domain (margin) before producing a physical system. Finally, as the linear time-invariant system was the foundation in operation of the controller cases in this research, more future study on combining the disturbance observer or a sliding mode observer is worth undertaking to overcome the non-linearity, parametric uncertainties and external disturbances [24,28,62].

5. Conclusions

Although the LQR ensured finding the optimal controller with the minimum cost function, the Q and R matrix values affected the system robustness, as it moved the system poles to various placements to stabilise the system. This study provides ample evidence that selecting appropriate Q and R influences not only the cost function but also the energy consumption and the performance of the optimal controller, specifically when the exoskeleton is restricted by certain design aspects.

The most robust optimal controller in STS showed great potential for energy consumption and the highest number of times pursuing the STS activity, while it was less efficient for walking activity. However, the most effective way to reduce the energy consumption of a rehabilitation hip exoskeleton in any activity is studying the speed required for rehabilitation and adjusting the activity to a lower torque.

This method ensured that using the exoskeleton over a longer distance when walking, and a greater number of times for STS, showed good performance when there was less choice in finding suitable Q and R matrices.

Author Contributions: Conceptualisation, R.A. and N.B.; methodology, R.A. and N.B.; software, R.A.; validation, R.A.; formal analysis, R.A. and N.B.; investigation, R.A.; resources, R.A. and N.B.; data curation, R.A.; writing—original draft preparation, R.A.; writing—review and editing, N.B.; visualisation, R.A.; supervision, N.B.; project administration, R.A.; funding acquisition, R.A. and N.B. All authors have read and agreed to the published version of the manuscript.

Funding: This research received no external funding.

Institutional Review Board Statement: Not applicable.

Informed Consent Statement: Not applicable.

Data Availability Statement: The data generated in the simulation part are available on request from the corresponding authors.

Conflicts of Interest: The authors declare no conflict of interest.

Abbreviations

The following abbreviations are used in this manuscript:

SCI	Spinal Cord Injury
LQR	Linear Quadratic Regulator
SEA	Series Elastic Actuator
STS	Sit To Stand
LO	Luenberger Observer
PID	Proportional Integral Derivative
CGM	Classical Gain Margin
CPM	Classical Phase Margin
GM&PM-C	Gain and Phase Margin Combined
DM	Disk Margin
DGM	Disk Gain Margin
DPM	Disk Phase Margin
RMSE	Root Mean Square Error
DH	Denavit Hartenberg Parameters
DoF	Degree of Freedom
RoM	Range of Motion

References

1. Andersson, R.; Björsell, N.; Isaksson, M. Robots Are a Promising Investment to Fight Pandemics. In Proceedings of the 8th International Conference on Signal Processing and Integrated Networks (SPIN), Noida, India, 26–27 August 2021; pp. 458–463.
2. Buongiorno, D. Advanced Control Strategies ForNatural Human-Exoskeleton Interaction. Ph.D. Dissertation, Sant' Anna School of Advanced Studies, Pisa, Italy, 2017.
3. Pons, J.L. *Wearable Robots: Biomechatronic Exoskeletons*; John Wiley & Sons: Hoboken, NJ, USA, 2008. [CrossRef]
4. Turja, T.; Saurio, R.; Katila, J.; Hennala, L.; Pekkarinen, S.; Melkas, H. Intention to Use Exoskeletons in Geriatric Care Work: Need for Ergonomic and Social Design. *Ergon. Des.* **2020**, *30*, 13–16. [CrossRef]
5. Patel, D.R.; Neelakantan, M.; Pandher, K.; Merrick, J. Cerebral Palsy in Children: A Clinical Overview. *Transl. Pediatr.* **2020**, *9*, S125–S135. [CrossRef] [PubMed]
6. Volpicelli, G.M. Review: Sportsmate 5 Fitness Exoskeleton | WIRED. WIRED.com. Available online: https://www.wired.com/review/sportsmate-5-fitness-exoskeleton/ (accessed on 10 July 2022).
7. Yuzbasioglu, N. Elderly to Make Up 22% of World Population by 2050. 2021. Available online: https://www.aa.com.tr/en/life/elderly-to-make-up-22-of-world-population-by-2050/2379462 (accessed on 10 July 2022).
8. Andersson, R.; Björsell, N. The MATLAB Simulation and the Linear Quadratic Regulator Torque Control of a Series Elastic Actuator for a Rehabilitation Hip Exoskeleton. *IRCE 2022, in press.*

9. İşcan, M.; Eken, H.; Vural, B.; Yılmaz, C. Design and Control of an Exoskeleton Robot: A Matlab Simscape Application. *J. Therm. Eng.* **2018**, *4*, 1867–1878.
10. Chen, B.; Ma, H.; Qin, L.Y.; Gao, F.; Chan, K.M.; Law, S.W.; Qin, L.; Liao, W.H. Recent Developments and Challenges of Lower Extremity Exoskeletons. *JOT* **2016**, *5*, 26–37. [CrossRef] [PubMed]
11. Hsu, S.H.; Changcheng, C.; Lee, H.J.; Chen, C.T. Design and Implementation of a Robotic Hip Exoskeleton for Gait Rehabilitation. *Actuators* **2021**, *10*, 212. [CrossRef]
12. Budhota, A.; Chua, K.S.G.; Hussain, A.; Kager, S.; Cherpin, A.; Contu, S.; Vishwanath, D.; Kuah, C.W.K.; Ng, C.Y.; Yam, L.H.L.; et al. Robotic Assisted Upper Limb Training Post Stroke: A Randomized Control Trial Using Combinatory Approach Toward Reducing Workforce Demands. *Front. Neurol.* **2021**, *12*, 804. [CrossRef]
13. Huo, W.; Mohammed, S.; Moreno, J.C.; Amirat, Y. Lower Limb Wearable Robots for Assistance and Rehabilitation: A State of the Art. *IEEE Syst. J.* **2016**, *10*, 1068–1081. [CrossRef]
14. Moreno, J.C.; Brunetti, F.; Navarro, E.; Forner-Cordero, A.; Pons, J.L. Analysis of the Human Interaction with a Wearable Lower-Limb Exoskeleton. *Appl. Bionics Biomech.* **2009**, *6*, 245–256. [CrossRef]
15. Long, Y.; Du, Z.J.; Wang, W.; Dong, W. Development of a Wearable Exoskeleton Rehabilitation System Based on Hybrid Control Mode. *Int. J. Adv. Robot. Syst.* **2016**, *13*, 1729881416664847. [CrossRef]
16. Ibrahim, H.A.; Ammar, H.H.; Shalaby, R. Modeling and Control of a Novel Design of Series Elastic Actuator for Upper Limb Rehabilitation. *Math. Model Eng. Probl.* **2022**, *9*, 85–93. [CrossRef]
17. Yu, H.; Huang, S.; Chen, G.; Pan, Y.; Guo, Z. Human-Robot Interaction Control of Rehabilitation Robots with Series Elastic Actuators. *IEEE Trans. Robot.* **2015**, *31*, 1089–1100. [CrossRef]
18. Zhang, J.F.; Yang, C.J.; Chen, Y.; Zhang, Y.; Dong, Y.M. Modeling and Control of a Curved Pneumatic Muscle Actuator for Wearable Elbow Exoskeleton. *Mechatronics* **2008**, *18*, 448–457. [CrossRef]
19. Ding, Y.; Kim, M.; Kuindersma, S.; Walsh, C.J. Human-in-the-Loop Optimization of Hip Assistance with a Soft Exosuit during Walking. *Sci. Robot.* **2018**, *3*, eaar5438. [CrossRef] [PubMed]
20. Kang, I.; Hsu, H.; Young, A. The Effect of Hip Assistance Levels on Human Energetic Cost Using Robotic Hip Exoskeletons. *IEEE Robot. Autom. Lett.* **2019**, *4*, 430–437. [CrossRef]
21. Tu, X.; Li, M.; Liu, M.; Si, J.; Huang, H. A Data-Driven Reinforcement Learning Solution Framework for Optimal and Adaptive Personalization of a Hip Exoskeleton. In Proceedings of the 2021 IEEE International Conference on Robotics and Automation (ICRA), Xi'an, China, 30 May–5 June 2021; pp. 10610–10616. [CrossRef]
22. Zhang, Q.; Nalam, V.; Tu, X.; Li, M.; Si, J.; Lewek, M.D.; Huang, H. Imposing Healthy Hip Motion Pattern and Range by Exoskeleton Control for Individualized Assistance. *IEEE Robot. Autom. Lett.* **2022**, *7*, 11126–11133. [CrossRef]
23. He, H.; Kiguchi, K. A Study on EMG-Based Control of Exoskeleton Robots for Human Lower-Limb Motion Assist. In Proceedings of the 2007 6th International Special Topic Conference on Information Technology Applications in Biomedicine, Tokyo, Japan, 8–11 November 2007; pp. 292–295. [CrossRef]
24. Mefoued, S.; Belkhiat, D.E.C. A Robust Control Scheme Based on Sliding Mode Observer to Drive a Knee-Exoskeleton. *Asian J. Control* **2019**, *21*, 439–455. [CrossRef]
25. Franks, P.W.; Bryan, G.M.; Martin, R.M.; Reyes, R.; Lakmazaheri, A.C.; Collins, S.H. Comparing Optimized Exoskeleton Assistance of the Hip, Knee and Ankle in Single and Multi-Joint Configurations. *Wearable Technol.* **2021**, *2*, e16. [CrossRef]
26. Mahdavian, M.; Arzanpour, S.; Park, E.J. Motion Generation of a Wearable Hip Exoskeleton Robot Using Machine Learning-Based Estimation of Ground Reaction Forces and Moments. In Proceedings of the 2019 IEEE/ASME International Conference on Advanced Intelligent Mechatronics, Tokyo, Japan, 8–11 November 2007; pp. 796–801. [CrossRef]
27. Ajayi, M.O. Modelling and Control of Actuated Lower Limb Exoskeletons a Mathematical Application Using Central Pattern Generators and Nonlinear Feedback Control Techniques. Ph.D. Dissertation, Tshwane University of Technology, Pretoria, South Africa, 2016.
28. Ugurlu, B.; Sariyildiz, E.; Kansizoglu, A.T.; Ozcinar, E.; Coruk, S. Benchmarking Torque Control Strategies for a Torsion-Based Series Elastic Actuator. *IEEE Robot. Autom. Mag.* **2021**, *29*, 85–96. [CrossRef]
29. Pratt, G.A.; Williamson, M.M.; Dillworth, P.; Pratt, J.; Wright, A. *Stiffness Isn't Everything.* In *Experimental Robotics IV*; Springer: Berlin/Heidelberg, Germany, 2005; pp. 253–262. [CrossRef]
30. Sun, J.; Guo, Z.; Zhang, Y.; Xiao, X.; Tan, J. A Novel Design of Serial Variable Stiffness Actuator Based on an Archimedean Spiral Relocation Mechanism. *IEEE/ASME Trans. Mechatron.* **2018**, *23*, 2121–2131. [CrossRef]
31. Pratt, G.A.; Williamson, M.M. Series Elastic Actuators. *IEEE Int. Conf. Intell. Robot. Syst.* **1995**, *1*, 399–406. [CrossRef]
32. Ham, V.R.; Sugar, T.G.; Vanderborght, B.; Hollander, K.W.; Lefeber, D. Compliant Actuator Designs: Review of Actuators with Passive Adjustable Compliance/Controllable Stiffness for Robotic Applications. *IEEE Robot. Autom. Mag.* **2009**, *16*, 81–94. [CrossRef]
33. Song, P.; Yu, Y.; Zhang, X. A Tutorial Survey and Comparison of Impedance Control on Robotic Manipulation. *Robotica* **2019**, *37*, 801–836. [CrossRef]
34. Zhang, Q.; Sun, D.; Qian, W.; Xiao, X.; Guo, Z. Modeling and Control of a Cable-Driven Rotary Series Elastic Actuator for an Upper Limb Rehabilitation Robot. *Front. Neurorobot.* **2020**, *14*, 13. [CrossRef]
35. Chen, J.; Fan, Y.; Sheng, M.; Zhu, M. Optimized Control for Exoskeleton for Lower Limb Rehabilitation with Uncertainty. In Proceedings of the 2019 Chinese Control And Decision Conference, Nanchang, China, 3–5 June 2019; pp. 5121–5125.

36. Yildirim, M.C.; Sendur, P.; Kansizoglu, A.T.; Uras, U.; Bilgin, O.; Emre, S.; Yapici, G.G.; Arik, M.; Ugurlu, B. Design and Development of a Durable Series Elastic Actuator with an Optimized Spring Topology. *J. Mech. Eng. Sci.* **2021**, *235*, 7848–7858. [CrossRef]

37. Khamar, M.; Edrisi, M.; Zahiri, M. Human-Exoskeleton Control Simulation, Kinetic and Kinematic Modeling and Parameters Extraction. *MethodsX* **2019**, *6*, 1838–1846. [CrossRef]

38. Piatt, J.A.; Nagata, S.; Zahl, M.; Li, J.; Rosenbluth, J.P. Problematic Secondary Health Conditions among Adults with Spinal Cord Injury and Its Impact on Social Participation and Daily Life. *J. Spinal Cord. Med.* **2016**, *39*, 693–698. [CrossRef]

39. Mao, Y.R.; Wu, X.Q.; Li Zhao, J.; Lo, W.L.A.; Chen, L.; Ding, M.H.; Xu, Z.Q.; Bian, R.H.; Huang, D.F.; Li, L. The Crucial Changes of Sit-to-Stand Phases in Subacute Stroke Survivors Identified by Movement Decomposition Analysis. *Front. Neurol.* **2018**, *9*, 185. [CrossRef]

40. Chen, B.; Zhong, C.H.; Ma, H.; Guan, X.; Qin, L.Y.; Chan, K.M.; Law, S.W.; Qin, L.; Liao, W.H. Sit-to-Stand and Stand-to-Sit Assistance for Paraplegic Patients with CUHK-EXO Exoskeleton. *Robotica* **2018**, *36*, 535–551. [CrossRef]

41. Mak, M.K.Y.; Levin, O.; Mizrahi, J.; Hui-Chan, C.W.Y. Joint Torques during Sit-to-Stand in Healthy Subjects and People with Parkinson's Disease. *Clin. Biomech.* **2003**, *18*, 197–206. [CrossRef]

42. Grimmer, M.; Seyfarth, A. Mimicking Human-Like Leg Function in Prosthetic Limbs. In *Neuro-Robotics*; Springer: Dordrecht, The Netherlands, 2014; pp. 105–155.

43. Pai, Y.C.; Rogers, M.W. Speed Variation and Resultant Joint Torques during Sit-to-Stand. *Achieve Med. Rehabil.* **1991**, *72*, 881–885. [CrossRef]

44. Grimmer, M. Powered Lower Limb Prostheses. Ph.D. Thesis, Technische Universität, Darmstadt, Germany, 2015.

45. Yildirim, M.C.; Kansizoglu, A.T.; Sendur, P.; Coruk, S.; Ugurlu, B. Development of a High Torque-to-Mass Ratio Series Elastic Actuator. In Proceedings of the 2018 International Conference on Robotics and Automation (ICRA), Brisbane, Australia, 21–25 May 2018. [CrossRef]

46. Ellis, G.; George, H. *Observers in Control Systems: A Practical Guide*; Academic Press: Cambridge, CA, USA, 2002.

47. Duriez, T.; Brunton, S.L.; Noack, B.R. *Machine Learning Dynamics and Taming Nonlinear Control—Turbulence*; Fluid Mechanics and Its Applications; Springer International Publishing: Cham, Switzerland, 2013; Volume 33. [CrossRef]

48. Skogestad, S.; Postlethwaite, I. Multivariable Feedback Control—Analisys and Design. *Int. J. Robust Nonlinear Control* **2001**, *1*, 585.

49. Ogata, K. *Modern Control Engineering*, 5th ed.; Pearson: London, UK, 2017.

50. Doyle, J.; Francis, B.; Tannenbaum, A. *Feedback Control Theory*; Macmillan Publishing Co.: London, UK, 1998.

51. Åström, K.J. *Model Uncertainty and Robust Control*; Valencia Polytechnic University: Valencia, Spain, 2000; pp. 63–100.

52. Palomba, I.; Wehrle, E.; Carabin, G.; Vidoni, R. Minimization of the Energy Consumption in Industrial Robots through Regenerative Drives and Optimally Designed Compliant Elements. *Appl. Sci.* **2020**, *10*, 7475. [CrossRef]

53. Van Kammen, K.; Boonstra, A.; Reinders-Messelink, H.; Den Otter, R. The Combined Effects of Body Weight Support and Gait Speed on Gait Related Muscle Activity: A Comparison between Walking in the Lokomat Exoskeleton and Regular Treadmill Walking. *PLoS ONE* **2014**, *9*, e107323. [CrossRef] [PubMed]

54. De Leva, P. Adjustments to Zatsiorsky-Seluyanov's Segment Inertia Parameters. *J. Biomech.* **1996**, *29*, 1223–1230. [CrossRef]

55. Lee, C.; Kwak, S.; Kwak, J.; Oh, S. Generalization of Series Elastic Actuator Configurations and Dynamic Behavior Comparison. *Actuators* **2017**, *6*, 26. [CrossRef]

56. Malzahn, J.; Roozing, W.; Tsagarakis, N. The Compliant Joint Toolbox for MATLAB: An Introduction with Examples. *IEEE Robot. Autom. Mag.* **2019**, *26*, 52–63. [CrossRef]

57. Oh, S.; Kong, K. High-Precision Robust Force Control of a Series Elastic Actuator. *IEEE/Asme Trans. Mechatron.* **2017**, *22*, 71–80. [CrossRef]

58. Munadi; Nasir, M.S.; Ariyanto, M.; Iskandar, N.; Setiawan, J.D. Design and Simulation of PID Controller for Lower Limb Exoskeleton Robot. *AIP Conf. Proc.* **2018**, *1983*, 060008. [CrossRef]

59. Seiler, P.; Packard, A.; Gahinet, P. An Introduction to Disk Margins [Lecture Notes]. *IEEE Control Syst.* **2020**, *40*, 78–95. [CrossRef]

60. Glad, T.; Ljung, L. *Control Theory: Multivariable and Nonlinear Methods*; Taylor & Francis: Abingdon, UK, 2000. [CrossRef]

61. Lee, J.; Edgar, T.F. Multiloop PI/PID Control System Improvement via Adjusting the Dominant Pole or the Peak Amplitude Ratio. *Chem. Eng. Sci.* **2006**, *61*, 1658–1666. [CrossRef]

62. Bergmann, L.; Liu, L.; Ngo, C.; Misgeld, B.; Leonhardt, S. Zero-Torque Control of an Exoskeleton Joint Actuator Using a Disturbance Observer. *AIP Conf. Proc.* **2019**, *2140*, 020004. [CrossRef]

MDPI AG
Grosspeteranlage 5
4052 Basel
Switzerland
Tel.: +41 61 683 77 34

Applied Sciences Editorial Office
E-mail: applsci@mdpi.com
www.mdpi.com/journal/applsci